"Richard Wrangham has written a brilliant and honest book about humanity's central contradiction: that we are capable of mass murder but live in societies with almost no violence. No other species straddles such a wide gap, and the reasons are staggeringly obvious once Wrangham lays them out in his calm, learned prose. This book is science writing at its best: lucid, rational, and yet deeply concerned with humanity."

—Sebastian Junger, author of *Tribe*

"This will prove to be one of the most important publications of our time. Fully supported scientific information from many directions leads us to a new and compelling analysis of our evolutionary history. Every page is fascinating, every revelation is unforgettable. It will change how we see ourselves, our past, and our future."

—Elizabeth Marshall Thomas, author of *The Hidden Life of Dogs*

"This is the most thought-provoking book I have read in years. In clear, elegant prose, drawing on riveting data and vivid scenes gathered from species all over the world, renowned anthropologist Richard Wrangham examines the issues most central to human morality. *The Goodness Paradox* is a breakthrough that deserves careful reading, thoughtful consideration, and lively debate among all those who care about our evolutionary history and the future of human morality."

—Sy Montgomery, author of *How to Be a Good Creature*

ALSO BY RICHARD WRANGHAM

Catching Fire: How Cooking Made Us Human

Demonic Males: Apes and the Origins of Human Violence
(with Dale Peterson)

The Goodness Paradox

The Goodness Paradox

The Strange Relationship Between Virtue and Violence in Human Evolution

RICHARD WRANGHAM

Pantheon Books
New York

 Published in the United States by Pantheon Books, a division of Penguin Random House LLC, New York, and distributed in Canada by Random House of Canada, a division of Penguin Random House Canada Limited, Toronto.

Pantheon Books and colophon are registered trademarks of Penguin Random House LLC.

www.pantheonbooks.com

Library of Congress Cataloging-in-Publication Data
Name: Wrangham, Richard W., [date] author.
Title: The goodness paradox : the strange relationship between virtue and violence in human evolution / Richard Wrangham.
Description: First edition. New York : Pantheon Books, 2019.
Includes bibliographical references and index.
Identifiers: LCCN 2018028837. ISBN 9781101870907 (hardcover : alk. paper).
ISBN 9781101870914 (ebook).
Subjects: LCSH: Human evolution. Human behavior. Aggressiveness.
Classification: LCC GN281.4 .W73 2019 | DDC 155.9—dc23 |
LC record available at lccn.loc.gov/2018028837

Jacket design by Kelly Blair

Printed in the United States of America

First Edition
2 4 6 8 9 7 5 3 1

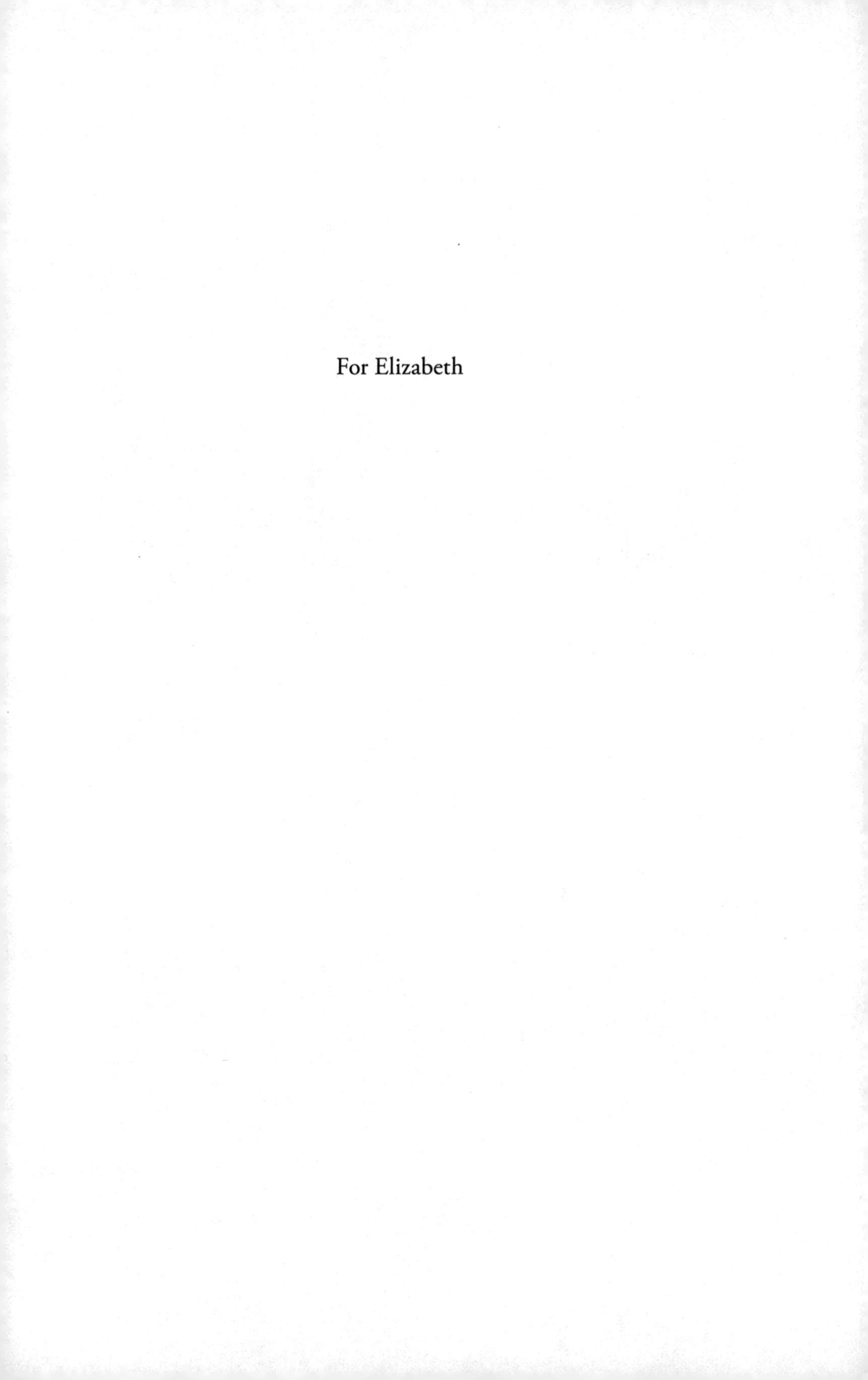

For Elizabeth

Contents

Preface

At the start of my career, I would have been surprised to learn that fifty years later I would be publishing a book about humans. In the 1970s I was privileged to be a graduate student working in Jane Goodall's research project on chimpanzees in Tanzania. Spending whole days trailing individual apes in a natural habitat was a joy. All that I wanted to do was study animal behavior, and in 1987 I launched my own study of wild chimpanzees in Uganda's Kibale National Park.

My bucolic research was disturbed, however, by discoveries that were too intriguing to ignore. Chimpanzees exhibited occasional episodes of exceptional violence. To shed an evolutionary light on this behavior, I compared chimpanzees with their sister species, bonobos. In the 1990s, research on bonobos was beginning in earnest. Chimpanzees and bonobos were proving to be an extraordinary duo, bonobos being much more peaceful than the relatively aggressive chimpanzees. In various collaborations that I describe in this book, but most particularly with Brian Hare and Victoria Wobber, my colleagues and I concluded that bonobos had diverged from a chimpanzee-like ancestor by a process that was strongly akin to domestication. We called the process "self-domestication." And since human behavior has often been considered similar to the behavior of domesticated animals, the insights from bonobos suggested lessons for human evolution. The key fact about humans is that within our social communities we have a low propensity to fight: compared to most wild mammals we are very tolerant.

I was acutely aware, however, that, even if humans are in some ways notably unreactive, in other ways we are a very aggressive species. In 1996, in a book called *Demonic Males: Apes and the Origins of Human Violence,* Dale Peterson and I described evolutionary explanations for similarities in aggression between chimpanzees and humans. The pervasiveness of violence in human society is inescapable, and the evolutionary theories explaining it seem sound. So how could our domesticated qualities and our capacity for terrible violence be reconciled? For the next twenty years or so, I grappled with this question.

The resolution that I describe in the following pages is that our social tolerance and our aggressiveness are not the opposites that at first they appear to be, because the two behaviors involve different types of aggression. Our social tolerance comes from our having a relatively low tendency for reactive aggression, whereas the violence that makes humans deadly is proactive aggression. The story of how our species came to combine these different tendencies—low reactive aggression and high proactive aggression—has not been told before. It takes us into many corners of anthropology, biology, and psychology, and will undoubtedly continue to be developed. But I believe that it already offers a rich and fresh perspective on the evolution of our behavioral and moral tendencies, as well as on the fascinating question of how and why our species, *Homo sapiens,* came into existence at all.

Much of the material in this book is so new that it has been published only in scientific papers. My goal here is to make this richly technical literature and its far-reaching implications more accessible. I approach the topic through the eyes of a chimpanzee-watcher who has walked, watched, and listened in many habitats of East and Central Africa. Those of us privileged to have spent days alone with apes have felt touched by Pleistocene breezes. The romance of the past, the story of our ancestors, is a thrill, and innumerable mysteries remain for future generations seeking the origins of the modern mind in deep time. Enlarged understanding of our prehistory and of who we are will not be the only reward. Dreams inspired by the African air can yet generate a stronger and more secure view of ourselves, if we open our minds to worlds beyond those that we know well.

The Goodness Paradox

Introduction

Virtue and Violence in Human Evolution

Adolf Hitler, who ordered the execution of some eight million people and was responsible for the deaths of many millions more, was said by his secretary Traudl Junge to have had an agreeable, friendly, and paternal manner. He hated cruelty to animals: he was a vegetarian, adored his dog Blondi and was inconsolable when Blondi died.

Pol Pot, the leader of Cambodia whose policies killed maybe a quarter of his country's people, was known to his acquaintances as a soft-spoken and kindly teacher of French history.

During eighteen months in prison, Joseph Stalin was always amazingly calm and never shouted or swore. In effect, he was a model gentleman-inmate, not obviously the kind of person who would later annihilate millions for political convenience.

Because seriously evil men can have a gentle side, we hesitate to empathize with their kindness for fear of seeming to rationalize or excuse their crimes. Such men remind us, however, of a curious fact about our species. We are not merely the most intelligent of animals. We also have a rare and perplexing combination of moral tendencies. We can be the nastiest of species and also the nicest.

In 1958, the playwright and songwriter Noël Coward captured the strangeness of this duality. He had lived through the Second World War, and the bad side of human nature was fully obvious to him. "It is hard to imagine," he wrote, "considering the inherent silliness, cruelty and superstition of the human race, how it has contrived to last as long as it has. The witch-hunting, the torturing, the gullibility, the massacres, the intolerance, the wild futility of human behaviour over the centuries is hardly credible."[1]

And yet most of the time we do wonderful things that are the very opposite of "silliness, cruelty and superstition," depending as they do on reason, kindness, and cooperation. The technological and cultural marvels that distinguish our species are made possible by these qualities, in combination with our intelligence. Coward's examples still resonate.

> Hearts can be withdrawn from human breasts, dead hearts, and, after a little neat manipulation, popped back again as good as new. The skies can be conquered. Sputniks can whizz round and round the globe and be controlled and guided . . . and *My Fair Lady* opened in London last night.

Heart surgery, space travel, and comic opera all depend on advances that would have amazed our distant ancestors. More important from an evolutionary point of view, however, they also depend on capacities for a quite exceptional ability to work together, including tolerance, trust, and understanding. Those are some of the qualities that cause our species to be thought of as exceptionally "good."

In short, a great oddity about humanity is our moral range, from unspeakable viciousness to heartbreaking generosity. From a biological perspective, such diversity presents an unsolved problem. If we evolved to be good, why are we also so vile? Or if we evolved to be wicked, how come we can also be so benign?

The combination of human good and evil is not a product of modernity. To judge from the behavior of recent hunter-gatherers and the records of archaeology, for hundreds of thousands of years people have shared food, divided their labor, and helped the needy. Our Pleistocene ancestors were in many ways thoroughly tolerant and peaceful. Yet the same sources of evidence also indicate that our forebears practiced raiding, sexual dominance, torture, and executions with varieties of cruelty that were as abominable as any Nazi practice. Certainly nowadays, a capacity for great cruelty and violence is not particular to any one group. For a variety of reasons, a given society might have experienced exceptional peace for decades even as another might have suffered spasms of exceptional violence. But this does not suggest any differences in the innate psychology of people throughout time and world over. Everywhere humans seem to have had the same propensity for both virtue and violence.

Babies show a similar contradiction in their tendencies. Before infants can talk, they will smile and chuckle and sometimes help a friendly adult in need, an extraordinary demonstration of our innate predisposition to trust one another. At other times, however, those same bighearted offspring will scream and rage with sublime self-centeredness to get their way.

There are two classic explanations for this paradoxical combination of selflessness and selfishness. Both assume that our social behavior is hugely determined by our biology. Both also agree that only one of our two notable tendencies is the product of genetic evolution. They differ, however, in which side of our personality each regards as fundamental—our docility, or our aggressiveness.

One explanation posits that tolerance and docility are innate to humanity. According to this view, although we are essentially good, our corruptibility stands in the way of our living in perpetual peace. Some religious thinkers blame supernatural forces such as the devil or "original sin" for this state of affairs. Secular thinkers, by contrast, might choose instead to imagine evil as engendered by societal forces such as patriarchy, imperialism, or inequality. Either way, it is assumed that we are born good but are susceptible to corruption.

The other explanation claims that it is our bad side that is innate. We are born selfish and competitive, and we would continue in the same vein were it not for efforts at self-improvement informed by civilizing forces, which might include enjoinments from parents, philosophers, priests, and teachers, or the lessons of history.

For centuries, people have simplified their understanding of a confusing world by adopting one or the other of these opposed views. Jean-Jacques Rousseau and Thomas Hobbes are classic icons for the alternatives. Rousseau has come to stand for humanity's being instinctively nice, Hobbes for humanity's being naturally wicked.[2]

Both positions have some merit. There is plenty of evidence that humans have innate tendencies for kindness, just as there is for our having spontaneously selfish feelings that can lead to aggression. No one has found a way to say that one kind of tendency is more biologically meaningful or evolutionarily influential than the other.

The intrusion of politics makes the debate all the harder to settle, because when these abstractly theoretical analyses become arguments with societal significance, both sides tend to harden their stance. If you are a Rousseauian, your belief in essential human goodness probably marks you out as a peace-loving, easygoing crusader for social justice with faith in the masses. If you are a Hobbesian, your cynical view of human motives suggests you see a need for social control, cherishing hierarchy and accepting the inevitability of war. The debate becomes less about biology or psychology and more about social causes, political structures, and the moral high ground. Prospects for simple resolution duly recede.

I believe that there is an escape from this morass about the fundamental nature of humans. Rather than needing to prove that either side is wrong, we should ask whether the debate makes sense at all. Babies point us in the right direction: the Rousseau and Hobbes perspectives were both right as far as they went. We are naturally good in the way that Rousseau is said to have claimed, *and* we are naturally selfish, much as Hobbes argued. The potential for good and evil occurs in every individual. Our biology determines the contradictory aspects of our personalities, and society modifies both tendencies. Our goodness can be intensified or corrupted, just as our selfishness can be exaggerated or reduced.

Once we acknowledge that we are at once innately good and innately bad, the sterile old argument gives way to fascinating new problems. If Rousseauians and Hobbesians are both partly right, then what is the source of our strange combination of behavioral tendencies? We know from the study of other species, particularly birds and mammals, that natural selection can favor a wide range of inclinations. Some species are relatively uncompetitive, some relatively aggressive, some both, some neither. The combination that makes humans odd is that we are both intensely calm in our normal social interactions, and yet in some circumstances so aggressive that we readily kill. How did this come to be?

Evolutionary biologists follow a principle crisply stated by the geneticist Theodosius Dobzhansky in a 1973 address to the National Association of Biology Teachers: "Nothing in biology makes sense except in the light of evolution." How evolutionary theory is best used, however, is a subject of debate. A key question for this book is: what is the significance of primate behavior?

A traditional view holds that animal and human mentality differ so widely that primates are irrelevant for the science of human nature.[3] Thomas Henry Huxley was the first evolutionary biologist to challenge that position. In 1863 he argued that apes provide rich clues to the origins of human behavior and cognition: "I have endeavoured to show that no absolute structural line of demarcation . . . can be drawn between the animal world and ourselves." Huxley anticipated his opponents' objections. "On all sides I shall hear the cry—'The power of knowledge—the conscience of good and evil—the pitiful tenderness of human affections, raise us out of all real fellowship with the brutes.'"[4] That kind of skepticism is understandable and has not completely disappeared. In 2003, the evolutionary biologist David Barash argued "it is highly questionable whether human beings carry a significant primate legacy at all when it comes to behavior."[5]

There are also variations of behavior galore due to culture. One society is peaceful, another violent. One reckons clan membership down the female line, one down the male line. Some have strict rules about sexual behavior, while others are lax. The diversity can seem so overwhelming as to render uniformity irrelevant for comparison with other species. After a detailed survey of hunter-gatherer behavior, the anthropologist Robert Kelly abandoned the notion that human behavior can be characterized as having any particular form. "There is no original human society, no basal human adaptation," he wrote in 1995. "Universal behaviors . . . never existed."[6]

In short, the idea that human behavior is so infinitely variable that our species has no special features in common with nonhuman primates is understandable. Two strong arguments stand against it, however.

On the one hand, human variation is limited. We really do have characteristic forms of society. Nowhere do people live in troops, as baboons do, or in isolated harems, like gorillas, or in entirely promiscuous communities, like chimpanzees or bonobos. Human societies consist of families within groups that are part of larger communities, an arrangement that is characteristic of our species and distinctive from other species.

Yet, on the other hand, in many ways humans and primates really do behave alike. The evolutionist Charles Darwin early on observed similarities in the expressions of emotion in humans and other animals, such as the "bristling of the hair under the influence of extreme terror" or "uncovering of the teeth under [the influence] of extreme rage." This "community of certain expressions," he wrote, is "rendered somewhat more intelligible if we believe in their descent from a common ancestor."[7]

The fact that we share smiles and frowns with our cousin primates is intriguing, but even that observation seems relatively trivial compared with the discoveries about chimpanzee and bonobo behavior that began in the 1960s, and continue to accumulate. Chimpanzees and bonobos are the two ape species most closely, and equally, related to humans. They present an astonishing pair. They look so similar to each other that they were not recognized to be separate species for years after both were known. Each of the two sister species shares extensive behavioral similarities to humans. Yet they are in many ways social opposites.

Among chimpanzees, males are dominant over females, and they are relatively violent. Among bonobos, females are often dominant over males, violence is muted, and eroticism is a frequent substitute for aggression. The behavioral distinctions between the two eerily echo competing social stances in the modern human world: the divergence of male and female interests, for example; or between hierarchy, competition, and power on the one hand, and egalitarianism, tolerance, and negotiated settlement on the other. The two species conjure such different visions of the essential ape that their opposition has become something of a battleground in primatology, each supposed by different schools to represent our ancestral lineage better than the other. As we will see, the notion that either chimpanzees or bonobos, but not both, point to human behavioral origins is not very helpful. A more intriguing goal is to understand why the two species are similar to humans in their different ways. Their behavioral contrasts are of a piece with the central question animating this book: why are humans both highly tolerant, like bonobos, and highly violent, like chimpanzees?

Chapter 1 launches the investigation by documenting behavioral differences among humans, chimpanzees, and bonobos. Decades of research suggest how species differences in aggression can evolve. Aggressiveness was once thought of as a tendency running from low to high along one dimension. But we now recognize that aggression comes in not one but two major forms, each with its own biological underpinnings and its own evolutionary story. As I show in chapter 2, humans are positively dualistic with respect to aggression. We are low on the scale of one type (reactive aggression), and high on the other (proactive aggression). Reactive aggression is the "hot" type, such as losing one's temper and lashing out. Proactive aggression is "cold," planned and deliberate. So our core question becomes two: why are we so lacking in reactive aggressiveness, and yet so highly proficient at proactive aggressiveness? The answer to the first explains our virtue; the answer to the second accounts for our violence.

Our low tendency for reactive aggression gives us our relative docility and tolerance. Tolerance is a rare phenomenon in wild animals, at least in the extreme form that humans show. It is found, however, among domesticated species. In chapter 3, I consider the similarities between domesticated animals and humans, and show why an increasing number of scientists believe that humans should be regarded as a domesticated version of an earlier human ancestor.

One of the exciting aspects of the biology of domesticated animals is that researchers are beginning to understand puzzling similarities that occur among many unrelated species. Why, for example, do cats, dogs, and horses frequently sport white patches of hair, unlike their wild ancestors? Chapter 4 explains new theories linking the evolution of physical features such as these to changes in behavior. Humans exhibit enough such features to justify calling us a domesticated species. That conclusion, which was first intimated more than two hundred years ago, creates a problem. If humans are like a domesticated species, how did we get that way? Who could have domesticated us?

Bonobos suggest a solution. In chapter 5, I review the evidence that bonobos, like humans, show many of the features of a domesticated species. Obviously, bonobos were not domesticated by humans. The process happened in nature, unaffected by human beings. Bonobos must have self-domesticated. That evolutionary transformation seems likely to be widespread among wild species. If so, there would be nothing exceptional in the self-domestication of human ancestors. In chapter 6, I therefore trace the evidence that *Homo sapiens* have had a domestication syndrome since their origin, about 300,000 years ago. Surprisingly few attempts have been made to explain why *Homo sapiens* arose, and as I describe, even the most recent paleoanthropological scenarios have not addressed the important problem of why selection should have favored a relatively tolerant, docile species with a low tendency for reactive aggression.

How self-domestication happened is in general an open question, with different answers expected for different species. Clues come from the way that aggressive individuals are prevented from dominating others. Among bonobos, aggressive males are suppressed mainly by the joint action of cooperating females. Probably, therefore, bonobo self-domestication was initiated by females' being able to punish bullying males. In small-scale societies of humans, females do not control males to the same extent as they do among bonobos. Instead, among humans, the ultimate solution to stopping male aggressors is execution by adult males. In chapters 7 and 8, I describe the use of execution in human society to force domineering men to conform to egalitarian norms, and I explain why I believe self-domestication through the selective force of execution was responsible for reducing humans' reactive aggression from the beginnings of *Homo sapiens*.

If genetic selection against reactive aggression indeed occurred through self-domestication, we should expect human behavior to share aspects of the behavior of domesticated animals beyond reduced aggression. In chapter 9, I examine this proposition. I emphasize that the proper comparison is not between humans and apes, because too many evolutionary changes have occurred in the seven million years or so since we had a common ancestor. Instead, the proper comparison is between *Homo sapiens* and Neanderthals, whom I take as a stand-in for our pre-*sapiens* ancestors. Chapter 9 reviews evidence that *Homo sapiens* had a more highly elaborated culture than Neanderthals. The difference, I propose, is plausibly due to *Homo sapiens*'s having lost more of the aggressiveness of the common ancestor than Neanderthals did.

A low propensity for reactive aggression enhances the capacity for tolerant cooperation, but it is not the only contributor to human social virtue. Morality is also vital. Chapter 10 asks why our evolved moral sensibilities often make people afraid of being criticized. Sensitivity to criticism, I conclude, would have promoted evolutionary success thanks to the emergence of the same new social feature that was responsible for self-domestication: a coalition able to carry out executions at will. Our ancestors' moral senses helped protect them from being killed for the crime of nonconformity.

The ability of adults (and particularly men) to conspire in the act of capital punishment is part of a larger system of social control using proactive aggression that characterizes all human societies. Chapter 11 discusses how humans echo chimpanzee behavior in this regard but have elaborated it far beyond chimpanzee style. Since proactive aggression is complementary to reactive aggression (rather than its opposite), proactive, planful aggression can be positively selected even while reactive, emotional aggression has been evolutionarily suppressed. Humans can therefore use overwhelming power to kill a selected opponent. This unique ability is transformative. It has led our societies to include hierarchical social relationships that are far more despotic than those found in other species.

A familiar and important occurrence of proactive aggression is in war, so, in chapter 12, I consider some of the ways in which the psychology of aggression influences warfare. Although contemporary war is much more institutionalized than most prehistoric intergroup violence, our tendencies for proactive and reactive aggression both play important roles, sometimes promoting and sometimes interfering with military goals.

Chapter 13 assesses the paradox that virtue and violence are both so prominent in human life. The solution is not so simple or morally desirable as we might wish: humans are neither all good nor all bad. We have evolved in both directions simultaneously. Both our tolerance and our violence are adaptive tendencies that have played vital roles in bringing us to our present state. The idea that human nature is at the same time both virtuous and wicked is challenging, since presumably we would all wish for simplicity. F. Scott Fitzgerald wrote, "The test of a first-rate intelligence is the ability to hold two opposed ideas in the mind at the same time, and still retain the ability to function." "I must hold in balance," he continued, ". . . the contradiction between the dead hand of the past and the high intentions of the future." I like Fitzgerald's thought. The moral contradictions of our ancestry should not prevent us from reaching a realistic assessment of who we are. When we do that, high hopes are still possible.[8]

1

The Paradox

I STARTED THINKING ABOUT the biological roots of peacefulness several decades ago, in a remote part of the Democratic Republic of the Congo. Later, the Congo would suffer badly, but in 1980, when Elizabeth Ross and I began a nine-month honeymoon in the Ituri Forest, all was quiet.

We were there as part of a two-couple team. Our job was to document the lives of two societies living alongside each other, Lese farmers and Efe foragers. Small villages of Lese farmers dotted the vast Ituri plain, some as far as two days' walk from their neighbors. Efe pygmies occupied the same area. When the Efe could find roots and fruits, they camped deep in the forest. In leaner times, they settled on the edge of a familiar village. Efe women then worked in the Lese gardens in exchange for cassava, bananas, or rice.

We lived in leaf-roofed, mud-walled huts in our own little clearing near a Lese village. We could not speak their first language, KiLese, but we knew enough KiNgwana, a version of Swahili, to have cheerful conversations with them. The Ituri people knew little of the outside world. Their economy was based largely on barter. Such things as nuclear bombs, soda cans, and electricity were outside their experience.

The living quarters of both the Efe foragers and the Lese farmers were small and dark and were hardly used at all by day. So life between dawn and dusk was public, which meant that, throughout the daylight hours, we could record behavior openly. We watched and listened and followed. We shared their food and joined in their activities. As a behavioral biologist who had studied chimpanzees and seen their intemperate aggression to one another, I was alert to the possibility of important events such as the sight of clenched fists or the drawing of bows. Having grown up in the sleepy British countryside where even a raised voice, let alone public fighting, was a rarity, I had wondered if aggression would be more obvious in this distant Congolese settlement.

It was wonderful to be able to see so much social interaction, but on the aggression score, there was little of interest. Even when dozens of people competed for meat around the carcass of an elephant, there was nothing more than occasional raised voices. Once, I met three men in fighting gear, loincloths furled around their hips, en route to the chief's village. They had heard that their teenage sisters had been taken to a feast by the chief's kin, and the three men were intent on preventing their sisters from being debauched. They proved able to rescue the girls without violence. We heard of an Efe man who hit his wife with a flaming log. Doubtless a few other incidents were hidden behind whispers and mud walls, but the only injuries we saw were the results of accident and disease.

Our Ituri companions had some of the toughest lives on earth. Living off what they grew, hunted, or found in an unproductive forest, they faced routine food shortages, poverty, discomfort, and desperate illnesses virtually unaided by modern medicine. Their cultural practices often seemed to make life harder. They beautified girls by crudely chipping their teeth. They remembered their own grandparents as cannibals. Pictures on the side of cans of meat showed smiling people, so the Lese teased us with the idea that Europeans who ate canned food were cannibals, too. Funerals brought arguments over the value of the dead: had a woman produced enough children to be worth her bride price of seven chickens? Even the most understandable misfortunes were attributed to witchcraft, a daily source of irrational threat. In many ways, the Ituri was a place where anything might happen.[1]

Beyond the practical hardships and weird superstitions, however, the Lese and Efe were entirely familiar in their basic psychology. Illogical beliefs, poverty, and strange medical practices took different forms in rural England and the Congo, but they were present in both. At heart, the Ituri people were hauntingly like the villagers in my native England—loving their children, quarreling over their lovers, worrying about gossip, looking for allies, jockeying for power, trading information, fearing strangers, planning parties, embracing ritual, ranting at fate—and very, very rarely getting into fights.

Obviously, humans can be more or less violent depending on social context. The Congo had a central government, and although the Ituri peoples were in many ways independent of it, they were not totally isolated. Perhaps the calm of the Lese and Efe was a result of the civilizing influences of cultural development that ultimately emanated from Kinshasa, the distant capital city. There was a police force, for instance. The police were mostly male kinsmen of the *chef de la localité.* They used their status less to uphold the law than to exploit the villagers. On the rare occasions when they toured the neighborhood, a few policemen would arrive in a village after a walk of a few hours. They never brought food. So they would find a trivial excuse for fining an unfortunate host a chicken, eat it that night, and then stay on for as many days as they could continue extracting meals. The mundane corruption was of course resented, so the police were not much respected. Even so, in theory their occasional presence, with their supposed ties to a larger state apparatus, might have tempered spontaneous expressions of anger. One could argue, therefore, that modern societal influences had reduced the level of aggression in the Ituri.

To find out whether the same gentleness holds when a group is truly independent of any governing body, we need to consider a society without a police force, a military, or any other presiding coercive institution.

New Guinea is one of the few places where small-scale societies have been documented while living in true political anarchy, free from even the remotest interference of a state. Its cultures are particularly informative because they show how people behave when living with a constant threat of being attacked by neighboring groups.

Anthropologist Karl Heider visited one such society. In March 1961, he took off in a small plane from the north coast of New Guinea, rose toward the heart of the island, came to a high mountain barrier, found a pass free of cloud, and saw the Grand Valley of the Baliem River opening up green and broad beneath. U.S. soldiers had discovered this hidden world when they crash-landed there in 1944. On finding fifty thousand Dani farmers living as if in the Stone Age, the soldiers had innocently named it Shangri-La—after the valley invented by James Hilton in his 1933 novel, *Lost Horizon,* a fictional Utopia. The peaceful appearance of the Dani's fertile land was in some ways deceptive, however. This was no paradise. It was a hotbed of war.[2]

The Dani had one of the highest killing rates ever recorded. Sometimes Heider found small groups of men setting off on raids to ambush an unsuspecting victim. Occasionally, there were battles: in the no-man's-lands between villages, minor skirmishes could dissolve into larger chaos, with up to 125 villagers killed at a time. In a macabre index of the bloodletting, fallen warriors were commemorated by the removal of a finger from girls as young as three years old; among the Dani there were hardly any women with intact hands. Heider's figures indicated that, if the rest of the world had been like the Dani, the twentieth century's sickening 100 million war deaths would have ballooned to an unthinkable 2 billion.[3]

Yet, when Heider wrote a book about the Dani, his subtitle was *Peaceful Warriors.* His phrase draws attention to the essential human paradox. Beyond the intermittent mayhem, in the calm of ordinary life, Shangri-La really was a fair name for the Grand Valley. The Dani raised their pigs and tubers with a farmer's typical steadiness. Heider wrote of the people's low-key temperament, gentle demeanor, and rarity of anger. These were pacific, caring individuals who lived in systems of mutual dependence and support. When Dani households were not lulled by easy conversation, he said, they rang with song and laughter. Restraint and respect marked their daily interactions. As long as they had no war, the Dani were in many ways ordinary rural folks leading calm, thoroughly unaggressive lives.[4]

The Dani proved typical of the remote New Guinea highlands in combining peace within the group with homicide of outsiders. Another New Guinea group, the Baktaman, occupied the headwaters of the Fly River. Every Baktaman community resisted trespass, often with violence. Territorial conflicts were so severe that they caused a third of the community's deaths. Yet, within the villages, violence was severely controlled and "killing denied to be conceivable."[5] It was the same in the basin of the Tagari River, in west central Papua New Guinea, where the Huli terrorized their enemies but had no violence within their own villages.[6] Those New Guinea peoples were rapidly changed by contact with missionaries and the state. But before governments intervened, the people showed something very important: even among people with continuous war, there was a huge distinction between "peace at home" and "war abroad."

Only a few other sites have offered the same opportunities as New Guinea for research into independent societies unaffected by a state. The anthropologist Napoleon Chagnon studied a remote population of the Yanomamö people of Venezuela for some thirty years, from the mid-1960s.[7] He found a similarly stark contrast. Despite a high rate of lethal violence in interactions among villages, within villages—even among these people, whom Chagnon characterized as "fierce"—family lives were "very tranquil" and episodes of aggression were largely regulated into formal duels.[8]

The anthropologists Kim Hill and Magdalena Hurtado documented intergroup fighting among the Aché hunter-gatherers of Paraguay shortly after an Aché group had settled into a mission station. The Aché reported having previously used their hunting weapons of bows and arrows to shoot at strangers on sight. The result was a significant death rate. But during seventeen years of working with the Aché, including treks in the forest for weeks at a time, Hill and Hurtado never observed so much as a scuffle within the group.[9]

In earlier centuries, the Age of Exploration brought European travelers into contact with independent small-scale societies in various parts of the world, including the Americas. Marc Lescarbot, a lawyer, writer, and poet, was an early example. He spent a year living with Mi'kmaq Indians in eastern Canada in 1606–1607. He was open about their perceived faults, such as gluttony, cannibalism, and cruelty to prisoners, but equally clear about their virtues. There was little fighting, he said. "As for justice, they have not any law . . . but that which Nature teacheth them—that one must not offend another. So have they quarrels very seldom." Lescarbot's observations proved highly influential in that his writing gave rise to the notion of the Noble Savage, a symbol of innate goodness that became popular in the nineteenth century in Britain. Nowadays the idea of the Noble Savage is often associated with Rousseau, but Rousseau never used the phrase; he did not think of humans as charitably as he is usually assumed to have done. Indeed, to judge from the ethnomusicologist Ter Ellingson's history of the Noble Savage concept, Rousseau's cynical views of human nature meant that he would probably not be regarded as a "Rousseauian" today![10]

Lescarbot was only one of many who were impressed by the internal peacefulness of small-scale societies. By the end of the seventeenth century, according to Gilbert Chinard, "hundreds of voyagers had noted in passing the goodness of primitive peoples." Their "goodness," however, was applied only to people of the same society.[11] In 1929, the anthropologist Maurice Davie summarized a consensus understanding that remains true today: people were as good to members of their own society as they were harsh to others.

> There are two codes of morals, two sets of mores, one for comrades inside and another for strangers outside, and both arise from the same interests. Against outsiders it is meritorious to kill, plunder, practise blood revenge, and steal women and slaves, but inside the group none of these things can be allowed because they would produce discord and weakness. The Sioux must kill a man before he can be a brave, and the Dyak before he can marry. Yet, as Tylor has said, "these Sioux among themselves hold manslaughter to be a crime unless in blood revenge; and the Dyaks punish murder. . . . Not only is slaying an enemy in open war looked on as righteous but ancient law goes on the doctrine that slaying one's own tribesmen and slaying a foreigner are crimes of quite different order."[12]

The distinction between how people behave in warfare compared with at home is all too familiar to soldiers in industrial nations. The Spanish Civil War in 1936 was typically brutal. George Orwell was a volunteer who spent his weekdays experiencing the horrors of the front lines, then returned to his wife at weekends. The change of atmosphere was "abrupt and startling." In Barcelona, only a short train-ride away from the mayhem, "fat prosperous men, elegant women, and sleek cars were everywhere." In Taragona, "the ordinary life of a smart seaside town was continuing almost undisturbed."[13]

From the Ituri Forest and New Guinea highlands to everywhere else in the world, the same pattern emerges. Whether or not their lives are consumed by war beyond their settlements, people can be strikingly peaceful when at home. My experience in the Congo seems to be the norm for our species.

From a comparative perspective, the rate of physical aggression among humans "at home" may be low, although from a moral perspective, it is still higher than most of us would wish. The evolutionary psychologist Steven Pinker, among others, has documented a decline in the probability of dying from violence within many countries over the past millennium, a trend for which we should all be grateful. Undoubtedly, life for millions of people would be more pleasant if the rate continued to decline.[14]

Nevertheless, from an evolutionary perspective, the human rate of physical aggression within social communities is already strikingly low. Because chimpanzees are one of humanity's two closest relatives, they provide an illuminating comparison. Chimpanzees are not like people. A day spent with wild chimpanzees gives you a good chance of seeing chases, sometimes some hitting, while hearing fearful screams. Every month, you are likely to see bloody wounds. The primatologists Martin Muller, Michael Wilson, and I quantified the difference between an ordinary group of chimpanzees and a particularly disturbed population of Australian Aboriginals who had recently stopped hunting and gathering. Among the Australians, social disintegration and alcohol were considered responsible for raising the likelihood of physical aggression to especially toxic levels. However, even in this comparison with an unusually violent group of humans, the chimpanzees were several hundred to a thousand times more aggressive. The difference in the frequency of fighting between humans and chimpanzees is enormous.[15]

Bonobos are the other species most closely related to humans, a similar-looking ape with a well-deserved reputation for being much more peaceful than chimpanzees. They are not unaggressive, however. A recent long-term field study found that wild male bonobos were aggressive at about half the rate of chimpanzees, while female bonobos were more frequently aggressive than female chimpanzees. So, although male bonobos are less violent than male chimpanzees, the rates of aggression in both these species of great ape are far higher than the rates among humans. Overall, physical aggression in humans happens at less than 1 percent of the frequency among either of our closest ape relatives. Compared to them, in this respect, we really are a dramatically peaceful species.[16]

The idea that humans are in general exceptionally peaceful within their home communities is an important claim that should be examined carefully. The statistics about fighting overall seem indisputable. School shootings may often be in the news in the United States, but their frequency is low compared to violence among chimpanzees and bonobos. Still, what about domestic violence?

Even within a famously pacific group such as Botswana's !Kung San hunter-gatherers (now more often called the Ju/'hoansi), domestic violence has been recorded often. Furthermore, this form of aggression may have been systematically underreported. Early voyagers and anthropologists tended to be men from patriarchal societies. Wife beating tends to happen in private and can therefore escape the attention of anthropologists. Does the frequency of men's aggression toward women undermine the proposition that humans are exceptionally nonviolent in their home community lives? With regard to male violence against females, how do humans compare with other primates?[17]

Certainly, wife beating—or, more generally, intimate-partner violence—is a common human phenomenon. In 2005, the World Health Organization's Multi-Country Study on Women's Health and Domestic Violence produced detailed data from twenty-four thousand women in ten countries.[18] Physical violence by partners included slapping, shoving, punching, kicking, dragging, beating, choking, burning, and using or threatening the use of a weapon. In cities, the proportion of women who said that they had experienced physical violence by their partners averaged 31 percent, from 13 percent in Japan to 49 percent in Peru. In rural areas, the rates were higher, averaging 41 percent. Between 50 and 80 percent of the intimate-partner violence was considered "severe." These rates appear to be slightly above those in the United States, where, in more than nine thousand detailed interviews, 24 percent of women reported severe physical violence inflicted by an intimate partner.[19] With such high reported rates, the conclusion of WHO researchers Christina Pallitto and Claudia García-Moreno is not surprising: "There is a clear need to scale up efforts across a range of sectors, both to prevent violence from happening in the first place and to provide necessary services for women experiencing violence."[20] Adding sexual violence to physical violence worsens the picture. A 2013 WHO study found that the proportion of women who had experienced either physical or sexual violence averaged 41 percent in cities and 51 percent in rural areas of their ten focal countries. The equivalent figure in the United States was 36 percent.[21]

So it is undeniable that, as deplorable as it is, violence against women is widespread the world over. Some 41 to 71 percent of women have been beaten by a man at some time during their lives. Yet this range is low compared with the incidence among our closest animal relatives. One hundred percent of wild adult female chimpanzees experience regular serious beatings from males.[22] Even among bonobos, whose females are routinely higher-ranking than males, males attack females rather often. In subgroups averaging nine individuals, the primatologist Martin Surbeck saw a male bonobo physically attacking females every six days on average.[23] If that rate had applied to the Efe foragers and Lese farmers of the Ituri Forest in the Congo, Elizabeth and I would have expected to see (or at least hear about) wife beatings several hundred times during our nine months there. But we saw none, and there were only occasional stories of beatings.

Aggressive behaviors of men toward women seem likely to be particularly prevalent in small-scale societies that celebrate the importance of men at war. Certainly, there are dramatic accounts of males being domineering and bullying toward females in societies such as the Sambia of New Guinea[24] or the Yanomamö of Venezuela,[25] both of which groups were studied in a time and place where violence between villages was a significant reality. Yet, once again, even though the rate and intensity of male violence toward females were probably as high in those contexts as in any human society, it paled in comparison to the rates among our primate kin. It is understandable why Elizabeth Marshall Thomas titled her book about the !Kung *The Harmless People,* Jean Briggs called hers about the Inuit *Never in Anger,* and Paul Malone called his book about the Penan people of Borneo *The Peaceful People.*[26]

Domestic violence is abhorrent and should always be taken seriously. It is a fact, however, that humans are less aggressive than our close relatives, even when we include the perpetual threat that men hold for women.

War is an altogether different matter. Events in the Democratic Republic of the Congo illustrate the contrast between home tranquillity within communities and violence with outsiders. Following the 1994 genocide of the Rwandan Tutsi and the arrival of Hutu militias in the Congo, the Ituri Forest became a killing ground. The Ituri peoples suffered through the First and Second Congo Wars, from 1996 to 2008. Forest life was a nightmare. Roaming military groups used their power to kill and rape ordinary villagers. In the Ituri and surrounding areas of eastern Congo, at least five million are thought to have died, and many hundreds of thousands were raped.[27]

War can vanish from a society for decades at a time, but when it starts up again, the numbers show that humans kill one another at rates higher than chimpanzees or any other primate. Among small-scale societies such as hunter-gatherers and horticulturalists, Lawrence Keeley found that the kill rate from violence between groups was not only higher than in primate populations. It was also higher than the rates recorded in Russia, Germany, France, Sweden, and Japan from 1900 to 1990, despite those countries' immense losses from two world wars.[28] Scholars debate how accurately Keeley's data represent a long-term average, but the numbers certainly show that the rate of killing people in other groups among small-scale human societies has often been unpleasantly high.[29]

High rates of killing or other forms of violence are not inevitable, and there is much variation among societies and over time. But overall tendencies are clear: compared with other primates, we practice exceptionally low levels of violence in our day-to-day lives, yet we achieve exceptionally high rates of death from violence in our wars. That discrepancy is the goodness paradox.

2

Two Types of Aggression

IN THE AFTERMATH of the Second World War, an important question was how to control excessive aggression. In 1965, the physiologist José Delgado put his life on the line to demonstrate a breakthrough in understanding. Alone in a bullring except for a bull charging straight at him, Delgado carried no red cape, no blades, and nothing else that might ordinarily be considered a defensive weapon. But he did have a radio transmitter. He had prepared for this moment with a laboratory operation. His patient was an adult male "fighting bull," a type bred for its aggression and feared by even the most courageous matadors. Delgado had implanted an electrode into its brain, placing the tip very precisely into the bull's hypothalamus, with wires leading to the surface of the bull's skull. He had made sure that he could control the electrode's activity using a radio signal.

Now was the moment of truth.

When the bull was released into the bullring, he saw Delgado, and charged at once. You can see the interaction on YouTube. The snorting animal closes fast. Delgado, seemingly crazy, stands his ground. He presses a button.

The bull stops, and Delgado walks away.

Delgado's work was part of a wave of enthusiasm for the idea that violent tendencies might be controlled through biological science. A neurobiologist working with animal aggression, he thought that experiments like those that he conducted with the fighting bull might have wider implications. He fantasized about being able to "psychocivilize" people by "the use of implantable brain electrodes that could be modulated via remote control."[1] That suggestion led nowhere, but Delgado's stunt shows that as long ago as 1965 there was a growing scientific understanding of the neural underpinnings of aggression. Since then, we have learned much more.

Aggressive behavior includes a rich and complex set of biological abilities and emotions. Some people are much more aggressive than others. How people express their aggression also varies. Some are confrontational, some are passive-aggressive, some are gossips. There is so much diversity that we might come to assume there are no simple ways meaningfully to categorize types of aggression.

Since the 1960s, however, many different scientific efforts to understand the biology of aggression have converged on the same important idea. Aggression, meaning a behavior intended to cause physical or mental harm, falls into two major types, so distinct in their function and biology that from an evolutionary viewpoint they need to be considered separately. I use the terms "proactive" and "reactive aggression," but many other word pairs connote the same sense: cold and hot, offensive and defensive, or premeditated and impulsive—all refer to the same core distinction.[2]

Reactive aggression is a response to a threat. It is the type of aggression that Delgado's bull was showing, and that most of us are very familiar with. It is readily seen in sports matches when players lose their temper with one another or with the referee. It is the type featured prominently in textbooks about animal behavior, perhaps illustrated by an account of Siamese fighting fish or rutting red deer. It is shown more by men than by women, and is associated with high levels of testosterone.[3] Mostly people do not show reactive aggression very frequently or intensely compared with other animals, but unfortunately there are exceptions. Here is a sad example.

In October 2015, on the day he died, sixteen-year-old Bailey Gwynne was at school in Aberdeen, Scotland, sharing a packet of biscuits with a group of boys. A smaller boy accepted one, ate it, and then asked for a second. Gwynne refused, called the boy a fat c—, and turned away. The frustrated boy countered with "Your mum's a fat bitch." The exchange of insults was enough. Gwynne turned back and squared up. Both young men started throwing punches. Gwynne was bigger than his antagonist, got him in a headlock, and hit him repeatedly against a wall. The smaller boy pulled out a small knife and stabbed Gwynne in the chest. Gwynne collapsed.

The killer was distraught with remorse. "That was my fault," he told the head teacher as Gwynne lay bleeding. Gwynne died a few minutes later. As he was being handcuffed, the teenager told the police, "It was just a moment of anger." "I didn't mean to," he said later, "but I stabbed him." At his trial, he was convicted of culpable homicide (a Scottish equivalent of manslaughter) and sentenced to nine years in detention.[4]

Petty insults had flashed to lethal aggression too quickly for the fighters to have second thoughts. Gwynne's death illustrates the tragic miscalculation of costs and benefits that is typical of an escalated "character contest" or "honor killing," a classic high-level form of reactive aggression. Character contests often stem from arguments in bars. Two men whose inhibitions have been loosened by alcohol start calling each other hostile names. They move outside to fight, one pulls a weapon, and suddenly a shouting match becomes serious. When the criminologist Marvin Wolfgang conducted the first major study of the reasons for murder in the United States, in 1958, he found that, during a four-year span in Philadelphia, character contests were responsible for 35 percent of the city's homicides, the largest category of any type of murder. Similar frequencies have been found elsewhere.[5]

Reactive aggression is variously described as hostile, angry, impulsive, affective, or "hot." It always comes with anger, and often with a loss of control, such as losing one's temper. It is a response to provocations such as a perceived insult, embarrassment, physical danger, or mere frustration. In the state of intense arousal that is typical of reactive aggression, a fighter easily lashes out at anyone around him or her. Reactive aggressors have no goal beyond getting rid of the provoking stimulus, which is often the person responsible for an insult.[6]

Just as some people are more prone to reactive aggression than others, species also vary. Most, such as chimpanzees or wolves, are more prone to reactive aggression than humans are. The pattern has been well studied in animals. Reactive aggression is especially prominent among males fighting over status or mating rights. Normally, fights among animals end without injury, but if the stakes are high the competition can be severe. During a study of rut fighting among male pronghorn antelope, 12 percent of conflicts over mating rights to estrous females led to the death of one or both males.[7] In various populations of red deer, 13 to 29 percent of male deaths came from rut fighting. There are many similar statistics. They suggest that, if men were as ready to commit reactive aggression as those rutting ungulates, the annual death toll from character contests among U.S. men would rise from its current estimate of fewer than ten thousand to more than one hundred thousand.[8]

Compare reactive aggression with violence that is coolly planned, such as David Heiss's behavior in killing Matthew Pyke. Heiss lived near Frankfurt, Germany, far from Pyke's home in Nottingham, England. Both were in their early twenties when they met online in 2007 on a gaming forum run by Pyke's girlfriend, Joanna Witton. Heiss developed a crush on Joanna. He decided he had to meet her. In 2008, Heiss arrived without warning at Joanna and Pyke's Nottingham apartment to declare his passion. Unfortunately for Heiss, she wanted nothing to do with him. But Heiss was not to be put off. He stayed in England for a month, leaving love notes and stalking her.

Heiss returned to Germany but after a few weeks his obsession led him back to Nottingham. Joanna again rejected him, and again he left. By September 2008, Joanna had announced that she planned to marry Pyke. That was the trigger. Heiss left Germany for Nottingham once more, this time having prepared an alibi, forged a suicide letter from Pyke, and armed himself with a knife. He approached the couple's apartment, watched Joanna leave for work, and rang the doorbell. When Pyke answered, Heiss immediately attacked him. He inflicted eighty-six wounds. Pyke wrote the name of his killer in his own blood as he lay dying. As a strategy to win a mate, Heiss's behavior was a pathetic failure. Heiss was sentenced to a minimum of eighteen years in prison. But as a strategy to remove his rival, it worked.[9]

Heiss's actions are a classic example of proactive aggression. Proactive aggression is also characterized as premeditated, predatory, instrumental, or "cold." Unlike reactive aggression, it involves a purposeful attack with an external or internal reward as a goal, rather than an effort to remove a source of fear or threat. It is the calculated act of the professional assassin, as seen in the purposeful flying of an airplane into a populated building, the deliberate driving of a hired truck into a crowd of innocents, or the carefully planned activities of a typical school-shooter. It need not involve outward anger or other expressed emotions at the time of the act, though of course emotions are involved in deciding what to do—in fact, as we will see, emotions tend to be particularly strong in proactive murderers, to judge from their brain activity.

The goals of premeditated violence may be something concrete, like money or power or a mate, or more abstract, like revenge, self-defense, or merely the keeping of a promise. Much of what humans do in war is premeditated, for example, such as raiding in a surprise attack. The high frequency of war deaths tells us that humans, like chimpanzees, exhibit high levels of premeditated aggression compared with most species. We are excellent planners, hunters, raiders, and, when we want to be, killers. The anthropologist Sarah Hrdy noted that to pack hundreds of chimpanzees into close quarters on an airplane would be to invite violent chaos, whereas most human passengers behave sedately even when they are crowded. As Dale Peterson observed, however, intense screening is needed to ensure that a secret enemy will not carry a bomb on board. The contrast illustrates the difference between our low propensity for reactive aggression and our high propensity for proactive aggression.[10]

When a violent action is criminal, the perpetrator may or may not be insane. In Heiss's deluded mind, the goal of killing Pyke appears to have been to win Joanna Witton, or perhaps to punish her for her wrong choice. Normally, however, the actor is not legally insane. This makes sense because proactive aggression involves various high-level cognitive abilities, including devising and conforming to a purposeful plan and focusing attention on a consistent target. The behavior is self-rewarding rather than serving to remove an aversive stimulus: the killer is pleased at achieving a goal. Proactive aggression can be triggered by a wide variety of motivating factors, including desire for money, power, control, or sadistic fantasy, some of which will seem extreme to ordinary people.[11] Yet successful aggressors initiate action only when they perceive they are likely to achieve their goals at appropriately low cost.[12]

Individuals who have a greater tendency for proactive aggression than others have characteristic social emotions. They tend to have reduced emotional sensitivity, feel less empathy for their victims, and experience less remorse over their actions.

The proportion of murders (as opposed to war deaths) that is proactive or reactive has not been well characterized but overall reactive murders are found to be more common. Whereas Marvin Wolfgang had found that 35 percent of Philadelphia murders resulted from character contests, he and criminologist Franco Ferracuti concluded that "probably less than 5 per cent of all known killings are premeditated, planned and intentional."[13] Those percentages sum to only 40 percent, leaving it unclear how many of the remaining 60 percent of homicides were proactive or reactive. The unclassified homicides fell into categories such as "domestic quarrel" (14 percent of homicides), "jealousy" (12 percent), and "money" (11 percent).[14] Certainly some of the remaining 60 percent were proactive since they included revenge. Criminologist Fiona Brookman reported revenge as a motivation in 34 percent of British homicides in which both killers and victims were men, and since revenge always includes a component of planning, it can be considered proactive.[15] A second reason for thinking that the 5 percent figure is too low is that because proactive murderers have had time to plan events carefully, they presumably get away with their crimes relatively often. The proportion of unsolved murders can be high. Data from the U.S. Federal Bureau of Investigation suggest that at least 35 percent of murderers in the U.S. are never brought to justice. For these reasons proactive killing probably occurs at a higher frequency than 5 percent of murders.[16]

Even so evolutionary scientist Johan van der Dennen found that reactive, not proactive, aggression is responsible for most homicides. A survey of crime in seventeen U.S. cities supported Wolfgang's findings from Philadelphia by attributing most killings to trivial disagreements: "Altercations appeared to be the primary motivating forces both here and in previous studies."[17] Character contests were again frequent. In the words of a Dallas homicide detective: "Tempers flare. A fight starts, and somebody gets stabbed or shot. I've worked on cases where the principals had been arguing over a ten-cent record on a juke box, or over a one-dollar gambling debt from a dice game." The greater frequency of altercations than planned attacks does not appear to result merely from the presence of modern weapons such as guns: in thirteenth- and fourteenth-century Oxford, England, as van der Dennen noted, most homicides were again found to be spontaneous.[18]

Proactive and reactive killings are explained in different ways. Because proactive homicides are deliberate, they are more easily understood. As illustrated by the case of David Heiss, the proactive murderer's goal is preordained and makes sense to him or her even if they are deluded about it being a good idea. Reactive killing is harder to explain because the intensity of the fight is often out of all proportion to the provocation, and the killing is often accidental. The killer is typically remorseful, and is regularly caught and punished, as the young killer of Bailey Gwynne was. Evolutionary psychologists Margo Wilson and Martin Daly argued that most killings that result from trivial altercations reflect a drive to maintain status, a drive that would have been adaptive in a world with no alcohol and less effective weapons, but that today is no longer adaptive because it leads to the aggressor's becoming homicidal.[19] Criminologists Kenneth Polk and Fiona Brookman argue that fights over status are especially frequent among lower-working-class and underclass men because their material resources are scarce, making honor all the more important. Daly and Wilson similarly show that reactive aggression is frequent in populations where incomes are highly unequal.[20] Reactive aggression has also been argued to be relatively easily provoked in "cultures of honor," where honor is valued highly for cultural reasons, such as in the American South.[21] While reactive aggression is thus subject to various economic and cultural influences, both killer and victim tend to lose, suggesting that these murders normally just happen to get out of hand: they are assaults "gone wrong." Even when the fatal outcome is an accident, however, the intensity of these reactive fights reminds us of the perceived importance of honor or respect.

—

Reactive and proactive aggression differ not only in their frequency and explanation but also in how they are viewed by the public and the law. Because proactive aggression involves a deliberate choice, we tend to judge those who are guilty of it more harshly than those who commit acts of reactive aggression. Think of William Penn, a celebrated Quaker. When Penn founded Pennsylvania in 1682, he was a pacifist averse to the death penalty. But even to his sympathetic mind, proactive homicide deserved the ultimate fate. According to his Pennsylvania statutes of 1682 and 1683:

> If any person . . . Shall wilfully or premeditatively kill another person . . . Such person Shall, according to the Law of God Suffer Death.[22]

Cold planning makes the killing heinous. When the statute was discussed in 1705, Penn's attorney general insisted that murders that did not involve premeditation were more excusable. The death penalty should apply, he thought, only to homicides that were planned:

> The Act against murder, whereby whoever shall wilfully or praemeditatively kill another person . . . shall suffer death, I think it unreasonable, for that willfull killing may be in a sudden affray, therefore it should not be wilfully or praemeditatively, but wilfully and praemeditatively.[23]

Queen Anne supported the notion, and the law was duly enacted. For a time, murders earned the death penalty only if they were premeditated.

Killing that resulted from an act of passion was more forgivable. A charge of murder could be reduced to one of voluntary manslaughter if the act involved a loss of self-control brought about by a "reasonable" provocation, such as finding a spouse committing adultery, or learning that one's child was being sexually abused. Reactive violence after such a provocation came to be viewed so sympathetically that in extreme cases the guilty party could be let off altogether.

Take the case of Eadweard Muybridge. In 1874, Muybridge was a well-known photographer who had recently married a woman twenty-one years younger than himself, Flora Shallcross Stone. Among his many pioneering achievements, Muybridge invented a system for photographing the motion of animals, such as a running horse. He was often away on assignment, and during those times a dashing young drama critic, Major Harry Larkyns, sometimes escorted Flora. One day, Muybridge was in the home of Flora's midwife when he saw a photograph of their baby son. He turned it over and saw the words "Little Harry" in Flora's handwriting. Muybridge exploded in anger and pressured the midwife to tell him more. She reluctantly showed him love letters from Flora to Larkyns.

Next day, Muybridge made plans. He first organized his business affairs with his professional associate. He then made his way from San Francisco by ferry, train, and an eight-mile horse-and-buggy ride to a ranch in Napa Valley where Larkyns was staying. Muybridge knocked at the door and called for Larkyns. When Larkyns arrived, Muybridge is said to have announced: "Good evening, Major, my name is Muybridge and here's the answer to the letter you sent my wife." He fired at Larkyns with a single shot from his Smith & Wesson No. 2 six-shooter. Larkyns died at once. At the trial, Muybridge pleaded insanity, but his own testimony showed so clearly the deliberate nature of his actions that the judge was in no doubt of the appropriate verdict. Stating that the crime was premeditated and the accused was sane, the judge told the jury to convict Muybridge of murder.

The jury, however, ignored the judge's instructions. They considered Muybridge's violence to be the uncontrollable result of strong emotions produced by the supposition of his wife's adultery, and, that being so, they found him guilty only of justifiable homicide. He was immediately set free. Muybridge was roundly cheered as he left the court. Reactive aggression, as the public judged Muybridge's act to have been, could be forgiven. Muybridge was lucky to have lived when he did: the judicial system is less generous now. He was reportedly the last beneficiary in California of a finding of justifiable homicide.[24]

The conflict between the judge's and the jury's opinions in the Muybridge case shows that it can be difficult to decide whether a physical attack is proactive or reactive. Currently, the U.S. legal system applies four criteria for a homicide to be judged as voluntary manslaughter (which conforms to reactive aggression) rather than murder (usually proactive aggression), but it allows various interpretations.

> (1) There must have been a reasonable provocation.
> (2) The defendant must have in fact been provoked.
> (3) A reasonable [person] so provoked would not have cooled off in the interval of time between the provocation and the delivery of the fatal blow.
> (4) The defendant must not in fact have cooled off during that interval.[25]

The words may be clear, but their meaning relies on subjective decisions. What makes a provocation "reasonable"? Some people would think that the discovery that one's spouse has been adulterous is a reasonable provocation for killing the rival, as Muybridge's jury did. Others would disagree. And how short must the interval be between the provocation and the killing, so that the killer has not cooled off? As the psychologists Brad Bushman and Craig Anderson reported, some U.S. states consider a murder to be premeditated "if the killer thinks about the act ahead of time even for a 'matter of seconds.'" Thus a rape victim who killed her assailant during the attack would be regarded as more forgivable than one who did so one minute later. A victim who killed a day later would likely be considered to have acted with premeditation. Her violence would therefore sustain a heavier punishment, given that it was judged more deliberate. The law may long have appreciated the greater role of free will in premeditated violence, but it has not found a way to define a universally acceptable definition separating murder from manslaughter—or, in the terms discussed here, proactive from reactive violence.[26]

Although the law and the public have long recognized the important distinction between acts of premeditation and responses to provocation, the difficulties of drawing a sharp boundary may have contributed to the idea that aggressiveness falls along a single scale, from low to high. It has taken scientific approaches from several directions to pin down the differences between proactive and reactive aggression. In the middle years of the twentieth century, studies of child development, criminology, clinical psychology, and animal behavior were all developing a version of the proactive-reactive division. By 1993, when the psychologist Leonard Berkowitz summarized the phenomenon in a book called *Aggression: Its Causes, Consequences, and Control,* the dual nature of aggression had become clear.[27]

Berkowitz labeled the types of aggression "reactive" versus "instrumental," and applied them to all categories of conflict, not just to homicide. The terms "reactive" and "proactive" were first paired in the study of children in the 1980s. Reactive aggression is an immediate response to an imminent threat that involves anger, fear, or both. It starts with arousal of the sympathetic nervous system, producing the fight-or-flight response: adrenaline is released, the heart beats faster, glucose is mobilized, pupils dilate, the mouth dries up, and nonessential processes such as digestion are inhibited. In contrast, proactive violence is not immediately preceded by any equivalent arousal. There is no instantaneous threat to respond to. Proactive aggression is characterized by the presence of a deliberate plan and frequently by an absence of emotion at the time of the assault.[28]

The distinction between proactive and reactive aggression is useful forensically for understanding behavior from children's aggression through the crimes of homicide (including sexual homicides and mass murder), stalking, and domestic violence. The psychologist Reid Meloy reports that the majority of spousal batterers are readily classified as either a predatory (proactive) or an impulsive (reactive) type. Predatory batterers are more violent in general, more concerned to dominate and control their partners, and more likely to be violent, particularly when the spouses talk back. In contrast, it is when spouses try to withdraw from an argument that impulsive batterers are most likely to lose control. Such distinctions help indicate risk factors for physical danger, improving the ability to identify likely repeat offenders, or pointing to appropriate medication for regulating aggression.[29]

Biological mechanisms are thus important for specifying differences between reactive and proactive aggression. A main focus of research on the biological bases of aggression has been on murderers. In 1994, the neurocriminologist Adrian Raine led the first study assessing whether the brain activity of convicted murderers vary according to whether the crime was proactive or reactive. Raine was impressed by the personality differences among murderers. Randy Kraft was a computer consultant with an IQ of 129 who picked up young men for sex in the twelve years up to 1983. He was so controlled and careful in how he drugged his victims and disposed of their bodies that he is believed to have killed at least sixty-four times before being caught by chance for driving under the influence. Kraft falls into the "proactive"-killer category. Antonio Bustamante was an impulsive small-time criminal who was taken by surprise during a burglary and beat an eighty-year-old man to death with his fists. Bustamante was disorganized and ineffective. When he tried to cash the traveler's checks that he had stolen, the old man's blood was still on them, and when he was arrested, his clothes were also bloody. The unplanned murder that Bustamante committed was obviously a crime of reactive aggression.[30]

Raine's interest in brain difference of proactive and reactive murderers focused on the role of the prefrontal cortex. The cerebral cortex is a thin layer of tissue three millimeters (0.12 inches) thick that covers the surface of the brain, including the many folds lying between ridges. It is involved in higher cognitive functions, such as thought and consciousness. The part of the cortex lying at the front of the brain, above one's eyes, is called the prefrontal segment. The prefrontal cortex is particularly responsible for the control of emotions—in other words, the inhibition of their emotional expression. Reactive aggression can be considered a failure of control (or inhibition) of emotions like fear and anger. Raine asked himself a straightforward question. Do impulsive (reactive) murderers, who tend to have little control of their emotions, tend to have less neural activity in their prefrontal cortex than other people? He speculated that this would be the case.[31]

Raine conducted his studies in California prisons using PET scans, which measure the rate at which different parts of the brain use glucose—basically, how hard those parts of the brain are working. He scanned the brains of forty men accused of a murder (even though some had not been found guilty at the time of the study). To characterize the accused as proactive or reactive aggressors, two members of Raine's team examined each man's criminal history, psychological and psychiatric evaluations, interviews with lawyers, newspaper reports, and medical records.

In one way the reactive and proactive murderers proved indistinguishable. Compared with nonmurderers, all of the accused had high neural activity in subcortical parts of the brain, including the limbic system, which is a brain network that processes emotional responses. This finding suggests that all of the accused murderers tend to experience particularly strong emotions. Just as Raine had anticipated, however, the brains of the accused murderers differed, depending on whether their murderous acts were characterized as reactive or proactive. The reactive murderers had less activity in their prefrontal cortex, the inhibitory part of the brain. The difference contributes to an explanation of why some people are more vulnerable to committing a crime of impulsive violence: they find it hard to control themselves.

Raine's data were collected long after the murders had happened. That meant that the brain activity differences that he found could not be attributed to the excitement of the murderous moment. Instead, the observed levels of brain activity were characteristic of those individuals. Some individuals are simply more emotionally reactive than others.

Subsequent research has refined our understanding of cortical control of emotional impulses by using information from people diagnosed as psychopaths. Unlike impulsiveness, which is more associated with reactive aggression, psychopathy is associated more with proactive aggression. So psychopaths provide an opportunity to understand some of the characteristics that predispose people to proactive aggression.[32]

Psychopathy is found worldwide. According to a standard rating scale devised by criminal psychologist Robert Hare, psychopaths tend to exhibit twenty features, including superficial charm, frequent lying, sexual promiscuity, and a low threshold for boredom. They are insensitive both to what others think and to what others feel. This can work in their favor at least in the short term, because despite their arrogance, ambition, and readiness to be deceptive, their confidence can make them attractive. They show less evidence of empathy than ordinary people, and they tend to feel less guilt or remorse. This lack of concern makes them relatively more liable to be aggressive. Finally psychopaths are prone to try to get what they want, regardless of the means necessary. In sum, psychopaths are self-centered and uncaring people with impaired moral judgments. Not surprisingly, they are likely to be criminally delinquent, and they are also predominantly male.[33]

A survey in the U.K. found psychopathy in less than 1 percent of the household population, which is probably the approximate proportion worldwide. Psychopathy is more common among men and younger adults than among women and the middle-aged or elderly. Psychopaths show more violent behavior than others. In the U.K., psychopathy was also associated with suicide attempts, imprisonment, drug dependence, antisocial personality disorder, and homelessness. Of the many features contributing to psychopathy, a lack of conscience is found to be particularly important.[34]

Stepping back from psychopaths for a moment, we can look across species to garner insights into the function of a brain area that distinguishes psychopaths from other humans. The limbic system is a subcortical series of small structures deep in the brain that are connected with one another and are heavily involved in the production of emotional responses such as anger, anxiety, fear, and pleasure. In keeping with their stronger emotional reactions, wild mammals tend to have a larger limbic system than domesticated mammals. A well-studied part of the limbic system is the amygdala, a pair of almond-sized regions. An amygdala bigger than normal is associated with more fearful and aggressive reactions across individuals, and larger amygdalae are more typical of a wild than a domestic animal.[35]

Psychopaths appear particularly fearless, a trait that seems to be supported by their tendency to have an amygdala that brain imaging reveals as smaller, sometimes deformed, and less active than in other people. Low amygdala activity is particularly notable when psychopaths are engaged in behaviors such as moral decision-making, fear recognition, and social cooperation: psychopaths tend to have relatively low emotional responses to circumstances that lead most people to feel empathy or fear. Low fear and low empathy both support proactive aggression. Thus reduced amygdala activity may underlie some individuals' low fear and empathy and contribute to an explanation of those individuals' readiness to engage in proactive aggression.[36]

The neurobiology of proactive aggression has not been intensely studied in humans because of the difficulty of performing ethically appropriate experiments on the brain, but a recent approach suggests fascinating opportunities. A team led by neurobiologist Franziska Dambacher found that, somewhat like Delgado's experiment with bulls, they were able to reduce aggression in men. Instead of reactive aggression, the aggression that Dambacher's team reduced was proactive. Happily, Dambacher's method did not require brain surgery. The team stimulated neural activity in a particular part of the prefrontal cortex (right dorsolateral) using a novel method called anodal transcranial direct current stimulation (tDCS). They assessed proactive aggressiveness by allowing the subjects to blast noise at a supposed competitor, and then measured the volume and duration of these noises. Their experiment found that in men (but not women) proactive aggression was predictably reduced by tDCS.[37]

It is no surprise that differences in aggressive behavior are associated with differences in neural activity. Overall the evidence shows what we would expect of the amygdala and prefrontal cortex. Part of the amygdala's function is involved in feeling negative emotions (such as fear); in psychopaths who tend to commit proactive aggression, the amygdala underperforms. The prefrontal cortex is involved in controlling impulses, processing rewards and punishments, and planning; in people likely to commit reactive aggression, the prefrontal cortex underperforms. Studies of anatomy and brain activity in both the amygdala and the prefrontal cortex are still at a relatively early stage but they have given us a start in understanding the distinctive neurobiology of the two types of aggression.

The better we understand the biological bases of reactive and proactive aggression in humans, the better chance we have of reducing aggression. The neural circuits in the prefrontal cortex that regulate reactive aggression are enhanced by a high turnover of the neurotransmitter chemical called serotonin. Individuals with low concentrations of brain serotonin are therefore more liable to impulsive violence. As a result, psychiatric patients with a history of excess reactive aggression can be helped by taking "selective serotonin reuptake inhibitors" (SSRIs), drugs that increase serotonin concentration.[38] By contrast, no successful psychopharmacological interventions have been found to influence proactive aggression in humans.[39]

The regulatory action of serotonin depends not only on its concentration but also on the density of the relevant kind of serotonin receptors. People with a high level of impulsivity (who are therefore liable to reactive aggression) tend to have unusually high densities of a particular kind of receptor (the 5-HT_{1A} receptor) in parts of the prefrontal cortex associated with impulse control. Sex steroids (such as androgens and estrogens) also regulate the serotonin system. Men with low brain serotonin are more likely to be aggressive if they produce a high ratio of testosterone to the stress hormone, cortisol. Women show changes in the distribution of 5-HT_{1A} receptors associated with changes in levels of circulating hormones across phases of the menstrual cycle. Severe cases of premenstrual syndrome, which can involve increased irritability and aggressiveness, can be helped by SSRIs. Again, these pharmacological interventions that affect serotonin levels reduce reactive, but not proactive, aggression.[40]

Evidence from humans indicates how proactive and reactive aggression are organized differently in the brain but does not reveal the fine details. Animal studies are needed for that; they reveal the specific neural pathways that control proactive and reactive aggression.

The first hint that animal aggression falls into two distinct types that are based on subtle differences in brain activity came before World War II. Cats whose hypothalamus was stimulated in slightly different locations were found to produce predictably different behaviors. The hypothalamus is a small part of the nervous system near the base of the brain that influences hormone production in the entire body, through its physical connections to a small gland just outside the brain, the pituitary. Researchers found that the behavioral response to stimulation of the hypothalamus with an electrode depends on precisely where the electrode reached. Stimulating one location within the hypothalamus led to a "quiet biting attack," directed at mice present in the same cage, which, as detailed below, is a form of proactive aggression. Stimulating another location in the hypothalamus produced a form of reactive aggression called "defensive aggression," which was expressed toward other cats or the human experimenters.

"Quiet biting attacks" in cats were correctly regarded as feeding behaviors: they were part of a hunting sequence. For this reason, the different behaviors produced by stimulation of different areas of the hypothalamus were not thought of as alternative types of aggression. Researchers regarded the difference as a contrast simply between feeding and fighting.[41]

The view that "quiet biting attacks" were limited to feeding behaviors changed, however, after parallel studies were conducted on rodents. Rats and mice were key experimental animals because, in both animals, individuals sometimes show proactive aggression toward members of their own species. Rats have been the principal study species. They stalk and attack and sometimes even kill other rats. The same area of the hypothalamus proved to be involved in the control of "quiet biting attack" aggression in rats as it was in cats. But whereas the cats directed their attacks at prey (the mice), rats sometimes directed the attack at other rats; this time the "quiet biting attack" was not classified as a feeding behavior. Given that it was directed at members of their own species, "quiet biting attacks" were rightly judged to be proactive aggression.

These animal studies reveal some details of the neurobiological bases of proactive and reactive aggression. In both cats and rats, the specific area of stimulation within the hypothalamus determined what kind of aggressive behavior was produced. Stimulation of the mediobasal part of the hypothalamus produced "defensive" reactive aggression; activation of the lateral part of the hypothalamus produced "quiet biting attack" proactive aggression. This was a remarkable and surprising breakthrough. A minuscule shift in the location of an electrode produced a radical difference in the kind of aggression expressed; and the differences were closely congruent among very distantly related species of mammals, namely cats and rodents.

A similar contrast emerged in another deep region of the brain. The periaqueductal gray is a control center at the base. Activation of the dorsal side of the periaqueductal gray produced reactive aggression; activation of the ventral side produced proactive aggression.

What is the relationship between reactive and proactive aggression? Do they work in concert, such that increasing one tends to increase the other (mutual facilitation)? Or do they work in opposition, so that expressing one kind of aggression suppresses expressions of the other type (reciprocal inhibition)? Addressing the relationship between proactive and reactive aggression within an individual provides insights into the evolutionary function of the two types of aggressive behavior. Probing such questions, a team led by Hungarian neuroscientist Jozsef Haller found a provocative difference between cats and rats.

In cats, the two hypothalamic regions have connections that appear to allow for reciprocal inhibition—that is, increasing activity in one of the areas suppresses activity in the other. As we have seen, "quiet biting attacks" in cats are a hunting (feeding) behavior, whereas in rats the same behavior is directed against other rats. Haller's group proposed that when a cat is fighting (reactive aggression), the nerves coming from the mediobasal part of the hypothalamus inhibit firing of nerves in the lateral part of the hypothalamus, suppressing "quiet biting attack" aggression. As a result, the cat cannot fight and hunt simultaneously—a useful adaptation to avoid the confusion of trying to do two incompatible things (fighting and feeding) at the same time. In rats, by contrast, connections between the mediobasal hypothalamus and lateral hypothalamus are minimal, so there is less inhibition of one type of aggression by expression of the other type. This means that if a rat initiates a premeditated, proactive "quiet biting attack" but finds the victim fighting back, the attacker can instantly respond with reactive aggression. Thus in rats, the absence of reciprocal inhibition allows proactive and reactive aggression to occur at the same time without one inhibiting the other.[42]

When a planned attack turns into a fight, human aggressors would benefit from an ability to readily adapt by producing reactive "defensive" aggression. Accordingly a similar lack of reciprocal inhibition seems likely in humans, more similar to rats rather than cats. If Haller's proposal applies to humans, I would expect that our species will prove to have few nerve connections between the mediobasal and lateral hypothalamus, given that, when a planned attack turns into a fight, human aggressors readily adapt by producing reactive aggression.

The animal evidence that reactive and proactive aggression are produced by different neural pathways indicates how different species might be adapted toward greater or lesser tendencies to produce each type of aggression. In a parallel way, studies of humans show differences in brain activity that suggest why some individuals are more or less likely to commit acts of proactive or reactive aggression.

Not all of the biological underpinnings of aggression result from genetic makeup; accidents during life can be important for the individual. Eadweard Muybridge shows how a propensity for reactive aggression can be increased by an event that has taken place relatively early in life, well before the committing of a crime. In 1860, when he was thirty, Muybridge and seven other passengers were in a carriage traveling down a mountainside in Texas, and the driver lost control of his horses. They crashed into a tree at speed. The coach was smashed to pieces, one man was killed, and everyone was injured. Muybridge landed on his head. He lost consciousness and was later unable to recall the accident. He experienced double vision, loss of taste, and loss of smell, symptoms consistent with damage to the prefrontal cortex. He took months to recover.

At Muybridge's murder trial, fifteen years later, a series of witnesses testified that his personality had changed severely following his accident. He had become eccentric and irritable. He was so socially disinhibited that he posed nude in front of his camera, and he exhibited such uncontrolled outbursts of emotion that they were written up in newspapers. The neuropsychologist Arthur Shimamura concluded from Muybridge's symptoms that his excessive emotionality arose from damage to the orbitofrontal cortex, a part of the prefrontal cortex involved in decision making. His particular history of neural injury had apparently left him exceptionally unable to control his impulses, including those for reactive aggression.[43]

Studies of twins provide the critical information about genetic influence. The key is that identical (monozygotic) twins share 100 percent of their genetic material, whereas fraternal (dizygotic) twins share only 50 percent, like any pair of siblings. So if monozygotic twins are more similar than dizygotic twins in some feature, their greater similarity is due to their being more similar genetically. The strength of genetic influence on a feature can be assessed by how much more similar to each other monzygotic twins are compared with dizygotic twins.

Environmental influences are not easy to account for, because "environment" is a broad concept that can include social responses as well as the physical world. A particularly tricky problem is that when identical twins live together, the fact that they look alike can encourage other people to respond to them in very similar ways. Identical twins who live together can therefore experience environments that are more similar to each other's than the environments experienced by fraternal twins who do not look so alike. For this reason, the best research is restricted to the relatively rare cases of twins who are separated from each other early in life and grow up in different families. The twenty-year Minnesota Twin Family Study, which collected data from 1936 to 1955, followed by decades of publication, led the way by being particularly effective at finding that kind of rare cases: twins reared apart. The researchers found that gene similarities influenced many characteristics, from intelligence, religiosity, and happiness to the way children hold their bodies when they stand.[44]

A 2015 survey of all available twin studies of aggressiveness, of which there had been forty in the preceding five years alone, found that the genetic heritability of aggressive behavior was typically in the range of 39 to 60 percent, averaging 50 percent. This means that, in those environments, genetic and socialization influences were roughly equally important in shaping individuals' aggression. Interestingly, the same does not apply to some closely similar behaviors, such as rule breaking. Differences among children in the tendency to break rules, as opposed to being aggressive to one another, have been found to come almost entirely from socialization.[45]

Researchers have only occasionally considered proactive and reactive aggression separately. Most studies of aggression have been conducted with boys. To measure aggression, researchers ask parents, teachers, and/or the boys themselves to complete a questionnaire. Acts of aggression judged to be proactive include "he threatens and bullies other kids" or "he damages or breaks things for fun." Examples of reactive aggression are "he damages things when he is mad" or "he gets mad or hits others when they tease him."[46]

A recent study compared twins in 254 monozygotic (identical) pairs and 413 dizygotic (fraternal) pairs who had lived in the same family from birth to twelve years of age. In this case, children's aggressiveness was rated by their teachers. Genetic factors accounted for 39 to 45 percent of the variance in proactive aggression scores, and for 27 to 42 percent of the variance in reactive aggression scores. This study's conclusion that genetic heritability was somewhat higher for proactive than reactive aggression is the latest of several that have found similar results. Based on this early work, proactive aggression might therefore prove to be more strongly influenced by genes than reactive aggression, but for the moment all we can say is that both types show important genetic influences.[47]

Twin studies show the strength of genetic heritability, but they do not identify which genes are important. For the most part, despite extensive research effort, we know little about the influences of specific genes for aggression. This is not surprising. Genetic contributions have their effects through multiple biological systems, such as the stress response, the anxiety circuit, the serotonin-neurotransmitter pathway, and the dynamics of sexual differentiation. Hundreds or thousands of genes can influence complex patterns of behavior. To isolate the effect of just one of the roughly twenty thousand genes in the human genome is very challenging, because it requires immense sample sizes, normally tens of thousands of individuals. Even if researchers get access to large numbers of genotypes, they would find it hard to characterize aggressive tendencies in a systematic way in so many people.[48]

Nevertheless, a few useful clues have been generated, such as how genetic factors influence reactive aggression via their effect on serotonin activity. A classic case is the MAOA gene, which lies on the X chromosome, a sex chromosome. Since boys have only one X chromosome, they have only one variant of the gene. This means that the effects of a rare variant are never masked by a counterpart, as they normally are in females (since they have two X chromosomes), or as they would be if the gene were on a nonsex chromosome. Furthermore, aggression in boys is easier to measure than it is in girls, since boys are more obvious about it. Most studies have therefore focused on boys.

The normal MAOA gene encodes an enzyme called monoamine oxidase A, which degrades serotonin and two other neurotransmitters, dopamine and norepinephrine. A family of variant genes is called the low-activity MAOAs (abbreviated to MAOA-L). The enzymes produced by these variants are relatively inefficient at degrading neurotransmitters. The variant forms of the gene therefore interfere with normal serotonin metabolism. Disruption to the serotonin system is expected to cause individuals to lose emotional control more easily, take more risks, and be more reactively aggressive.

A 2014 survey of thirty-one studies found a small but consistent tendency among males who carry MAOA-L to express antisocial behavior at relatively high rates. Something similar has even been detected experimentally. The political scientist Rose McDermott and her colleagues tested whether the presence of this gene would increase the amount of unpleasantly hot chile sauce that subjects would administer to those who had apparently wronged them; it did, and more so when the subjects were more strongly provoked.[49]

As a result of such studies, MAOA-L has sometimes been called the warrior gene. The nickname is unfortunate, because many people who carry the gene show no tendency to be more aggressive than the rest of the population. And even in those who have the gene, its effects interact with experience. Thus, the typical effects of MAOA-L on antisocial behavior are more predictable when childhood experience is taken into account: an adult carrier of MAOA-L is more likely to express violent behavior if he was physically maltreated when young. By contrast, there is no indication that MAOA-L is associated with tendencies for proactive aggression or psychopathy.

The interaction between child abuse and the MAOA-L gene is a reminder that genetic influences never take place in a vacuum. The environment in which a young human grows up is likely to affect all genetic influences on behavior. Mostly, individual gene differences are only weakly predictive.

Comparable cautions apply to brain activity. After Adrian Raine discovered differences in the brain activity of men accused of proactive and reactive murders, he had his own brain assessed. His PET scans placed him closer to the proactive murderers and psychopaths than to the comparison group of men not accused of murder. The result intrigued him. "When you have a brain scan that looks like a serial killer's it does give you pause," he said. He reflected on other similarities he shared with psychopaths, such as having a low heart rate. He decided he was lucky to have been nudged into his path as a researcher. He could easily have been a criminal. Genes can influence behavior; they rarely determine it.[50]

Aggressive behavior is influenced by genes; proactive and reactive aggression are controlled by different neural pathways and have different scores of heritability; and certain genes promote reactive but not proactive aggression. In the future, we can expect that twin studies, adoption studies, and studies of the genes themselves will increasingly specify separate risk factors for reactive and proactive aggression. For the moment, we can say that the two kinds of aggressive behavior represent sufficiently contrasting emotional and cognitive reactions that they are subject to different biological underpinnings.

Reactive and proactive aggression are therefore expected to be capable of evolving separately from each other. One Russian experiment found exactly that result with rats. Selection of Norway rats for tameness toward people led to a reduction in reactive aggression, supported by elevated levels of serotonin. There was no change, however, in the rats' propensity for proactive aggression.[51]

Given that humans, compared with other species, are low on reactive aggression and high on proactive aggression, the question is why we have this mixture. We start our search for an answer by addressing our low propensity for reactive aggression. Among animals that are (like humans) unusually docile, many are domesticated. We need to find out what happens when a species is domesticated.

3

Human Domestication

DOMESTICATION IS NOT the same as tameness. A wild animal can sometimes be tamed, but that does not make it domesticated. Raymond Coppinger could have told you that.

Coppinger was a dog-sled racer and research biologist who bred dogs and understood them as well as anyone in the world. In 2000, his friend Erich Klinghammer, director of Wolf Park, Indiana, invited him into the cage of a captive wolf. Coppinger hesitated. "I don't know much about tamed wolves," he said. Klinghammer reassured him. His wolves were progeny from several generations of captive breeding, far removed from the wild. They had been hand-raised by human "puppy parents" since they were ten days old. Even as adults, they were still handled every day. They were used to being on leashes, and were as tame as wolves could be. "Just treat them like dogs," Klinghammer said.

So Coppinger did. He joined Klinghammer and the wolves inside their pen. Saying something like "Good wolf" to an adult female called Cassi, he patted her on her side.

In Coppinger's words:

> That was when she became all teeth. Not a nip, but a full war—a test of my ability to stay on my feet and respond to Erich's excited command, "*Get out, get out!* They'll *kill* you!" Note the wording "They'll kill you!" I had a blurred vision of a collection of wolves

gathering, and a wolf tugging on my pants as Cassi focused on my left arm.

"Why did you hit her?" Erich said later, almost too softly to be heard over my pounding heart.

"It wasn't hitting! I was patting her! You said treat them like dogs and I pat dogs and if I do some social misconduct with a dog, I don't get my head bit off, and why is it that all you people who socialize wolves have those nasty scars!" I said in a single breath while applying a tourniquet to the mangled arm of my goose-down jacket. Never again did I think that tame wolves could be treated like dogs.[1]

Wolves are different from dogs. However much you tame a wolf, it will not become domesticated. After years of behaving well, a wolf can suddenly and unpredictably forget its training. You should not trust wild animals, because they are all too reactively aggressive. Domesticated animals, in contrast, have changed genetically from their wild ancestors; they are less easily stimulated into producing reactive aggression.

The issue is not how well an animal can learn. Chimpanzees are as smart as any animal, and when they have good relationships with particular people they can behave as well as Klinghammer's wolves do with Klinghammer. Take conservationist Karl Ammann. He and his wife Kathy have had a chimpanzee called Mzee, a refugee from the bushmeat trade, in their Kenyan home for twenty years. Karl reports that even as a young adult, Mzee always slept in their bed. In fact, Mzee did not like to go to sleep unless he lay between Karl and Kathy, holding their hands.

I met Mzee when I stayed with the Ammanns a few years ago. He behaved well enough, but there was an edge to our interactions. At one point, over breakfast, he and I reached for the jug of orange juice at the same time. He grabbed my hand as I held the jug, and he squeezed. Ouch. "You first!" I squeaked, and was still rubbing my fingers back to life after he had finished his drink. Karl and Kathy were wonderful in their interactions with him, but it would take a lot of training for most people to be able to live safely with him—let alone sleep with him.

Mzee had a terrific relationship with the Ammanns because of their devotion to him. The same is true of many other apes who have lived closely with people, such as the gorilla Koko with Penny Paterson, or the chimpanzee Washoe with Roger Fouts. Adult great apes, however, should never have the freedoms that we give to well-trained dogs, because it is not safe. In a sympathetic account of how the psychologist Roger Fouts and his team worked with the linguistically skilled chimpanzee Washoe, the animal trainer Vicki Hearne noted the need for leashes, a tiger hook, and a cattle prod. Such precautions might have saved Charla Nash from losing her sight, face, hands and some of her brain to the chimpanzee Travis. Travis was a thirteen-year-old veteran of TV shows who was treated like a member of the family by his owner, Sandra Herold. One day, when her friend Charla was holding one of Travis's toys, he attacked her and inflicted appalling injuries.[2]

Whether reared in human homes or studied all their lives by people who love them deeply and thoughtfully, chimpanzees cannot be trusted not to use their strength in aggression—even when they understand the rules perfectly well. Lucky ones, like Washoe, end up with chimpanzee companions and a space designed to make them feel at home. The unlucky spend their adult lives in solitary imprisonment. Either way, chimpanzees' poor self-control forces us to use barriers. As Hearne noted, only with domesticated animals can we have relationships of easy trust.

Humans' place in this tamed/domesticated dichotomy is clear. We are calm compared with a typical wild animal—more like a dog than a wolf. We can look each other in the eye. We do not lose our tempers easily. We normally control our aggressive urges. In primates one of the most potent stimulus for aggression is the presence of a strange individual. But child psychologist Jerome Kagan reports that in hundreds of observations of two-year-olds meeting unfamiliar children, he has never seen one strike out at the other. That willingness to interact peacefully with others, even strangers, is inborn. Like domesticated animals, humans have a high threshold for producing reactive aggression. In this respect humans resemble domesticated animals much more closely than we resemble wild animals.[3]

The notion that humans are a domesticated species is at least as old as ancient Greece. Back in those days, more than two thousand years ago, the idea came in two versions. One was that domestication is a universal human characteristic and the other, unfortunately more prominent, was that human groups vary in levels of domestication. Theophrastus was Aristotle's successor as leader of the Peripatetic School in Athens. He took the view that domestication is a human universal. If only everyone had listened to him. The other view was Aristotle's, and when his notion resurfaced in the nineteenth century it led to trouble. Aristotle regarded most humans, such as the Greeks and Persians, whom he knew best, as so much less aggressive than wild animals that he put them in the same category of tameness as horses, cattle, pigs, sheep, goats, and dogs. On the other hand, he considered hunter-gatherers to be wild, hence undomesticated. So Aristotle regarded some humans as being more domesticated than others.

His disdain foreshadowed Nazi justifications for violence toward people whose degree of domestication was supposedly inferior to their own.

Two thousand years later, the topic of human domestication re-emerged in the intuitions of an influential early anthropologist, Johann Friedrich Blumenbach. Born in Germany in 1752, Blumenbach spent his working life in Göttingen. He showed his promise early, publishing his Ph.D. thesis in fifteen pages at the age of twenty-three, titled *On the Natural Variety of Mankind.* He became a professor of medicine at the age of twenty-four and spent the rest of his life researching how humans fitted into the natural world. A series of exciting discoveries was beginning to turn biology from medieval ignorance into a realistic assessment of what it means to be human. Blumenbach never lost interest in the goal of understanding humans as animals.

His contributions were huge. The great systematist Linnaeus had claimed that orangutans and humans were the same species. Blumenbach showed that they were different. Blumenbach distinguished chimpanzees and orangutans from each other, and he became the naming authority for the chimpanzee (*Pan troglodytes* [Blumenbach 1775]). With regard to humans, he was fascinated by population differences.

He produced a classification of races that included "Caucasian," a term he invented. For that reason, he is sometimes criticized today as if he had been an early racist. But he was thoroughly anti-racist. He insisted that all human populations were equally intelligent, and he declared that slavery was a wrong. In the words of the paleontologist Stephen Jay Gould, "Blumenbach was the least racist, most egalitarian, and most genial of all Enlightenment writers on the subject of human diversity." Blumenbach died in 1840 covered in scholarly glory, a towering figure sometimes memorialized as the father of anthropology.[4]

Despite all the reverence, Blumenbach had one big idea that no one took seriously. He was convinced about a special feature of humans, about which he could not have been more explicit. "Man," he wrote in 1795, "is far more domesticated and far more advanced from his first beginnings than any other animal." In 1806 he explained that our species' domestication was due to biology: "There is only one domestic animal . . . (domestic in the true sense, if not in the ordinary acceptation of this word) that also surpasses all others in these respects, and that is man. The difference between him and other domestic animals is only this, that they are not so completely born to domestication as he is, having been created by nature immediately a domestic animal." His view was equally clear in 1811. "Man is a domesticated animal . . . born and appointed by nature the most completely domesticated animal . . . the most perfect of all sorts of domestic animals that have been created."[5]

Blumenbach's confident judgment was not widely appreciated. A problem was race. Blumenbach had echoed Theophrastus by applying his concept of domestication to humans as a species, not just to some populations. That was too much for the intelligentsia of the day. To Blumenbach's critics, the world was full of uncivilized peoples, who were therefore also undomesticated. Some humans were domesticated, others were not, according to the critics.[6]

The uncivilized people that Blumenbach's opponents cited included two kinds. One group was the "savages" who were being discovered by Europeans around the globe, but few of the skeptics had direct contact with those distant populations. A second group of uncivilized people were children who had been found living alone in the woods of Europe.

The latter were more scientifically useful, because individuals could be produced and studied.

In 1758, the biologist Carl von Linné, also known as Linnaeus, had included these "wild children" in the tenth edition of his great taxonomic book about biological diversity, *Systema Naturae.* His method of incorporating them shows how chaotic the understanding of biology was then. Instead of recognizing them for what they were—namely, unfortunate outcasts with diminished mental or physical abilities—he treated them as if they represented a subspecies of humanity. He named them *Homo sapiens ferus.* Linnaeus was the scientific idol of the day, an ultimate authority forty-five years older than Blumenbach. When he implied that the "wild children" came from a "wild population," most people seem to have just assumed he was right. To Blumenbach, however, the idea made no sense.[7]

Blumenbach mounted a daring challenge to the great Linnaeus by researching the latest example of a "wild child." Peter of Hameln was reckoned to be about twelve years old when he was discovered in Germany in 1724. The boy was obviously uncivilized. He ate forest plants, had no language, and sometimes slept on all fours. He had no shame: he was unembarrassed about performing bodily functions in front of others. King George I of England acquired Peter and passed him around to intellectual circles, including the poet Alexander Pope and the writer Jonathan Swift. Peter was one of only a handful of wild children documented since the fourteenth century, and since his case was relevant to the topical nature-nurture question, he became a *cause célèbre*—"more remarkable than the discovery of Uranus," said the linguist Lord Monboddo, who took a close interest in the boy.

Monboddo aligned himself with Linnaeus by assuming that Peter had come from a wild population. Jean-Jacques Rousseau agreed. These and other intellectuals extolled Peter "as a specimen of the true natural man." Monboddo wrote, "I consider (Peter's) history as a brief chronicle or abstract of the history of the progress of human nature, from the mere animal to the first stage of civilized life."[8] The claims of Linnaeus, Monboddo, Rousseau, and their friends seem astonishing to us now. Those leading scholars apparently regarded Europe's wild forests as big enough to hide whole populations of undomesticated humans whose existence was revealed only by the occasional emergence of individuals like Peter. Their acceptance of the idea of *Homo sapiens ferus* had an obvious implication: as a member of a feral population, the fact that Peter was thoroughly uncivilized contradicted Blumenbach's claim of universal human domestication. Blumenbach responded to these challenges by conducting more detective work than his opponents. He showed that Peter was not a wild child at all, but a damaged boy who had had a hard life. He had once lived with his father, but when a stepmother moved in, the boy had been beaten and forced out. The reason Peter did not talk was that he was mentally handicapped: that was why his father's new wife objected to taking care of him. Despite his difficulties, Peter was clever enough to have survived in the woods on his own for a year. He was a disabled child who had had a traumatic upbringing, but he had come from an ordinary village home, not from a supposed population of wild people.[9]

Blumenbach's discoveries showed that "wild children" were not in fact "wild." "Wild children" were a myth, as irrelevant to theories about human nature as they were to the question of human domestication. If they were the best evidence of the existence of natural, undomesticated humans, there was a larger lesson to be drawn. Blumenbach closed his account of Peter unambiguously. "Above all, no originally wild condition of nature is to be attributed to Man, who is born a domestic animal."[10]

The "wild children" distraction was settled. Linnaeus and Monboddo had gotten it wrong. The refutation of "wild children" as aboriginal humans seemed to leave the field clear for Blumenbach's claim that humans, being naturally well behaved, should indeed be regarded as a domesticated species.

There was a second difficulty, however, which would prove to be a much more serious obstacle, so great that for a century Blumenbach's big idea was not taken seriously. The question was: how could the domestication of humans have happened? In the case of farmyard animals, humans were obviously responsible for the domesticating. But if humans were domesticated, who was responsible? Who could have domesticated our ancestors?

Even Blumenbach had no answer. He attempted only one suggestion, which he hid away in a footnote in his *Contributions to Natural History.* He attributed his possible solution to "a very profound psychologist" whom he did not name. The solution invoked divinity. "There must have been in the primitive world a class of higher existences on earth, to whom man acted as a sort of domestic animal." A vanished superhuman species had domesticated humans? The odd idea was ignored.[11] Blumenbach's failure to explain how humans became domesticated apparently did not matter to him personally, because he thought like a creationist: he simply accepted the nature of humanity without being concerned with questions about how it came to be. He died in 1840, nineteen years before Charles Darwin introduced the theory of evolution by natural selection in his *On the Origin of Species.* Blumenbach seems to have been comfortable with the idea that humans had been a domestic species since our creation.

Darwin was the next great thinker about the place of humans in nature, and, unlike Blumenbach, he was disturbed that the idea of human domestication came without any explanation for how it had happened. Darwin of course assumed that humans had evolved, and he knew that, in order to make his arguments persuasive, it was critical to show how evolution worked. In his two-volume book *The Variation of Animals and Plants Under Domestication* (1868), he had not considered whether humans might be domesticated. In his 1871 book on human evolution, *The Descent of Man, and Selection in Relation to Sex,* by contrast, Darwin contemplated Blumenbach's proposition. If humans really were domesticated, he wanted to know how and why.

Darwin agreed that "civilized men . . . in one sense are highly domesticated," and that "man in many respects may be compared with those animals which have long been domesticated."[12] His expertise in domestication, however, soon led to a problem. In his *Origin of Species,* he had launched his thesis about the evolution of wild animals by discussing the evolution of domestic pigeons. He had argued that evolution happens because of selection favoring the breeding of some individuals over others. He saw selection as occurring by nature, in the case of wild animals, or by humans, in the case of domesticated animals. By analogy to the farmyard, he insisted that the evolution of a more domesticated kind of human must depend on someone's doing the selecting. He could not see how that could have happened.[13]

Darwin cited only one effort to select for particular traits in a human population. The episode was barbaric, a telling illustration of the difficulty of imagining artificial selection in our own species.

From 1713 until his death in 1740, King Frederick William I of Prussia, an arrogant and domineering man who was also a drunkard, wanted to make his Potsdam Guards regiment the most impressive in the world. To do so, he paid a thousand recruiters to roam fifteen European countries to capture the tallest men and bring them to Prussia. The king committed vast sums to the effort. Rewards for the recruiters were so high that all scruples were abandoned. Choice individuals could be kidnapped. Their protecting guardsmen might be killed. One way and another, twelve hundred recruits were obtained as the initial group.

The tall soldiers were intended to be the pride of the army, but because they were there against their will they had to be imprisoned. Protests led to severe punishment. Disobedience was rewarded with torture. As many as 250 absconded each year, only to have their noses and ears sliced off and be condemned to a life sentence in Spandau Prison if they were caught. Regimental morale plummeted. Suicide notably reduced the ranks. Nevertheless, by the time the King died, he had created three battalions totaling 3,030 tall men.

The difficulties of building up his collection of "giants" were so great that the king turned to artificial selection as an alternative method of arranging for an army of tall soldiers.[14] He decided that if giants could not be easily recruited he had better breed them. Accordingly, his men searched peasant villages for tall women to marry his tall guardsmen, to whom they would be assigned as mating partners. The historian Robert Hutchinson described the system: The king "did violence to the virtuous wives and daughters of his peasantry. No consent was sought, no inquiries were made as to previous marriage relations. Every rule of decency and morality was ruthlessly violated, in face of the most stringent laws to the contrary. Stature alone was taken into account. . . . Here was a unique application of the divine right of kings."[15]

The Prussian king's experiment is said to have led to some unusually tall people in Potsdam, but overall it was a failure. It was deeply resented by both husbands and wives, and when the king died the experiment ended. Obviously, artificial selection of humans was an aberration if even a powerful monarch could not make it happen. Since the failed Prussian experiment was the only attempt Darwin knew of, and there was nothing like it in nature, he declared human domestication an impossible idea: "[Man's] breeding has not been controlled, either through methodical or unconscious selection. No race or body of men has been so completely subjugated by other men, that certain individuals have been preserved and thus unconsciously selected. . . ." For this reason, Darwin concluded, "man differs widely from any strictly domesticated animal."[16]

This might have ended the debate, but Darwin was always thorough, and he found a second difficulty with Blumenbach's idea. Unfortunately, in doing so, he amended Blumenbach's proposal. Blumenbach thought of domestication as Theophrastus had, a uniform phenomenon across the human species, equally applicable to all populations. But Darwin translated Blumenbach's concept into Aristotle's idea that some people are more domesticated than others. Ultimately, this twist was to set a terrible precedent, but the first result was mild: it merely led Darwin to a wrong conclusion. This happened as a result of his time with "wild" people.

In December 1832, during his voyage on HMS *Beagle,* Darwin had encountered hunter-gatherers in Tierra del Fuego, an island at the southern tip of South America, now split between Chile and Argentina. The meeting was striking because traveling with Darwin on the *Beagle* were three Tierra del Fuegians who had been taken to England on an earlier expedition, a young woman and two men. The captives came from the Yámana population (then called Yaghan). The captain of the *Beagle,* Robert Fitzroy, had kidnapped them in the hope that they would become effective missionaries when returned to their island after a spell in England. Darwin enjoyed interacting with them on the

Beagle. He was unprepared for how different the Yámana would be at home on Tierra del Fuego.

The *Beagle* spent several weeks at the island, giving Darwin ample chance to see the resident Yámana hunter-gatherers. He was shocked. "These were the most abject and miserable creatures I anywhere beheld . . . their skins filthy and greasy, their hair entangled, their voices discordant, their gestures violent and without dignity. Viewing such men, one can hardly make oneself believe they are fellow-creatures, and inhabitants of the same world. . . . Their skill in some respects may be compared to the instinct of animals; for it is not improved by experience. . . ."[17] The encounter seems to have led Darwin to explore the idea (which he wrongly attributed to Blumenbach) that humans varied in their degree of domestication. Suppose that "civilized" people can indeed be thought of as more domesticated than "wild" people, he said. If that is true, he surmised, the biology of civilized men should be more similar to that of domesticated animals.

Darwin knew that domesticated animals reproduce faster than their wild counterparts. If Blumenbach was right, "civilized men" would be expected to reproduce faster than "wild men." However, Darwin reported, the opposite was true: "wild men" reproduced faster than "civilized men." On this basis, Darwin concluded that the differences between "civilized men" and "wild men" were not due to a domesticationlike process that produced the former. There were other problems, too. Darwin knew that domesticated animals have smaller brains than their wild ancestors, whereas the size of the human brain and skull had apparently increased over time. So Darwin regarded the idea that humans are domesticated animals as a double failure. Not only was there no mechanism for human domestication, but also humans did not follow the patterns of domesticated animals. Darwin accordingly rejected Blumenbach's concept.

In the intellectual excitement following the publication of *On the Origin of Species,* Darwin was only one of many people thinking about the significance of evolutionary theories for human behavior. For the essayist Walter Bagehot, the contrast of human docility with the aggressiveness of wild animals was too fascinating to ignore. In an 1872 treatise on the theory of political evolution, which Darwin admired and annotated carefully, Bagehot wrote, "Man . . . was obliged to be his own domesticator; he had to tame himself." So Blumenbach's idea did not vanish entirely.[18] But Bagehot was a journalist. He did nothing with his speculation about human domestication. Perhaps Darwin's skepticism discouraged further thinking on the matter.[19] At any rate, for a few decades there was little mention of the idea.

For Darwin, the notion that some populations of humans were more domesticated than others seems to have been an intellectual analysis with no political implications. He was a committed abolitionist who wrote of his abomination of slavery, and in general (Fuegians aside) of his admiration for nonwhite peoples. Calamitously, however, when the idea of human domestication reappeared in the early twentieth century, it was not Blumenbach's "universal" version that became the focus. Instead, it was Darwin's (and Aristotle's) theory that different populations were domesticated to different extents. Human domestication came to be seen not only as a cause of racial differences but also as an index of human value: some races or populations were thought to be better than others, depending on how domesticated they were. The divisive potential of this idea became explosive in the hands of Nazis and their associates.

The trouble began with a 1914 essay titled "The Racial Characteristics of Man as a Result of Domestication." The author, the German anthropologist Eugen Fischer, argued that Aryans were superior because they were more domesticated than other races. According to Fischer, a semiconscious preference for blond hair and pale skin had led to the selective breeding of fine Aryan traits. In 1921, Fischer built on his essay by coediting a book on human genetics and "racial hygiene," together with Erwin Baur and Fritz Lenz. According to the historian Martin Brüne, this book provided the key scientific justification for Nazi eugenics. The authors all "supported the legalisation of sterilisation and dismantling of welfare institutions to reinstitute the laws of natural selection." Their recommendations were used in support of the anti-Semitic Nuremberg Laws passed in 1935. Jews were not the only targets. In 1937, Fischer studied six hundred children of African fathers living in Germany. His theories led to their being forcibly sterilized.[20]

Confusingly, a contrasting concept of the significance of human domestication proved to have an almost equally unfortunate conclusion, even though it took an opposite theoretical perspective. The Austrian ethologist Konrad Lorenz was a brilliant researcher on animal behavior who in 1973 would share a Nobel Prize for his pioneering studies. He worked with both wild and farmyard animals, but he was troubled by the latter, because they seemed poorly adapted to the hardships of life. The title of a notorious paper that he published in 1940 summed up his concern: "Disorders Caused by the Domestication of Species-Specific Behavior." He disdained Muscovy ducks as being squat, fat, ugly creatures compared with their clean-limbed wild ancestors, and he applied a parallel distinction to humans. Lorenz considered that, under the influences of civilization, humans had become overly domesticated, leading to unattractive, infantilized, and unviable people. So, unlike Fischer's group, who credited the fine qualities of Aryans to their being highly domesticated, Lorenz considered more highly domesticated populations to be a degraded version of the natural ideal.[21]

The conclusion of Lorenz's pseudoscience contrasted directly with Fischer's, but it justified his promoting a similarly nasty eugenics. For Lorenz, domestication was the source of degeneration. He therefore argued that civilization would decay "unless self-conscious, scientifically based race politics prevents it."[22] Lorenz is sometimes said to have perverted his biology in order to satisfy wartime Germany's Nazi masters, but, in spite of intense postwar criticism, his commitment to the degradation theory of domestication outlasted Hitler. In the 1970s, he told his biographer Alec Nisbett, "The great devil he was fighting then, and is still fighting, is the progressive self-domestication of humanity."[23]

So, regardless of how human domestication was valued, it had become a political weapon. If Fischer was right and domestication was a good thing, this justified oppression of the less domesticated. If Lorenz was right and it was a bad thing, this justified oppression of the more domesticated. Either way, theory was used for evil.[24]

Appalling though its association with eugenics had been, after World

War II the concept of human domestication slowly recovered its legitimacy. The problem had come from the eugenicists' claim that some populations were better than others, not from the idea that humans were domesticated. As long as no value judgments were assigned to population differences, even explicitly anti-racist scholars could be on board. Margaret Mead was a notable example, an icon of cultural relativity who felt comfortable writing in 1954, "Man is a domesticated animal."[25] Her adviser had been the so-called father of American anthropology, Franz Boas, a fieldworker and theorist much admired for his insistence on the psychological sameness of mankind. In 1934, Boas had written that humans as a whole had been domesticated, even approving of Fischer's ideas that races had been produced by domestication. "Man is not a wild form," Boas claimed, "but must be compared to the domesticated animals. . . . He is a self-domesticated being."[26]

Subsequently, human self-domestication in some form or other has been explored by evolutionarily minded scholars from an astonishing range of perspectives, including archaeology, social anthropology, biological anthropology, paleoanthropology, philosophy, psychiatry, psychology, ethology, biology, history, and economics. Everywhere, the essential rationale is the same. Our docile behavior recalls that of a domesticated species, and since no other species can have domesticated us, we must have done it ourselves. We must be self-domesticated. But how could that have happened?[27]

The idea of self-domestication was repeatedly reinvented as a way to understand human docility, but until recently the term was used only as a description. Some, such as Boas, said humans are *like* domesticated animals. Others, such as Blumenbach, said we actually *are* domesticated animals. But either way, their analysis stopped there. The reference to domestication was a teaser. It merely hinted that the biology underlying human docility might have parallels in animals. Unfortunately for those visionaries, the biology of domestication was obscure. Also, too little was known then about human evolution. The words of the evolutionary biologist Theodosius Dobzhansky made sense in 1962.

"The concept of human domestication," he wrote, "is too vague an idea at this time to be scientifically productive."[28]

The evidence is clearer now. It suggests that humans became less reactively aggressive and increasingly docile around the time of our becoming *Homo sapiens.* Critical clues come from comparisons with domesticated animals. In his 1868 book, *The Variation of Animals and Plants Under Domestication,* Darwin reported that there are various surprising biological markers of the domestication process other than docility. For instance, there is a strong tendency for domesticated mammals to have floppy ears. Some breeds of dogs, such as German shepherds, have pricked ears, but in many breeds the ears flop strangely downward, as they always do in puppies. Darwin found that, like dogs, every other kind of domesticated mammal includes some adults with floppy ears. This was amazing, because floppy ears are very rare in adult wild animals. Elephants were the only wild species that Darwin knew to have them. Making matters more mysterious, there was no obvious reason why docility should be linked to floppy ears. It was just something that happened.

Another example is white spots on foreheads, common in horses, cows, dogs, and cats but not in wild animals. It was the same story for curly tails, variable hair quality, and white feet. The reasons domesticated animals show these mysterious associations were entirely unknown then, and are only now beginning to be hinted at. But, regardless of its explanation, the list of traits associated with domestication, called the domestication syndrome, is useful, because it provides telling clues to the human past. Critically, the domestication syndrome includes changes to bones. Fossil bones allow archaeologists to recognize when species such as dogs, goats, and pigs became domesticated. So, as the archaeologist Helen Leach argued in 2003, they can do the same for humans.[29] Leach listed four characteristics of the bones of domesticated, but not wild, animals that are found in contemporary humans.

First, domesticates mostly have smaller bodies than their wild ancestors. After a domesticated variety has become established, artificial selection by humans can create deliberately large breeds, such as carthorses or great Danes, but the initial size drop is consistent. The effect is so predictable in dogs or herd animals like sheep and cattle that archaeologists use it as one of their main criteria to recognize when domestication of different species has occurred. Nowadays increases in the amount and quality of food mean that many of us humans are bigger than our ancestors were a few hundred years ago. However, further back in time, there were drops in height of humans in many places around the world. The drops occurred around the end of the Pleistocene, twelve thousand years or so ago. Smaller human bodies in the past are also indicated by changes in the relative thickness of bones. The limb bones of our ancestors are thicker, both at the ends and in the midshaft, compared with their length. Cross sections also show that limb bones formerly had thicker walls surrounding the marrow cavity. The thicker the bones, the more weight they carried. Based on bone thickness, humans have been gradually weighing less since the time of *Homo erectus,* about two million years ago, including an especially pronounced weight loss since the appearance of modern-looking *Homo sapiens.* Such changes are often summarized as humans' becoming less robust and more gracile.[30]

Second, the faces of domesticated animals tend to be shorter, projecting relatively less forward, than those of their wild ancestors. Teeth also become smaller, and jaws smaller still, trends that appear to be responsible for tooth crowding in early-domesticated dogs. Humans follow the same patterns. A study in the Sudan, where people lived continuously for the last ten thousand years, showed that faces became consistently shorter during that time. The trend began much earlier, however; for one, the first *Homo sapiens* had smaller faces than pre-*sapiens* species such as *Homo erectus.* Decline of tooth size has been noticed over the last hundred thousand years. Teeth diminished in size at a rate of approximately 1 percent per two thousand years until ten thousand years ago, when their decline accelerated to a loss of 1 percent of volume every one thousand years. The rates of decline were similar in many areas across Europe, the Middle East, China, and Southeast Asia.[31]

Third, differences between males and females are less highly developed in domesticated than in wild animals, always for the same reason: males become less exaggeratedly male. In ungulates such as cattle and sheep this change is seen in a reduction in the size of horns of domesticated species in comparison with their wild ancestors. In humans there is no evidence of any change in the relative height of males and females until recently. However during the last 35,000 years, according to anthropologist David Frayer, males have become more like females not only with respect to stature but also to the size of the face, the length of the canine teeth, the area of the chewing teeth, and the size of the jaws.[32] Further back in time, around 200,000 years ago, male faces were already becoming relatively feminized. An analysis by biologist Robert Cieri and colleagues showed that brow ridges above the eyes became less projecting in males, and the male face became shorter from the top of the nose to the upper teeth.[33]

Finally, domesticates have a strong tendency to have smaller brains than their wild ancestors, whether mammals or birds. The average reduction in brain volume for a given body weight is around 10 to 15 percent, but some level of brain reduction is found in every domesticated mammal except for laboratory mice. Although human brain size, as measured by the volume inside a skull (or cranial capacity), has steadily increased over the last two million years, the trajectory took a surprising turn around thirty thousand years ago, when brains started to become smaller. In Europe, modern brains are some 10 to 30 percent smaller than those of people living twenty thousand years earlier.[34] Strikingly, in domesticated animals the loss of brain size is not associated with any consistent reduction in cognitive ability. Indeed, the smaller-brained species sometimes outperform their bigger-brained ancestors. For a given body weight, the brains of guinea pigs, for example, are about 14 percent smaller than those of their wild ancestors, but guinea pigs learn to navigate mazes, and to make associations, and reversals of associations, faster. Domesticated, small-brained rats are similarly impressive in learning and memory tasks compared with their wild cousins. Reduced brain size is an interesting and surprising fact about most domesticated animals, but it is not a reason to think that they, or small-brained *Homo sapiens,* are cognitively impaired compared with their ancestors.[35]

Conventionally, these four kinds of changes in human fossils have been explained separately, often in ways unique to humans. Body-size reduction might be understood to result from climate change, or from a reduction in food availability, or from adaptation to new diseases. Facial reduction might have been the result of novel cooking methods, such as boiling, which made food softer. Reduced sex differences might have issued from the increased use of technology, with males no longer needing to rely on certain physical skills to be good hunters. Smaller brains might be explained as the result of lighter bodies, maintaining a consistent relationship between brain size and body size. But when we step back from each specific change, we see a bigger picture. The differences between modern humans and our earlier ancestors have a clear pattern. They look like the differences between a dog and a wolf.

The coincidence seems too strong to be due to chance. As we have seen, the idea that humans have been domesticated goes back at least two thousand years. Now we find that anatomical changes that came about during *Homo sapiens*'s history show a strong similarity to the anatomical changes experienced by dogs in their evolution from wolves. Half a million years ago, our ancestors were heavier-bodied, with more projecting faces, relatively bigger males, relatively masculine faces, and bigger brains. To extrapolate from domesticated animals, those characteristics probably indicate that our ancestors were less docile than we are today. They would have had a greater propensity for reactive aggression, losing their tempers more easily, quicker to threaten and fight one another. But somehow we became domesticated. Our social tolerance increased even as our bones changed. Our more wolflike behavior of the past became our more doglike behavior of the present.

Blumenbach would surely have been thrilled to know of such direct evidence in support of human domestication. He would have been even more delighted to realize what we now know: that the parallels between humans and domesticated animals occur not only in bones, but also in other aspects of biology.

4

Breeding Peace

THE DOMESTICATION SYNDROME was a surprise to Darwin. He never understood the reason for domesticated breeds of different mammalian species sharing a series of apparently unrelated traits. The puzzle has continued into the twenty-first century. The problem is that the various features of the domestication syndrome seem to be biologically unrelated to each other. What is the connection between reduced aggressiveness and smaller chewing teeth? No relationship is obvious. Animals do not fight with their chewing teeth, so there seems no reason for the teeth to become reduced in size, merely because reactive aggression is reduced. The same difficulty applies to the relationship between aggression and white patches of fur, floppy ears, or small brains. What could explain why domesticated animals tend to have such traits?

The most popular traditional hypothesis is one of "parallel adaptation." According to the parallel-adaptation hypothesis, reduced aggressiveness, white patches, floppy ears, and other features of the domestication syndrome are adaptations that evolved independently in response to a new context, namely living with humans. We will shortly see why the parallel-adaptation hypothesis can be confidently rejected, but one can see why scholars in search of a solution to the curious puzzle of the domestication syndrome might have thought it a reasonable solution. Short faces and smaller teeth could be adapted to a higher quality of food, so soft that it would take less chewing than the wild foods that domesticates' ancestors used to eat. Perhaps white patches of fur might have been favored because they allow humans to identify their animals, or because farmers found white "socks" on their animals cute. Floppy ears might, in theory, be permitted because excellent hearing ability is less important for domesticated than for wild animals. Assuming that smaller brains represent less ability to be alert to threats, that could be explained by their living in a relatively safe environment. Explanations such as these led to the idea that the entire domestication syndrome could in theory represent a series of parallel adaptations to the new, human-created, environment.[1]

This is the same kind of reasoning that has been applied to human evolution, as we saw earlier, based on the fact that humans have been acquiring an increasing number of cultural adaptations for the last half-million years. In animals, the parallel-adaptation hypothesis suggests that the domestication syndrome evolved as a series of independent adaptations to "living with humans." In humans, the parallel-adaptation hypothesis suggests that the domestication syndrome evolved as a series of independent adaptations to "increasingly sophisticated culture." Improvements in cooking techniques might have led to a higher-quality diet, and so to less intense chewing and a more gracile chewing anatomy. Better spears, longer-range bows and arrows, and the increased use of snares and nets could mean that there was less pressure on males to be physically developed for the hunt, leading to a more feminine physique. Like explanations for the domestication syndrome in animals, the parallel-adaptation hypothesis suggests that the domestication syndrome in humans has nothing to do with reduced aggression. Instead, it has emerged as a series of independent biological responses to humans acquiring various kinds of culturally inherited skills and tools.[2]

The hypothesis of parallel adaptation is appealing because it fits an important general expectation from evolutionary theory. Biologists ordinarily assume that features evolve because they are adaptations that promote an individual's ability to survive and reproduce. This adaptationist perspective is a core principle of evolutionary theory that, ever since Darwin, has successfully revolutionized our understanding of life.

Occasionally, however, the adaptationist perspective is wrong: biological features are not always adaptations. Nipples provide no benefit to males, yet mammals have maintained them since the origin of suckling around two hundred million years ago. In the growing embryo, the sequence of development responsible for female nipples, which are indeed adaptive, also leads to male nipples, which are not. So male nipples have no function, but the costs of getting rid of them are apparently higher than the costs of keeping them. Thus, they have survived intact and useless for millions of generations. They illustrate how an organism's ability to adapt can sometimes be constrained by its developmental program.[3]

Male nipples are the poster child for a nonadaptive feature. A second example that is often cited as a nonadaptive trait is the clitoris in birds. Most bird species have neither a clitoris nor a penis. Those that mate in water, however, tend to have both. Male ducks (also known as drakes) risk having their sperm washed away unless they deposit their sperm deep inside the female's body. Unlike most land birds, drakes have therefore evolved a penis, and the sequence of development responsible for the penis also generates, for the females, a clitoris. As with male mammalian nipples, therefore, the clitoris of ducks appears to be an evolutionary byproduct, a feature that is present because selection promotes a penis in males. Male nipples and female phalluses are fascinating cases, but in the sweep of life they are minor players. The fact that they are prime examples of features that persist despite not having a positive selective function implies that nonadaptive characteristics are rare phenomena.[4]

The results of a large-scale and long-term experiment by the Soviet geneticist Dmitri Belyaev therefore came as a surprise when it showed that the domestication syndrome is not a series of adaptations at all. Instead, it represents a series of nonadaptive responses provoked by the single major factor of being domesticated. The entire domestication syndrome, Belyaev discovered, is equivalent to the male nipple. The domestication syndrome is a stunning example that suggests a potentially massive biological role for nonadaptive byproducts of selection.

—

Belyaev's career serves to remind us how lucky most contemporary scientists are to work in safety.[5] In 1939, at the age of twenty-two, he had his first job in Moscow, at the Department of Fur-Bearing Animals of the Central Research Laboratory. An excellent scientist who fully understood the potential value of selective breeding for improving fur production, he was eager to advance the science. But at that time in the Soviet Union, it was dangerous to be a geneticist. Joseph Stalin had led the USSR since 1924, and Stalin regarded Western genetics as a pseudoscientific tool designed to promote anti-Soviet ideology.[6] Geneticists who did not follow the party line were consigned to labor camps or worse. Belyaev's own family had suffered from Stalin's paranoia. As a teenager, Belyaev had been inspired by the research of his elder brother Nikolai, who was a well-known geneticist. In 1937, Nikolai was arrested and shot without trial because of his interest in Western-style genetics. Intellectually honest genetics research did not become permissible until several years after Stalin died, in 1953.[7] In 1958, Belyaev joined the Institute of Cytology and Genetics of the Siberian Department of the Soviet Academy of Sciences at Novosibirsk. There he was able to keep hundreds of foxes and other animals in captivity and finally to test an idea he had long been drawn to.[8]

In general, Belyaev was interested in the domestication syndrome, and in particular, in the breeding rate of captive silver foxes, a subspecies of red fox that had been brought over from Prince Edward Island, Canada, in the 1920s. Silver foxes produce an unusual fur color that is a Siberian and worldwide favorite. By the time Belyaev began his research, families in thousands of small rural farms all over the Siberian countryside had been keeping silver foxes for up to eighty fox generations. These foxes were certainly captive, but no deliberate attempts had been made to domesticate them. Indeed, unlike most genuinely domesticated animals, the foxes had not increased their breeding rate beyond the level in the wild: they still bred only once a year. This was disappointing for the farmers, but the slow breeding rate was not particularly surprising, given that the silver foxes were to a large extent simply wild animals held captive.[9]

Belyaev's hypothesis was that selection purely for increased docility (a heritable trait) might produce the domestication syndrome—including a faster rate of reproduction. His great insight was that animals could not coexist with humans if they were so fearful that their reaction to people was spontaneous aggression. Accordingly, he imagined that the early stages of domestication would always involve selection in favor of the least reactively aggressive, most tractable animals. Whether a farmer did this consciously or unconsciously did not matter: most likely, it would happen simply because the more aggressive animals were more dangerous and difficult to deal with.

Although no one understood the details of how a tamer psyche evolved, Belyaev knew that it had to be a complicated process involving several biological systems. Anatomical changes would probably be needed in the brain, and beyond that would be differences in the production and control of hormones, neurotransmitters, and other physiologically active chemicals. The effects of such extensive biological changes could hardly be confined in their effects solely to reducing an animal's aggressive reactions. Furthermore, the earlier in an animal's life that these changes started happening, the more numerous and broader the consequences were likely to be. Since the biological control of aggression tends to use similar mechanisms in different mammals, genetic taming might cause different species to respond in similar ways. In short, animals selected to be docile could reasonably be expected to develop all sorts of other features as unrelated by-products of their reduced emotional reactivity, including—maybe—an accelerated breeding rate.

Belyaev had the courage to commit himself to this idea even though, for all he knew, decades might be needed to produce a result. In 1959, he began by selecting his initial population from among several thousand foxes at different fur farms. To find the calmest animals, experimenters approached each cage and tried to open it. Most foxes responded by growling and trying to bite. A small proportion, about one in ten, were less fearful and more friendly than others. They were chosen as the first breeders, a hundred females and thirty males.[10]

Once the foxes started breeding, Belyaev asked his team to assess the pups rather than the adults. Experimenters offered food to young pups while trying to stroke and handle them at the same time. The calmest pups tolerated this treatment without snarling. About 20 percent of females and 5 percent of males were judged to be especially tame, and they would be selected for the new mating pool. Year after year, Belyaev's team followed this protocol. In the first fifty years, they would test about fifty thousand fox pups, or a thousand per year. Some two hundred were chosen each year for breeding.[11] For comparative purposes, Belyaev also wanted to have an unselected line of foxes. He monitored a separate line of foxes bred in the normal way, without regard to how aggressive or docile they were.

The experiment struck gold sooner than expected. Within just three generations, some offspring in the experimental population were no longer showing aggressive and fearful responses. In the fourth generation, the experimenters were amazed to find a few pups approached by humans wagging their tails as if they were dogs. Unselected foxes had never wagged their tails.[12] The sixth generation marked the appearance of a "domesticated elite." Elite foxes not only wagged their tails, they even whimpered to attract attention, and they approached the experimenters to sniff and lick them. According to Belyaev's collaborator Lyudmila Trut, by the tenth generation these elite foxes made up 18 percent of the pups; by the twentieth, 35 percent; and by the thirtieth to thirty-fifth generations, 70 to 80 percent. Within a few years, the American Kennel Club applied to import domesticated silver foxes as pets.[13]

Impressive as the rate of change in tameness was, what really caught Belyaev's attention was the appearance of other traits that had not been targeted by the experimenters. By 1969, only ten years after selective breeding began, "a peculiar piebald spotting was first observed in a male fox."[14] In unselected foxes the "piebald spotting," which Belyaev called a "star mutation," was rare, but in the selected foxes it was relatively common, including being found "between the ears"—in other words, this was the forehead "blaze" of white often found on horses, or similar white patches commonly seen on cows, dogs, cats, and many other domesticated animals. The star mutation had never before been described in silver foxes, but it soon appeared in forty-eight separate families of foxes at the experimental farm. These represented only a minority of the families, but, in keeping with Belyaev's hypothesis, thirty-five of them included foxes "remarkable in their high degree of tameness." Docility, Belyaev concluded, had led to the star mutation.[15]

The appearance of the star mutation was dramatic because it provided the first experimental support for the idea that selection against aggression could generate features of the domestication syndrome that had no adaptive significance. Many such discoveries followed, including changes to the reproductive system. By 1962, 6 percent of the selected females were breeding not only in the summer but also in the spring and fall. By 1969, 40 percent of females were breeding three times per year, compared to once in the unselected lines. The change was evidently genetic, because it was concentrated in the same families. The shift from one to two annual mating cycles was not always beneficial: litters often failed. But even though there were no immediate practical benefits in terms of increased fur production, for Belyaev the theoretical triumph was clear. Selection for docility alone had led to a loss of breeding seasonality. It also led to foxes' reaching sexual maturity a month earlier than unselected foxes, having longer mating seasons, and producing bigger litters.

Other effects piled on. "Next came traits such as floppy ears and rolled tails similar to those in some breeds of dog," wrote Lyudmila Trut, who continued the work after Belyaev's death in 1985. "After 15 to 20 generations we noted the appearance of foxes with shorter tails and legs, and with underbites or overbites." These were all features of the domestication syndrome.[16] Bones of the domesticated lines also supported Belyaev's hunch. Skull shape changed. Trut reported in 1999 that the domesticated foxes had narrower skulls, with less cranial height, than the farm foxes. Much like the pattern in long-domesticated animals, "the skulls of males became more like those of females."[17]

Belyaev's gambit had paid off. Foxes selected for mating had been chosen not on the basis of any features of anatomy, color, or other external characteristics, but merely because they were friendlier when they were young. The result was not just a rapid increase in tameness in the selected population, but also a whole suite of incidental effects. The unselected foxes that Belyaev had allowed to breed in the traditional way, by contrast, had far lower rates of acquiring domestication-syndrome features. Belyaev repeated the experiment with American mink and rats. Similar results emerged. There was no doubt. Belyaev had discovered the selective force underlying the domestication syndrome: selection for reduced reactive aggression.

A century earlier, Darwin had puzzled over some strange co-occurrences—for example, that blue-eyed cats with white fur tend to be deaf—which suggested that natural selection could not always force animals into an optimal design. Darwin wondered if some features that were nonadaptive or even maladaptive (like deafness in cats) might be dragged into being by other, adaptive features that had a meaningful biological rationale. Belyaev had now shown this could happen. Selection merely for tameness can lead not only to the rapid evolution of friendly, attentive foxes, but also to a series of surprising and otherwise unrelated physical changes that are also found in a wide range of other domesticated animals and yet have no specific purpose in domesticated life.[18]

The Siberian experiment did more than just show that adaptive traits can drag nonadaptive traits into existence. It also showed for the first time exactly what the domestication syndrome is. It is not a series of adaptive features each sculpted by evolutionary pressures to respond to human environments. Instead, it is a largely useless series of traits that happen to signal an evolutionary event. The domestication syndrome shows that the species has recently undergone a reduction in reactive aggressiveness.

Belyaev's research evinced some immediate effects of selecting for tameness, but it did not show how long the domestication syndrome would last. This is critical. If features of the domestication syndrome are sufficiently maladaptive, natural selection would be expected to reverse them. So, in theory, although the domestication syndrome might be thrown up rapidly by selection for tameness, it could be lost equally quickly.

The history of domesticates suggests otherwise. According to a recent assessment, dogs began domestication more than fifteen thousand years ago, goats and sheep around eleven thousand years ago, followed within less than a thousand years by cattle, pigs, and cats. Other domesticates, such as llamas, horses, donkeys, camels, chickens, and turkeys, are thought to have evolved within the last five or six thousand years. These animals are recognizable as domesticates thousands of years ago because by then their bone anatomy had developed the features of the domestication syndrome that they have maintained or exaggerated ever since. Once the domestication syndrome has evolved, it can persist for thousands of generations.[19]

Admittedly, in most of these cases the newly evolved domesticated species continued to live under human control. Furthermore, they were also often subject to continuing selection for novel characteristics and increased tameness. We therefore need to ask whether the traits of the domestication syndrome are retained when the animals are returned to the wild and have to survive there. Even in the wild, it turns out, lineages of originally domesticated animals can thrive for many generations without reverting to their ancestral type.

The biologists Dieter Kruska and Vadim Sidorovich studied American mink descended from some eighty generations of breeding on Canadian ranches. In the eighteenth and nineteenth centuries, fur trappers caught large numbers of mink in the wild, but by the 1860s, creative pioneers had found that they could produce high-quality pelts cheaply from minks in captivity. So fur farms became a new industry, and the domestication syndrome ensued. In a typical manifestation of the syndrome, the captive-bred mink developed a shorter face and a brain 20 percent smaller than those of its wild relative of similar body weight.[20]

American mink were bred so successfully in Canada that they became the major source of farmed mink fur, and remain so even today. Europeans were so impressed by the commercial prospects, however, that they imported the American domesticates for breeding. Unfortunately, many escaped into the wild, where they thrived. By 1920, these feral American mink were spreading fast. Hundreds of thousands of the invasive species established themselves throughout continental and archipelagic Europe, including Norway, Italy, Spain, Britain, Ireland, Iceland, Russia, and Belarus. In Belarus, the success of the domesticated American mink caused a substantial decline in the populations of two native species of carnivores, the European mink and polecats. The American mink outcompeted their wild cousins.[21]

Although they were now in the wild, the newly feral mink retained the small brains and short faces of their domesticated forebears. Only about eighty generations had been needed to produce aspects of the domestication syndrome; after fifty generations in the wilds of Belarus, the mink had shown no reversal to wild anatomy. Small brains and short faces appeared to be just as well adapted to the woods and waterways of Europe as to Canadian cages.

The mink case is informative because the history of mink breeding is well known. Other mammals that thrive in the wild, despite having small and domesticated brains, include goats, pigs, cats, and dogs. Domestic goats on Chile's remote Juan Fernández Island have retained their small cranial capacities in the wild for about four hundred years. The most striking example of a domesticated species that has succeeded in the wild is the Australian dingo. Dingoes are descended by thousands of generations from domesticated dogs, but even after at least five thousand years back in the wild, they have brains no larger than those of dogs. Dingoes' brains have not reverted to being large and wolflike.[22]

The study of dogs is just as informative. Although many live closely with humans and rely on their owners for food and care, others differ from domesticated species by surviving and reproducing without any direct involvement with humans. These are the free-ranging dogs that cluster in such waste-ridden locales as dumps, slaughterhouses, fishing ports, or markets. The biologist Kathryn Lord and her colleagues tried to estimate their number. Worldwide, they assessed the dog population as between seven hundred million and one billion, of which they estimated that 76 percent breed independently. Again, these village dogs have not reverted to being wolflike. They have retained the domestication syndrome despite living in large feral populations independently of humans, as some may well have done from the earliest days of domestication.[23]

The success in the wild of small-brained, short-faced species such as feral mink and dingoes clearly undermines the notion that domestication is quickly reversed when the species returns to the wild. This raises a new puzzle. If traits of the domestication syndrome works well in the wild, in what way were the wild ancestors better adapted than the descendants of domesticates to the wild environment? If a small brain and short face are good enough for American mink on the loose in Belarus, why did the mink ancestors evolve a larger brain and longer face in the first place?

The answer is unknown. An intriguing possibility is that these features are adaptations to competition within the species, rather than adaptations to finding food or escaping predators. In other words, perhaps they represent the end of an evolutionary "arms race" in which smaller-brained, shorter-faced animals are ultimately outcompeted by those with larger brains and longer faces. This fits the fact that domesticated animals flourishing in the wild are found mostly in habitats where their wild ancestors are absent, such as the American mink in Europe, pigs on the Galapagos Islands, horses in the American Southwest, or dingoes in Australia. We might eventually expect a slow reversal to the wild traits, because animals with bigger brains are able to use them for whatever small advantage they provide—such as enabling them to be better primed for reactive aggression. But the domesticates living in the wild show that any such reversal is slow. The domestication syndrome's longevity is akin to that of the male nipple. Both can survive long periods even though there was never any adaptive reason for them.

The aggressiveness that was deliberately reduced by Belyaev was aggressiveness toward humans, which is the type that is always reduced in domesticated animals: without this reduction, humans would not be able to manage their animals efficiently. Whether domestication also leads animals to be less aggressive to members of their own species, as well as to humans, varies. To date, Belyaev and Trut's experiment has not produced evidence that domesticated foxes are unaggressive with one another. In many other species, however, domesticates are relatively nice to one another, compared with their wild ancestors.[24]

Cavies and guinea pigs are, respectively, the wild and domesticated versions of the same animal, native to Peru, Bolivia, and Chile. They allow a particularly strong comparison, because both can be easily kept in captivity under identical conditions. Cavies, or montane guinea pigs, are still abundant in the high Andes. Guinea pigs were definitely domesticated (for food) by forty-five hundred years ago, and perhaps as early as seven thousand years ago. Since guinea pigs can have as many as five generations in a year, it is possible that they have been domesticated for more than twenty thousand generations. Their domestication led to the typical syndrome, including relatively small brains, short faces, and white patches of fur.[25]

The biologists Christine Künzl and Norbert Sachser have shown that the behavior of cavies and guinea pigs toward group members of their own species differs markedly. Cavies are more aggressive toward one another, whereas guinea pigs not only show less aggression but are also more tolerant and friendly (for example, they groom or nudge one another), and more active in courtship. To examine the influence of living in captivity, investigators compared adult cavies trapped in the wild with those kept in cages for thirty generations. Males of the two groups of cavies showed no difference in behavior or stress response. So for males, the differences in aggressiveness between guinea pigs and cavies could be clearly attributed to genetic effects of domestication and not to the effect of merely being caged for many generations.[26]

Similar evidence that domesticates are less aggressive than their wild ancestors not only toward humans but also toward others of their own species has been found in dogs, rats, cats, mink, and ducks. Consider wolves and dogs. In general, wolves are more aggressive toward humans than dogs are. Wolves are also significantly more violent toward members of other packs than are dogs. Wolves are so innately aggressive to wolf "strangers" that the leading cause of death in the wild is being killed by other wolves, accounting for as much as 40 percent of adult mortality. Among packs of feral dogs, by contrast, killing a "stranger" from another pack has been described only once. Within packs, too, dogs appear to be more tolerant of one another than are wolves, as indicated by their sharing breeding opportunities more equally.[27]

That domesticated animals are restrained in their aggressiveness toward humans does not necessarily mean they are equally restrained in their aggressiveness toward one another. In general, however, animals that are managed by humans are doubtless often selected for being tolerant toward one another, since fights in the farmyard are costly to the farmer.

By showing that selection for docility toward humans can generate the domestication syndrome, Belyaev's study gave an invigorating new explanation for the otherwise puzzling distribution of many features. His speculation had been right. The evolution of tameness depends on changing a set of biological systems affecting emotional reactivity, and those systems have secondary effects on a series of other traits. White patches of fur happen to be one of those traits. So are short faces, smaller teeth, small brains, faster reproductive cycles, and floppy ears. His lovely discovery therefore raised a new question: how does taming lead to the domestication syndrome?

Two biological systems, it has been argued, influence every feature of the domestication syndrome, and therefore explain its very existence. These two systems are basic to the lives of every mammal, and closely related to each other. They are the pattern of neural-crest cell migration, and the hormonal control exerted by the thyroid gland.

Neural-crest cells are a leading candidate for involvement in many and possibly all of the traits of domestication.

Once upon a time, when you were about two to three weeks past conception, you were largely a single-layered sphere of cells surrounding an emptiness. Then you gastrulated, which means that some of your cells started shifting inward, creating multiple layers and turning you into a microorganism with inner and outer parts. This process of gastrulation marked the emergence of the four types of tissue that made up your little embryonic existence: ectoderm, mesoderm, endoderm, and neural-crest cells. Ectoderm, mesoderm, and endoderm are the outer, middle, and inner layers of cells, and they give rise to parts that maintain their same approximate location in our adult bodies, such as skin, muscle, and soft organs, respectively. The neural-crest cells are different.

Neural-crest cells are a unique kind of tissue. They appear in vertebrate embryos in a stripe along the back of the developing head and trunk, immediately under the epidermis (which will become the skin). Whereas most tissues develop slowly into their various destinies, soon after the neural crest is formed it vanishes as a separate entity, because its constituent cells leave their place of origin on the back of the embryo (the dorsal neural tube): the cells then separate from one another and spread around the embryo in groups. This uniquely penetrative system of migration means that even though the neural crest disappears very early in development, by the time individuals reach adulthood many organs in the body are at least partly derived from neural-crest cells.[28]

Some components of the embryo, such as the melanocyte system, are much more influenced by the neural crest than others are. Melanocytes are found in the lowest layer of the skin, where they produce the melanin pigments that give color to hair or skin. White patches in an animal's fur normally reveal the absence of melanocytes in that area. In domesticated species, many animals have a white tip to the tail, or white "socks" on their legs. The reason is that the neural-crest cells migrated so slowly, or were produced in such small numbers, that they never reached those terminal parts of the body. So the tail tip or ends of the legs lacked melanin, and ended up white-haired.

This simple dynamic also explains why domesticated animals often have the white "blaze" on their forehead, Belyaev's "star mutation." In their migration from the back of the head and trunk toward the forehead, neural-crest cells first move toward the mouth, and then upward until they meet, from each side of the body, above the eyes. If they do not quite finish their migration, the center of the forehead receives no melanocytes and therefore has no cells capable of producing pigment. Hence the blaze.

White patches of fur in domesticated animals thus show that the migration of neural-crest cells has suffered some kind of delay or reduction. In the case of foxes with piebald patches, Trut's team discovered that the neural-crest cells that were destined to become melanocytes, called melanoblasts, were delayed in their migration by one to two days.[29]

Much is known of the genetic systems that control how these cells' journeys take them to the specific destinations where they give rise to various distinct derivatives. The fact that a deficiency in neural-crest cell migration so clearly causes the white patches of the domestication syndrome raises the possibility that other aspects of the domestication syndrome are similarly linked to this very early change in embryonic development. In 2014, the biologists Adam Wilkins, Tecumseh Fitch, and I proposed exactly that idea. Fitch had visited Novosibirsk in 2002 and was impressed with the evidence linking neural-crest cell migration and white patches. With that starting-point, we considered the domestication syndrome as a whole and found an intriguing fit.[30]

Most traits of the domestication syndrome fall in line with the idea that they come from changes to neural-crest cell migration. Neural-crest cells give rise to the adrenal glands, whose reduced size and rate of hormone production are central to the reduction in emotional reactivity of domesticated animals, as Trut and her colleagues have emphasized.[31] Darwin had identified small jaws and short snouts (or flat faces) as characteristic of domesticates in general, and these traits were also found among the foxes and mink selected by Belyaev's group for tameness. The development of the jaws is well understood. They are derived from two pairs of primordial bone that develop after neural-crest cells arrive at the end of their migration. The number of neural-crest cells that reach the primordia determines the size of the jaw: the fewer the cells, the smaller the jaws.

The size of teeth is controlled by genes that are different from those that determine the size of the jaw, but, again, neural-crest cells are critical. About halfway through a human pregnancy, at around seventeen or eighteen weeks, neural-crest cells reach the fetus's tooth buds and are transformed into a cell type called odontoblasts. Odontoblasts form the outer surface of the living tissue, building the tooth from within by producing dentine during the tooth's growing phase. Smaller populations of neural-crest cells reaching the teeth thus produce smaller teeth.

Floppy ears provide an entirely different example of how neural-crest cells appear to influence a trait of domestication. Ears are floppy if the internal cartilage is too short, leaving the end part of the ear unsupported and liable to flop over. The sources of tissue are different for the supportive cartilaginous part of the ear and the ear as a whole, which is called the pinna. Both the cartilage and the pinna receive neural-crest cells, but they receive them from different parts of the neural crest. So animals with floppy ears appear to be those whose cartilage received relatively minor amounts of neural-crest cells. In thinking of humans as domesticated animals, it is disappointing that humans do not have floppy ears. Perhaps our ears are too small for a delay in neural-crest cell migration to matter.

Occasional genetic conditions in humans offer a different kind of support for the idea that these kinds of morphological changes are caused by reductions in neural-crest cell migration. Consider Mowat-Wilson syndrome, a rare condition apparently caused by a problem with neural-crest cell migration. Individuals suffering from Mowat-Wilson syndrome tend to have severe intellectual deficiencies, narrow jaws, and small ears. The syndrome is associated with mutations to a gene called ZEB2. ZEB2 mutations cause some neural-crest cells to stay at their site of origin, and this looks like the mechanism responsible for the small jaws and small ears of individuals with Mowat-Wilson syndrome. Suggestively, these individuals also tend to have "a happy demeanour with frequent smiling," the opposite of reactive aggressiveness.[32]

Reductions in the number of neural-crest cells and in their rate of migration are thus prime candidates for being root causes of the domestication syndrome. While such changes in the pattern of neural crest migration are mostly due to an individual's genes, environmental influences could also play a role. During the embryo's early development the hormone thyroxine is necessary for neural-crest migration. Thyroxine is supplied entirely by the mother's thyroid gland. Zoologist Susan Crockford suggested that the domestication syndrome could be caused partly by reductions in the mother's production of thyroxine.[33]

Although the hypothesis for the core role of neural-crest cell migration is a neat explanation for most parts of the domestication syndrome, until recently one important feature appeared to be an exception: the consistently small brain. Brain size is reduced in almost every one of the more than twenty species of animal that have been domesticated, whether birds (such as chickens, geese, and turkeys) or mammals (from rats to camels). In fact the only long-term domesticate that is an exception to this rule is the laboratory mouse, which has a brain no smaller than that of the wild house mouse from which it derives—

although possibly the ancestral house mouse is not truly wild, given its long association with humans.[34] In Belyaev's foxes, no reduction in brain size has yet been detected, although forty generations of selection reportedly led to a reduction in skull size. Different parts of the brain are not equally reduced through domestication. Kruska found that the largest reductions in the brains of domesticates occurred in areas concerned with sensory processing, especially auditory and visual, and in the limbic system, which is involved in emotion, reactivity, and aggressiveness. The corpus callosum, by contrast, a bundle of nerve fibers linking the left and right sides of the brain, maintains the same relative size in domesticated brains as in those of wild ancestors.[35]

The mechanisms of brain reduction have not been compared directly between wild and domesticated animals, but general principles regulating brain growth likely apply. Brain tissue itself is not derived from the neural crest, but neural-crest cells are essential for brain development. Proteins called growth factors, produced or regulated by migrating neural-crest cells, promote growth of the brain. For example, the facial neural-crest cells, which contribute to the growth of the face, produce agents, engagingly called Noggin and Gremlin, that influence the production of a vital protein called FGF8. The less FGF8 produced by the cranial neural-crest cells, the smaller the brain. Reductions in the rate of migration of neural-crest cells, or their number, are therefore likely to lead to a brain that grows more slowly and ends up smaller.[36]

The brain includes several structures of particular relevance to the development of reactive aggression. The foremost part of the developing brain (the telencephalon) is where the amygdala develops. Recall that the amygdala is critically engaged in promoting fearful responses (which can lead to reactive aggression). There are various hints that a smaller amygdala is associated with less fear and less aggression, including a woman whose amygdala was so badly damaged that she felt no fear at all, despite experiencing all other basic emotions. Whether the amygdala is reduced in size in domesticated animals has not been much studied, but in one mammal (the rabbit) and one bird (the Bengalese finch, a domesticated version of the white-rumped munia) the expected reduction in amygdala size has been found.[37]

Behind the telencephalon lies an area called the diencephalon, where the hypothalamus develops. Like the amygdala, the hypothalamus is a core part of the neural network that underlies reactive aggression (as well as proactive aggression). The hypothalamus also strongly influences the activity of the adrenal glands, and is involved in the regulation of females' estrous cycles and reproductive behavior.

Thus, in the brain as in the rest of the body, the traits of the domestication syndrome are likely to be affected by changes to the activity of neural-crest cells. Neural-crest cells have clear input into systems regulating stress, fear, and aggression, including the sympathetic nervous system and a set of brain structures regulating emotional responses. As we saw, neural crest development was first associated with the white patches typical of domesticated horses, dogs, cows, and other animals. We followed up our hypothesis that neural-crest cells are involved in many of the features of domesticates, finding support for links to such superficially unrelated traits as smaller jaws, smaller teeth, floppy ears, and even brain changes related to decreased reactive aggression.

A further critical test of our hypothesis is whether domesticated species show changes in genes that affect neural-crest migration. Since 2014, such effects have been found in six species (horses, rats, dogs, cats, silver foxes, and mink). Whether changes to neural-crest migration will be found in every domesticated species is uncertain, but at this early stage no exceptions have been found.[38]

Key to this growing understanding of the domestication syndrome, simply put, is juvenilization. As we will see in chapter 9, a reduction in reactive aggression is achieved most easily by selecting for juvenile traits, because juveniles are less reactively aggressive than adults. So the stress systems and brains of domesticated animals are expected to be like those of the juveniles of their wild ancestors, rendered juvenile in the domesticates partly by selection for reduced populations of neural-crest cells, leading to slow development of the relevant system.

The idea that neural-crest cells play a pervasive role in the domestication process is fascinating, because it offers a new kind of test for whether a species such as *Homo sapiens* has undergone selection against reactive aggression. A team from the neurobiologist Cedric Boeckx's lab considered this question by asking if humans and domesticated animals tend to share genetic evolution in parallel.[39] They listed a total of 742 genes that have been positively selected in *Homo sapiens* compared with two extinct human species, Neanderthals and Denisovans. Boeckx's team then noted all the 691 genes known to have been positively selected in four domesticated species (dogs, cats, horses, and cattle). They found a highly significant overlap in genetic changes between humans and the domesticated foursome. In all, forty-one of the genes that had been positively selected in humans had also been positively selected in the domesticated species. Although the biological roles of most of these overlapping genes are unknown, two cases seemed clearly related to domestication. For instance, the BRAF gene, which was positively selected in cats, horses, and *Homo sapiens,* has an important role in neural-crest development. Boeckx's team were clear about their conclusion. They considered their result to reinforce the hypothesis "that self-domestication took place in our species,"[40] meaning *Homo sapiens* specifically and no other species of *Homo.* This is a critical observation that we will come back to later. If the neural-crest hypothesis continues to be upheld, it will offer very precise ways to examine the evolutionary past of domestication and self-domestication.

Regardless of how accurate or complete the neural-crest hypothesis is, however, Belyaev's experiment has sharpened our understanding of the domestication syndrome. What he taught us is that the domestication syndrome is produced by selection against reactive aggression, not merely by living with humans. The implication is remarkable. It means that a domestication syndrome should be produced any time that reactive aggression is selected against. Since aggressiveness must sometimes be selected against in the wild, there should be many cases of domestication syndromes in the wild.

Until recently, however, no one had ever looked for such a case.

5

Wild Domesticates

Belyaev had asked if selection against reactive aggression would generate the domestication syndrome. When his bold intuition proved right, the discovery galvanized a research program at Novosibirsk that would last the rest of his life. Research continues there today, under the direction of his colleague Lyudmila Trut, comparing lines of less aggressive, more aggressive, and unselected animals. There were so many enthralling questions arising from Belyaev's findings that even now, after more than half a century, few domesticated species have been studied in detail. Silver foxes, rats, and mink are the main objects of research in Belyaev's former institute. Elsewhere, dogs, guinea pigs, mice, and chickens are the best-studied domesticated species. Some twenty others await investigation. Beyond the familiar domesticates, even more opportunities await, because a whole other class of animals begs for attention. Many wild species have experienced selection against reactive aggression.

The study of wild animals should yield fascinating results, given that the domestication syndrome has nothing to do with the specifics of how humans looked after their animals. If selection against reactive aggression were to produce the domestication syndrome in an animal living in nature without contact with humans, it would be a real-world example of domestication happening without a domesticator. It would strengthen the likelihood that humans, too, could have become domesticated even in the absence of Blumenbach's "class of higher existences on earth, to whom man acted as a sort of domestic animal."[1] Whether or not a domestication syndrome can be found in the wild is therefore a critical question both for understanding a potentially widespread evolutionary process and for helping to solve the puzzle of human docility.

Domestication is often understood to be a process that depends for its existence on humans' being the domesticators. The archaeozoologist Juliet Clutton-Brock, author of *A Natural History of Domesticated Mammals,* gives a typical definition: a domesticated animal, she wrote, is "bred in captivity, for purposes of subsistence or profit, in a human community that controls its breeding, its organization of territory, and its food supply." The origin of the word "domestication" tells the story even more directly. It comes from the Greek *domos* for "house." With this typical definition and etymology in mind, the idea that a wild animal exhibits a domestication syndrome would make no sense at all.[2]

But other definitions of domestication are independent of the farmyard by regarding domestication as a process predating human involvement. Entomologists apply the term "domestication" to ant agriculture, a system that goes back about 50 million years. Leaf-cutter ants of the American tropics rely on their relationship with fungi. Ants carry the fungi to new homes, give them fodder to grow on, and eat the product. The fungi have no independent existence. They live only with their ant hosts, and have evolved to be easy to harvest by producing special nutrient-rich swellings called gongylidia that their hosts distribute as food packages throughout the colony. These leaf-cutters, like a variety of other ants, have domesticated their food crops. Termites and bark beetles have also evolved primitive systems of agricultural domestication.[3]

In a very different context, the reason that Theophrastus, Blumenbach, Franz Boas, Margaret Mead, Helen Leach, and others called humans "domesticated" was that they considered humans to be "naturally" or "genetically" tame, and they needed a word to express that idea. Most of these writers gave no indication of concern with whether humans had ever been subject to any kind of controlled breeding. Blumenbach had referred lightly to the idea, but even for him the question of a domesticator was a side issue that he consigned to a footnote.

Obviously, the reason these writers called humans "domesticated" had nothing to do with their being "bred in captivity" or with "purposes of subsistence or profit. . . ." Instead, they wanted to emphasize their view that there is something special about human behavior that is shared with domesticated animals. This special quality is a social tolerance and low emotional reactivity to provocation. How that special quality originated was so immaterial to these writers that most did not even acknowledge the problem as worth mentioning.[4]

Admittedly, the fact that the word "domesticated" is used in various ways can be potentially confusing, but there is no other word that means "genetically tame." So in this book I follow Blumenbach and others by using the word "domesticated" to mean "tamed as a result of genetic adaptation" (as opposed to "tamed within a lifetime"). "Self-domestication" then refers to the process happening within a single species—that is, a reduction in a species's propensity for reactive aggression that has happened without having any other species responsible for facilitating the process. That is self-domestication.

So, could self-domestication have happened in the wild? Of course it could.

Aggressive behavior is old. It had certainly evolved by 540 million years ago, when the Cambrian "explosion" produced a host of new animal taxa—including most of the phyla found on earth today—such as insects and flatworms, many of which are quite violent toward members of their own species. Tendencies for reactive aggressiveness have doubtless waxed and waned ever since the Cambrian explosion. Currently, we have no reason to think that across animals as a whole, species are on average becoming either more or less aggressive. So around half of living species whose reactive aggressiveness differs from that of its immediate ancestor can be expected to have experienced an increase in aggressiveness, and the other half a decrease. These species that have experienced a decrease in reactive aggression would, accordingly, be cases of "self-domestication."

If Belyaev's insight applies in the wild, therefore, many animal species will have experienced not only a reduction in reactive aggression, but also a concomitant domestication syndrome—or, as we should call it for cases uninfluenced by a different species such as humans, a self-domestication syndrome.

Unfortunately, there has been very little effort made so far by scientists to test this hypothesis, partly because it is a new way of thinking, but even more because it is inherently hard to investigate. Even if self-domestication has been common, we might not be able to find evidence for it. For most species, the necessary evidence will be elusive because the behavior of the (normally extinct) ancestor is a crucial part of the equation but is extremely difficult to ascertain. If you want to test the idea that Asian elephants are self-domesticated (which they could well be), you need to know enough about a recent ancestor of living elephants to show that it was more aggressive than today's populations. But if the ancestor is extinct, very often we will have no fossils of it, and anyway its behavior will have left no trace. More often than not, we cannot know whether reactive aggression has decreased.[5]

But occasionally we get lucky, and one such piece of luck brings us amazingly close to our own evolutionary tree. It is the curious case of bonobos and chimpanzees. These two sister species of great ape are our closest living relatives; and the ancestor of bonobos appears to have been very similar to a chimpanzee.

Bonobos and chimpanzees look very similar. Both are black-haired apes that walk on their knuckles, weigh around 30 to 60 kilograms (65 to 130 pounds), with bigger males than females, and live in the moist rain forests of equatorial Africa. The easiest way to distinguish between the two is to note the bonobos' smaller head topped by hair with a central parting of hair. Bonobos are also unique for their pink lips, even though the rest of their face is uniformly dark. The two species live separated by the Congo River that winds around Africa's equator in the Democratic Republic of the Congo, chimpanzees on the right bank, to the north of the river, and bonobos on the left bank.[6]

Bonobos and chimpanzees share much of their social behavior. Both live in communities of a few dozen individuals, including more females than males. Community members inhabit a common territory, which they defend against encroachers from neighboring communities. Within the territory, they form changing subgroups (also known as parties) of a few individuals up to twenty or thirty or more. They travel alone at times. Sons never leave their native community but daughters mostly do. Around the time when a daughter reaches puberty, she tends to leave her mother and move to a different community, where she will spend the rest of her life. Females typically mate several hundred times before giving birth to each infant, which means that they normally mate often with every adult male in their community. Mothers look after their young essentially without any help from others.

So the two species are in many ways alike, which makes their few differences in behavior all the more striking. The feature that points most directly to bonobos' being self-domesticated is their relatively muted tendency for aggression. Bonobos are much less aggressive toward one another, and much less fearful of one another, than chimpanzees are. Zookeepers find bonobos easy to accommodate, because groups readily accept new individuals without any serious tension. Introducing chimpanzees to one another, by contrast, tends to be a painfully slow process, weeks or months of gradually familiarizing strangers with one another through wire mesh in order to minimize the risk of violence. Even after such cautious preparation, when chimpanzees who have not spent time together finally meet, they might easily fight. In the wild, too, long-term research has shown strong differences between chimpanzees and bonobos in every kind of competitive or aggressive behavior.[7]

Chimpanzee males fight often with other members of their community. Sometimes they fight over valuable foods, such as hunks of meat. Sometimes they fight over mating privileges. Mostly, however, they fight over nothing more than status. They regularly charge at one another in displays intended to demand clear expressions of subordinacy. If the target does not give any signal of submissiveness, fighting tends to erupt. Usually, the aggressor wins, the reluctant subordinate screams passionately, and everyone else in the party runs for cover.[8]

Males also commonly beat up on females, often in surprise attacks launched for no obvious reason. These bullying tactics can be sustained for many minutes. One attack at Kanyawara, in western Uganda, observed by the researcher Carole Hooven, lasted for a full eight minutes, during which the male grabbed sticks and beat the female intermittently with them, when he was not slapping, punching, and kicking her. The male's aim in such attacks is to intimidate a chosen female into readily acceding to his future demands for sex.[9] For each female, one male distinguishes himself from other males by being the one who most frequently attacks her. The tactic is often successful. Over subsequent weeks, a female's most frequent aggressor tends to be her most frequent sex partner, and eventually, even though she is likely to mate several times with every male in her community, he will be the most likely father of her next baby.[10] This stomach-churning practice is part of the reason why, as males become adult, they go through a ritual of beating up on every female. A male's ability to intimidate females is a vital component of his strategy for having as many offspring as possible.

Chimpanzee aggression within communities can be even more extreme. Infants less than a few months old have occasionally been killed. Adults of either sex can be responsible for killing an infant, though the killer is never the mother. When a coalition of adults unites in aggression, fights can also lead to the death of an adult male. The coalition collaborates in a frenzy of grabbing, hitting, and biting until the victim is overwhelmed and immobile, sometimes dead on the spot, and at other times crawling away to die within hours or days from the injuries inflicted during the attack.

Interactions between chimpanzee communities are never relaxed or friendly. Most encounters involve wary avoidance, sometimes accompanied by "shouting matches" when the parties from each community are separated by a large enough distance for individuals to call bravely at each other. Danger looms when parties get close (which can happen either by accident or design) and when one party has many more males than the other. The males in the larger party seek to press their advantage against the other community. Sometimes they catch and kill a helpless victim, whether an infant or an adult. Targets of between-community aggression like this are lucky to escape alive.[11]

The most extreme forms of chimpanzee violence are not frequent, and even the milder types need not occur every day. Nevertheless, long periods without emotional outbursts are rare. Particularly when fruits are abundant and subgroups are large, charges, fear screams, and beatings are an almost daily part of chimpanzee lives. Good feeding conditions can lead females to be sexually receptive and males to have abundant energy, a potent combination inviting the likelihood of raw male aggression.

Bonobos could hardly be more different from chimpanzees in terms of competition and aggression. Even though bonobo males can be ranked by dominance, competition among them rarely involves charging displays and includes no explicit signaling of status. Success in becoming high-ranking is rarely even dependent on interactions among the males themselves, because more important than their own fighting ability is the support given by mothers to their adult sons. The top-ranked males are mostly those with high-ranked mothers; when their mothers die, they are liable to fall in rank. Fighting over meat is infrequent, low-level, and more common among females than among males. Far less male intimidation of females occurs among bonobos than among chimpanzees. In one study bonobo females were more often aggressive toward males than vice versa. Male bonobos do not beat up on females, and in competition over food, females are more likely to win than males. No one has recorded any violent infanticides, nor any killing of adults, either within or between bonobo groups. Bonobos are certainly not free from disputes, and when groups from different communities meet, they sometimes fight to the point of inflicting bite wounds and scratches. But overall, the intensity of aggressive bonobo behavior is immensely less than among chimpanzees.[12]

The primatologist Isabel Behncke documented a telling example of just how different the behavior of bonobos can be from chimpanzees. Unlike neighboring communities of chimpanzees with their predictable hostility, subgroups from neighboring bonobo communities often enjoy one another's company to the point where they groom, have sex, and play together. Bonobo play includes a scary test of risk and trust between an adult and a juvenile that human observers call the "hang game." Sitting on a branch up to thirty meters (one hundred feet) high in a tree, the adult holds one of the juvenile's limbs and swings the willing plaything back and forth. The juvenile does not cling to the adult, which means that the adult holds the juvenile's fate in his or her hand: a drop would risk a severe wound or death. Yet the juvenile shows "visible joy," in Behncke's words, smiling from intense fun, apparently unaware of the very possibility that the adult might drop him or her. Astonishingly, the "hang game" happens even between bonobos of different communities. The trust exhibited among bonobos is clearly fostered by a degree of nonaggression and fear reduction that would be remarkable in almost any group-living species, and even more so in this close relative of the explosive chimpanzee.[13]

Similarly surprising, adult male bonobos from neighboring communities can play a "ball game" in which they chase one another slowly around a sapling, each trying to grab the testes of the one in front. Adult male chimpanzees occasionally play the same game with other males within their own community, the activity accompanied by excited faces and slow guttural chuckles. But the notion that any of them would play the game with a member of a different community would strike anyone who studies chimpanzees as ridiculous, given that the only relationship that has ever been seen between males from neighboring territories is instant hostility, leading to flight, shouting, or fight.[14]

Why should two species that look so much alike be so different in their intensity of aggression? Anatomy, ecology, and psychology all play a role. Male chimpanzees have one anatomical difference from bonobos that is clearly associated with higher aggressiveness: their dagger-like canine teeth are substantially bigger. Compared to the canines of bonobos, those of male chimpanzees are longer (or "taller" as the anatomists would say); their upper canines 35 percent taller and lower canines 50 percent taller. Females show a similar but slightly smaller difference (upper canine 25 percent taller, lower canine 30 percent taller in chimpanzees). Long canines are unnecessary for fruit-eating species like chimpanzees and bonobos, and are probably even a nuisance. But canines make strong fighting weapons, so the possession of longer canines in chimpanzees reveals an important aspect of this species' evolutionary history. Being a good fighter paid higher dividends in chimpanzees than bonobos.[15]

We can readily imagine that male chimpanzees learn the power their canines give them. Once their canines have erupted, which happens at about the age of ten years, the males are much more dangerous as opponents. Youthful acquisition of tall canines could therefore increase the willingness to fight. Something similar happens in humans. Boys who are big for their age learn as early as three years old that they can win in fights with smaller peers. Being rewarded for successful aggression, they end up being more aggressive throughout childhood. Large body size is even a risk factor for antisocial personality disorder. Just as being big affects the psychology of young boys, the longer and more dangerous canines of chimpanzees could in theory promote more aggressive behavior.[16]

We might expect that sex differences in size would also play a part in aggressive tendencies. In species in which males are bigger than females, such as gorillas or baboons, males tend to be more aggressive. Surprisingly, however, the little data available from the wild indicate that the difference in body weight between the sexes might be slightly less in chimpanzees (26–30 percent) than it is in bonobos (35 percent). Sheer body mass does not explain why male chimpanzees are more aggressive than male bonobos.[17]

A second plausible explanation for differences in aggression between bonobos and chimpanzees comes from their environments. Bonobos might have less to fight about, or might live in conditions that are less conducive to aggressive behavior. We will see shortly that differences in the food supply indeed appear to contribute to differences in aggression, although the effect is likely due to the effects of food type on the grouping patterns of chimpanzees and bonobos rather than to the benefits of winning access to contested resources.

Even more than anatomy and ecology, temperament is the key to understanding why bonobos are so peaceful compared to chimpanzees. In captivity, the two species have been studied in similar conditions, with the exigencies of the wild food supply entirely irrelevant. Much of this work has been conducted in sanctuaries where people dedicate their lives to looking after orphan apes that have been rescued from the bushmeat trade. Only one such sanctuary houses bonobos, the aptly named Lola ya Bonobo, meaning Bonobo Paradise, located in the Democratic Republic of the Congo near Kinshasa. Lola has been a haven for bonobos ever since Claudine André founded it in 1994. North of the Congo River, in the neighboring Republic of the Congo, orphan chimpanzees live in a similar home, the Tchimpounga Sanctuary, maintained by the United States' Jane Goodall Institute. Research at these two locations on dozens of bonobos and chimpanzees of all ages has done much to deepen our understanding of the two species.

The biological anthropologists Victoria Wobber and Brian Hare led the effort. Both were my graduate students. In 2005, I was delighted to accompany Hare and his wife, Vanessa Woods, on their first visit to Lola ya Bonobo. While designing experiments to study the behavior of a thoroughly non-murderous species, we occupied a guesthouse built by the former president, the murderous Mobutu Sese Seko.

Through systematic testing of dozens of Lola bonobos, as well as chimpanzees living in similar conditions in Uganda and the Republic of the Congo, Hare and Wobber confirmed and extended what we know about the two species' psychological differences. Outright reactive aggression is not easily studied in apes, because experimenters do not want individual apes to be stressed or wounded. But reactive aggression is closely related to emotional reactivity, which can be assessed by measuring social tolerance. In the first experiment designed to find out whether sanctuary bonobos were more tolerant than sanctuary chimpanzees, Hare simply put some pieces of banana in one or two small piles in an empty room, and then allowed two individuals to enter simultaneously through the same door. The species difference was beautifully clear. When two chimpanzees entered the room, normally only one individual fed: he or she dominated access to the food while the other went off on its own, despite showing obvious interest in the banana. But in the equivalent situation, bonobos showed none of the grabbing or monopolizing of food, nor sad withdrawals: two individuals would eat side-by-side without tension. Differences between the species were the same regardless of whether the test subjects were juveniles or adults, or males or females.[18]

A series of such studies all found the same kind of results. Bonobos shared food voluntarily, were more tolerant of others joining them at food, and were more skilled at tasks that demanded mutual tolerance, such as cooperating to get food that is out of reach. A particularly surprising finding came from studying whether bonobos preferred to eat alone or in the presence of another. Even when the potential companion belonged to a different social group, bonobos were so unaggressive that they voluntarily opened a door to allow the companion to join them and share their food pile. This left less food for the bonobo who opened the door, but that was not a concern. Company appeared to matter more than food.[19]

The contrasts in tolerance were supported by other differences. Bonobos were more playful and affiliative. When two bonobos were introduced into a room with food, a typical response was rushing toward each other for a sexual interaction before approaching the food. The interaction could vary from mild rubbing of each other's genitals all the way to full copulation, but, either way, the effect was the same as often seen in the wild: bonobos enjoy giving and receiving sexual pleasure, and they often use sex to relieve or deflect social tensions. After having sex together, whether in captivity or the field, bonobos fed next to each other easily. Chimpanzees never turned to play or sex in such tests.

In short, experiments with captive chimpanzees and bonobos have contributed a major part of the explanation for the pronounced species differences in frequency of reactive aggression. The differences are rooted in contrasting psychological tendencies. The greater tolerance shown by bonobos reflects a reduced emotional reactivity. Bonobos' placid nature indicates a lower propensity for reactive aggression than found in the boisterous, hot-blooded, charming, but dangerous chimpanzee. Neurobiologists are beginning to investigate how brain mechanisms contribute to these differences. In keeping with the behavioral results, relevant differences have been found in brain regions of the amygdala and cortex. Recall that higher levels of serotonin in the brain are associated with reduced reactive aggression. Strikingly, in bonobos, the amygdala contains twice as many serotonergic axons (nerves responding to serotonin) as it does in chimpanzees, suggesting one way in which bonobos have evolved a greater ability to regulate aggressive and fearful impulses. The biology of the brain appears, as expected, to be adapted to efficiently producing the kinds of emotional responses and social interactions that are characteristic of each species.[20] Genes must underlie the differences.

So how did this difference in psychology evolve?

Bonobos and chimpanzees have been separate species for at least 875,000 years, and possibly as long as 2.1 million years, based on our current understanding of genetic differences between species. That is, ancestors of chimpanzees and ancestors of bonobos diverged from their common ancestor species between roughly 0.9 and 2.1 million years ago. From that divergence until now, chimpanzees and bonobos evolved into the two species we observe today. An average chimpanzee generation is about 25 years, which indicates that the psychological and anatomical differences between bonobos and chimpanzees evolved during at least thirty-five thousand generations of separate evolution. If bonobos are self-domesticated, their ancestor must have been more aggressive than bonobos are. So a critical question for the self-domestication hypothesis is whether, thirty-five thousand generations ago or more, the common ancestor of chimpanzees and bonobos was more aggressive than bonobos are today.[21]

Behavior does not fossilize, and anyway no relevant fossils from that period have been found. However, bonobos have a series of anatomical distinctions from chimpanzees that are unique to their own lineage and that tell a story about their origins. Most notable is the feature that prompted the discovery of bonobos as a separate species from chimpanzees: the bonobo's juvenilized skull.

It took sharp eyes to realize that the skull of the bonobo was special. As early as 1881, a bonobo skull had arrived in the British Museum of Natural History, but no one noticed that it was different from that of a chimpanzee. From 1910 onward, more bonobo bones went to Belgium. Later on, Western scientists even had the chance to see a living individual. Prince Chim, a juvenile, was brought to the United States in 1923 before dying of pneumonia while in the care of the American primatologist Robert Yerkes. Yerkes assumed that Prince Chim was a chimpanzee, albeit one with a quite exceptionally delightful personality. Everyone else who met Chim thought the same, including one Harold Coolidge, a twenty-year-old student on his way to join an expedition in West Africa searching for primates. No one realized that Prince Chim belonged to an undescribed species. Dead or alive, bonobos stayed under the radar for almost fifty years after their 1881 introduction to scientists.[22]

The eventual breakthrough came out of nowhere. After his return from Africa, Harold Coolidge went to Tervuren, Belgium, in 1928 to measure gorilla skulls. Here, in his own words, is what happened when he visited the Royal Museum for Central Africa:

> I shall never forget, late one afternoon in Tervuren, casually picking up from a storage tray what clearly looked like a juvenile chimp's skull from south of the Congo and finding, to my amazement, that the epiphyses were totally fused. It was clearly adult. I picked up four similar skulls in adjoining trays and found the same condition.[23]

The skull bones were fused! So, even though it looked like a juvenile's skull, it had stopped growing. It had to be the skull of an adult. In juveniles of all mammals, flexible tissues called sutures link the growing bony plates of the skull. The sutures provide sufficient flexibility to allow the skull bones to move independently, accommodating the growing brain. Only when the brain has reached its maximum size do the sutures fuse, creating a stable structure. Coolidge mislabeled the sutures "epiphyses," but he understood perfectly the significance of what he was seeing.

Coolidge's observation meant that he was looking at a new variety of ape, similar in anatomy to a chimpanzee but differing by its adult skull being relatively small, rounded, and seemingly not fully developed, like the skull of a juvenile chimpanzee. Within days a German anatomist heard of Coolidge's finding and scooped him by quickly publishing an account of bonobos as a new taxon, but he made a mistake by calling them merely a subspecies. Coolidge had the last laugh by calling bonobos a full species. In 1933 the newly identified species was named *Pan paniscus* Schwarz 1929.[24]

As scientific scrutiny caught up with bonobos, the strange phenomenon of a juvenile-like skull in an adult body became more than than merely a distinction from chimpanzees. It also presented a way to reconstruct the evolutionary history of the newly named species. Chimpanzee and bonobo skulls are different now, but what was the skull of their common ancestor like? That is, since the divergence of the two species, in which species has the skull become more changed to a different form? The obvious possibility was that the bonobos' juvenile-like skull was an example of paedomorphism or adult retention of a characteristic found in the juvenile of its ancestor. If this skull was a case of paedomorphism, we could conclude that bonobos had evolved from a chimpanzee-like ancestor.

An alternative idea, however, also deserved consideration. The species difference could have come by peramorphism in chimpanzees rather than by paedomorphism in bonobos. Peramorphism refers to an adult characteristic being extended beyond its form in the ancestral species, and is thus the opposite of paedomorphism. If peramorphism explained the difference between bonobos and chimpanzees skulls, we could conclude that the bonobo skull morphology must have been the type found in the ancestor.

There is a way to tell whether the chimpanzee skull is peramorphic or, alternatively, the bonobo skull is paedomorphic: check the other apes. If the close relatives of bonobos and chimpanzees have a skull anatomy akin to that of bonobos, chimpanzees are the odd ones out; so the chimpanzee's skull would be peramorphic, whereas bonobos' would be relatively unchanged from the ancestral state. However, if the skulls of other apes are more like those of chimpanzees, the bonobo skull must be paedomorphic, a novel anatomy derived from an ancestor with a chimpanzee-like skull.

The answer is easy. The skulls of other apes, including gorillas, orangutans, and extinct species such as *Australopithecus,* are much more similar in these respects to skulls of chimpanzees than to those of bonobos. As the closest relative of bonobos and chimpanzees, gorillas provide the most relevant and revealing comparison. They follow the chimpanzee growth pattern so closely that they have been said to "resemble an 'overgrown' chimpanzee."[25] In short, bonobo skulls are paedomorphic; chimpanzee skulls are not peramorphic. The common ancestor from which both bonobos and chimpanzees descended most likely had a skull that looks like a modern chimpanzee skull. Bonobo skulls are the ones that changed; bonobos are the oddity.

Bonobos are oddities in other ways as well. Chimpanzees, gorillas, and orangutans all have limited estrus periods, and male dominance over females. Bonobos differ by having wildly extended estrus periods, and extensive female dominance over males. They have gone their separate way.

Why is it so important that we know that bonobo skulls changed so much? The skull houses the brain, and the brain directs behavior. Chimpanzee skulls retain the same ancestral style of cranial growth as gorillas and other great apes, suggesting that chimpanzee behavior has been relatively stable since they evolved from the last common ancestor of chimpanzees and bonobos. In contrast, bonobo skulls changed greatly, implying changes in bonobo brains and behavior. Identification of bonobos as a species separate from chimpanzees, and then establishing that bonobos are the species that changed more radically, indicates that the peaceability of bonobos is a new phenomenon, breaking away from the ancestral mold.[26]

Thanks to the fact that bonobos and chimpanzees are closely related yet differ greatly in their propensites for aggression and their skull morphologies, and they are blessed with a number of informative relatives, we can have considerable confidence that bonobo skulls, brains, and behaviors have changed more from their common ancestor than have those of chimpanzees. In other words, the low reactive aggression of bonobos is a newly evolved phenomenon. So we can entertain a strong prediction: Bonobos should show the domestication syndrome.

Brian Hare, Victoria Wobber, and I tested that prediction in 2012, and found the first evidence for the domestication syndrome in a wild species. The cranial anatomy of bonobos turns out to fit the domestication syndrome extraordinarily closely. To start with, bonobo brains (or cranial capacities) are smaller than those of chimpanzees. The reduction is especially marked in males, where it can reach as much as 20 percent. This echoes the decrease in brain size of almost every species of domesticated vertebrate compared with their wild ancestors. All of the other major cranial features of the domestication syndrome are present as well. The faces of bonobos are relatively short, projecting less than faces of chimpanzees. Bonobos have smaller jaws, and smaller chewing teeth. The skulls also show reduced exaggeration of maleness, with males more feminized than in chimpanzees, and sex differences smaller.[27]

These unusual features of bonobos have long been known, but they have not previously been discussed in relation to the theory of domestication. The physical anthropologist Brian Shea did as much as anyone to document the unique features of bonobo crania. He suspected that the key to understanding bonobos was the fact that the skulls of males and females were much more similar than they are in chimpanzees. "It seems likely," Shea wrote, "that the reduced sexual dimorphism in the facial region of [bonobos] is related to social factors such as lowered male-male and male-female aggression, increased female bonding, increased food-sharing, and perhaps aspects of sexual behavior." But how the morphological skull features are related to the behavioral tendencies remained unclear. And why should bonobos have smaller brains or smaller chewing teeth? They live in forests that are very similar to those occupied by chimpanzees a few miles away across the River Congo. The adaptive problems faced on either side of the river seem too similar to explain these important species differences in any simple way.[28]

In the light of self-domestication theory, the differences make sense. As in captivity, we can reconstruct that, when there was selection against aggression in bonobos, the domestication syndrome emerged. Small brains, short faces, small teeth, reduced sex differences, and a paedomorphic skull—all are features both of domesticated animals found in bonobos. Admittedly, bonobos do not have floppy ears or white spots in their coats. Perhaps thirty-five thousand generations have eradicated those common components of the self-domestication syndrome, or perhaps the bonobos never acquired them. The frequency of these characteristics varies among domesticated animals. Few cats are floppy-eared, and water buffalo rarely have white spots. Still, depigmentation does occur in bonobos. Most individuals sport around their lips a striking pinkness, a loss of pigmentation that could well be associated with a delayed migration of neural-crest cells similar to the process known to occur in domesticated species. And at their rear end bonobos feature a tuft of white hair like that of chimpanzees in infancy. Bonobos differ from chimpanzees in this regard by retaining the white tail tuft (paedomorphically) until adulthood.

Beyond anatomy, bonobo social behavior also remarkably fits the behavioral pattern Belyaev identified in domesticated silver foxes. In addition to reduced aggressiveness, there are two features of social behavior that are especially characteristic of domesticated animals: sex and play.[29]

Domesticated animals such as dogs and guinea pigs show a greater variety of sexual behavior than their wild counterparts. So do bonobos compared with chimpanzees. Homosexual behavior is a striking example. In young primates, males readily mount both males and females in a premature version of sexual copulation that includes no actual intercourse. As they mature, the males shift their mounting to females, so that during adulthood homosexual mounts are very rare. Chimpanzees follow this pattern, but bonobos exhibit extensive homosexual behavior as adults. If homosexual mounting found among adulthood is a retention from the juvenile period, as these observations across species suggest, it is paedomorphic.[30]

Adult homosexual behavior among bonobos is especially prominent among females. Among behavioral biologists it is prosaically called genito-genital rubbing, whereas in the Congo it is known as *hoka-hoka*. *Hoka-hoka* typically involves two females face-to-face, excitedly swinging their genitals side-to-side. The interaction sometimes ends in an orgasm-like pause including a tense face and contracted limbs. *Hoka-hoka* frequently follows social tension, such as when a subgroup of females finds an especially exciting food, or after a conflict between two females. If *hoka-hoka* is paedomorphic in chimpanzees, therefore, one should expect to see juvenile female chimpanzees engaged in similar behavior. They hardly ever do, but there are occasional reports of it.

A lovely example happened in Uganda's Kibale National Park in 1994, several years after I started studying chimpanzees there. Our research team had discovered a juvenile chimpanzee being kept illegally as a pet in a local village, presumably orphaned after her mother had been killed for meat. With permission from the appropriate authorities, we rescued her. We called her Bahati and tried to introduce her into our wild study community. Bahati was about five or six years old. Females normally join a new community when they are around twelve years old, so she was definitely younger than she should have been to try entering a new group. She was also in poor physical condition, since she had not been able to climb trees while she was kept in the village. For three weeks, the researchers Lisa Naughton and Adrian Treves camped with Bahati in the forest to help her build up her strength and re-adapt to forest foods.[31]

One day, a subgroup of our chimpanzee study community was nearby, and Lisa and Adrian led Bahati to them. The males were fascinated to meet the new juvenile. After some tense charging (which persuaded Bahati to stay close to Lisa and Adrian), some of the males approached her more gently and hugged her. It was a big relief for the humans that the wild chimpanzees were so welcoming to Bahati. Bahati seemed to feel the same. At any rate, while Lisa and Adrian watched nervously, Bahati went off with the wild chimpanzees to spend her first night away from humans since her capture, months earlier. She stayed with her new friends, day after day.

A few weeks later, I was filming Bahati as she traveled with some of our chimpanzees. By then, the males no longer showed much interest in her, but she had developed friendly relationships with others of her age. At one point, a female of her own age, Rosa, waited for Bahati. As Bahati approached, Rosa rolled onto her back and opened her arms to encourage the still-timid orphaned stranger. Bahati embraced her, the two young ones hugged, and they swung their pelvic regions against each other. I had never seen this behavior before in chimpanzees, but it was instantly familiar. It looked like *hoka-hoka* between female bonobos. The meaning seemed clear. The behavior of Bahati and Rosa was a rare chimpanzee juvenile act. Bonobos had extended and elaborated it paedomorphically into a characteristic feature of their adult social lives.

Like homosexual behavior, social play is seen more in domesticated species such as dogs than among wild ancestors such as wolves; and it is also found among young primates more than among adults. And again, as Isabel Behncke showed, like a domesticated species, adult bonobos play more than do adult chimpanzees. The primatologist Elisabetta Palagi made careful comparisons between bonobos and chimpanzees in captivity where the two species lived in similar conditions. Not only did adult bonobos initiate play and use play faces more often than adult chimpanzees, but, interestingly, bonobos also played more roughly. One might have expected rough play to be the choice of the more aggressive chimpanzees, but since roughness demands more tolerance from the partner, rougher play is explicable by the overall nonaggressiveness of bonobos.[32]

In bonobos, sex and play are often linked. In Behncke's words, "erect penises, playful intromission, and exploration of mature females' [sexual] swellings" are some of the many elements that can make up a bonobo play session, unlike anything seen in chimpanzees.[33]

The case for bonobos' being self-domesticated could hardly be stronger. Bonobos are unequivocally less aggressive than chimpanzees. Their common ancestor with chimpanzees is best reconstructed as having a chimpanzee-like skull, brain, and behavior. And the bonobos' differences from chimpanzees are characteristic of the domestication syndrome, whether anatomical (in the cranium) or psychological (in their sex and play). Those features of bonobos have not been explained by conventional adaptive logic. Thus no convincing reasons have been advanced using the parallel-adaptation hypothesis, which would state that bonobos' reduced aggressiveness, paedomorphic skull, short face, and small teeth have all evolved in parallel as a result of responding to a series of separate selective pressures. However, Belyaev found a pattern in domesticated animals that appears applicable to bonobos. Belyaev's rule, as we can call it, is that selection against reactive aggression leads to the domestication syndrome. That idea fits bonobos, because selection appears to have occurred against reactive aggression, generating features of the domestication syndrome. Seen in this light, many of the characteristics that so richly distinguish bonobos from chimpanzees did not evolve as adaptations. Instead, they evolved as incidental side effects of selection against reactive aggression. Genetic tests of this hypothesis will be rewarding, especially comparison of neural-crest genes in chimpanzees and bonobos.

The evidence offered by bonobos has exciting implications for vertebrate evolution. It suggests that in other species that have experienced a downward regulation of aggression, a self-domestication syndrome will likewise be found. If so, the male nipple of mammals might prove to be less unusual than it has seemed. Much as the male nipple results from a developmental constraint rather than being an evolved adaptation, so might short faces, small teeth, white patches, and other traits of the domestication syndrome also prove to have been produced by developmental constraints—even in wild animals. The bonobo evidence implies that reduced aggression has routinely led to incidental side effects.

It has a further implication of direct relevance to human evolution. If bonobos could be self-domesticated, the bonobo case supports the idea that humans could be self-domesticated too.

But what the evidence of self-domestication does not do is explain why bonobo aggression was reduced.

You might think that more aggressive individuals would always fare better in the competition for evolutionary success. In fact, of course, too much of anything is a bad thing. An animal that fights too often, or too intensely, wastes energy and takes unnecessary risks. The trick is to get the balance right, to fight in the right contexts and at the right level of intensity, and only when the payoff is worthwhile.

So what is it about bonobos that makes aggression less profitable for them than it is for chimpanzees? Among chimpanzees, males practice the most frequent and dangerous forms of violence, and yet male bonobos are relatively unimposing, so the question is really about males. Ultimately, bonobo psychology has evolved to the point where males show less interest in dominating others, whether female or male, than chimpanzees do. The deeper question is why, over evolutionary time, males with gentler, less aggressive proclivities tended to have higher reproductive success.

Female power is clearly an important part of the answer.[34] A male bonobo who confronts an adult female might well win if she is the only female in earshot. But female bonobos are rarely far from other females. The challenging male must expect that, if he makes a female scream, within seconds he may be confronted by a coalition of females ready to attack him, and so effective in doing so that his best response will be to run away. Female support for one another explains why males give up easily when competing with females over food, or why males rarely try to bully females, or why males do not, on average, outrank females. Coalitionary attacks need not be common. The primatologists Martin Surbeck and Gottfried Hohmann found that, although females in the wild can use coalitions well, they do so rarely, mostly when males threatened their young.[35] Despite their size disadvantage, females very effectively suppress bullying by males. Males seem to have learned where the ultimate power lies: numbers beat physical strength.

The reason that female bonobos can predictably present a united front appears mundane: they stay close to each other. Bonobo parties consistently contain a core of female associates and tend to have more females than males. Mostly the females are unrelated to each other by any ties of close kinship, because they are immigrants who joined the community as strangers when they were late juveniles or early adolescents. In the absence of kin ties, immigrants take several weeks or more of patiently following a group to be accepted. Eventually they engage in *hoka-hoka,* play, and grooming to become well integrated into the network of resident females. From then on they can count on support.[36]

Chimpanzee parties, by contrast, are numerically dominated by males. Females tend to travel alone or in smaller subgroups. It seems likely that because of this relatively dispersed way of life, female chimpanzees fail to gain confidence in each other's support against males. The only circumstances in which female chimpanzees have been seen to successfully mob aggressive males has been in those zoos where groups of females were alone together for months before males were introduced. Without males, chimpanzee females developed mutual trust. In the wild, most adult females apparently spend too little time together to learn to depend on each other.[37]

As we peel back the layers of bonobo evolution, we note that females can form their defensive coalitions because they form stable associations, and males are less aggressive because females render their aggression ineffective. But why are female bonobos more capable of forming stable associations than female chimpanzees? Animals are ultimately adapted to their environments, so the obvious place to start looking for an answer is in the special characteristics of the habitats of bonobos. In most respects, the habitats of chimpanzees and bonobos are very similar. The critical need for both species is access to rain forests or riverine gullies that harbor trees producing abundant fruit. Their habitudes differ in latitude, since chimpanzees live on average farther north and bonobos farther south. The Congo River, which separates the two species, curves so wildly, however, that in places chimpanzees live north, west, east, and even south of bonobos. Thus, in the equatorial region, the climate, soils, and forest types occupied by the two apes cannot be strictly distinguished. The forests vary on both sides of the river, and there is no evidence of any systematic difference in botanical structure or fruit production for the two ape species.

Nevertheless, a major zoological difference between their habitats affects the availability of the bonobos' food. Gorillas are found throughout the equatorial areas occupied by chimpanzees but are absent from bonobo habitats. The presence or absence of gorillas seems to account for a cascade of effects linking bonobo diet choices to their grouping patterns and social alliances and, ultimately, to their reduced aggressiveness. The cascade starts with gorilla competition for the foods eaten by chimpanzees. By contrast, bonobos, having no gorillas living in the same area, are relieved from that competition. So bonobos have more food choices than chimpanzees do.

Gorillas are the only other great ape in Africa. They resemble bonobos and chimpanzees in being largely confined to rain forests, and their diets are broadly similar, too: they eat fruits when they can find them easily, and leaves and stems when fruits are scarce. But gorillas are much larger than the other apes. Females gorillas are two to three times the weight of female bonobos and chimpanzees, and males are three to four times heavier, averaging around 170 kilograms (375 pounds). Gorillas' larger bodies mean that individuals find it harder to eat enough fruits during times when there are few productive trees. As a result, gorillas readily resort to a diet dominated by leaves and stems. Populations of gorillas high in the mountains even eat leaves and stems full-time, because at altitudes above 1,800 to 2,400 meters (6,000 to 8,000 feet) the climate is too cold to support more than the occasional edible fruit.[38]

The plant types preferred as food are mostly the same for all three African apes. They are the young leaves, basal stems, or growth tips of large fast-growing plants such as Zingiberaceae (gingers), Marantaceae (arrowroots), and Acanthaceae (acanthuses). These plants tend to occur in "meadows," often occupying gaps in the forest created by a treefall.

The ability of gorillas to specialize in eating these herbaceous foods, and their willingness to do so once fruits are scarce, appear to create a problem for chimpanzees. Every morning chimpanzees eat newly ripened fruits as their first main meal within a few minutes of leaving their night beds. They continue eating fruit until ripe fruits become too scarce to be found easily, which might be around midday. Then they resort to finding a patch of leaves or stems to eat. But if gorillas have been there first, the herbaceous food patch will be inadequate for the chimpanzees. So the chimpanzees have to search for other food patches. Mothers travel slowly because of the presence of their young, and cannot keep up with fast-walking males. Needing to eat for many hours each day, they disperse, often alone, to find the small patches of food that will give them the calories they need.

Bonobos, in contrast, without feeding competition from gorillas, have free rein to eat all of the ape foods that flourish in the environment. No other animal in the habitat of bonobos offers serious competition for eating these choice herbs, so bonobos can take the best of them. And that makes all the difference. An ape that can rely on cropping the herbaceous meadows can travel in relatively stable subgroups, slowly working their way from plant to plant. That is what gorillas do. The bonobos' daily access to "gorilla foods" appears responsible for subgroups' relative stability (as is also the case with gorillas) when compared with the shifting and smaller subgroups of chimpanzees.[39]

The chain of logic takes us to a final question. Why are there no gorillas in the bonobo habitats? While the ancient distribution of gorillas is unknown, we do know that there are no mountains on the south side of the Congo River, where bonobos live. On the north side of the river, where chimpanzees live, there are mountains to the west and the east. The western mountains of Nigeria, Cameroon, and Gabon are the centers of diversity of the western gorilla. The eastern mountains of the Democratic Republic of the Congo, Rwanda, and Uganda are the core areas of the eastern gorillas. Mountains are the sites where gorillas can survive when a hot, dry climate means that the flat lowlands lose their lush herbs.

If the lack of mountains on the south side is as significant as I think it is, we can reconstruct bonobo history as follows: The beginning of this history is undisputed, thanks to recent geological data. For longer than apes have been in existence the Congo River has been a barrier to animals moving south: maritime sediments show that the Congo has been pouring into the Atlantic Ocean for 34 million years. So the ancestors of chimpanzees, bonobos, and gorillas would always have lived north of the Congo River.[40]

Opportunities to cross to the south of the Congo River came during the Pleistocene era, which began about 2.6 million years ago and included cold, dry, Ice Age periods. Signs of low rainfall are found in the outflow area of the Congo River, in the form of marine sediments recording deposits of African dust. The deposits of dust are thought to coincide with reductions in the extent of forest, when the climate was dry. One such dry period occurred about 1 million years ago. The reduced rainfall probably caused the upper reaches of the Congo River to become sufficiently shallow that in a few places, even a nonswimming species like a great ape could make its way across. The ancestors of chimpanzees and bonobos duly crossed the river. They found an area much like the drier regions occupied by chimpanzees today. As long as there were riverine gullies where fruit trees could survive, the ancestral species would have thrived.[41]

Gorilla ancestors might have crossed, too. But even if they did, the south side of the Congo River was unsuitable for them. The lack of mountains meant that there were no areas sufficiently damp to house the moist herbs and stems that gorillas needed for food. So, even if gorillas did cross the Congo River a million years ago, they would have died out in that area not much later.

Then, perhaps after a few thousand generations, the rains came back, the river became a barrier once more, and the lowland habitats grew lush again. South of the river there was plenty of food for two species of ape, but there were no gorillas there to crop the abundant forest herbs. The only ape was the ancestor of chimpanzees, which now evolved to become the ancestor of bonobos. The proto-bonobos thrived, eating both tree fruits and the newly abundant high-quality herbs. Mothers shifted from a frequently solitary foraging style to begin travelling, gorilla-like, in stable, larger subgroups of individuals of both sexes sharing the herb meadows. Males who tried to bully the females could now be repulsed.

With females' power came a greater ability to choose the less aggressive males as mates. Females greatly extended their periods of sexual receptivity, and in so doing evolved a concealment of estrus. They could afford to have long periods of sexual attractiveness, since the presence of interested males was not a big problem in the big patches of gorilla food: there was little competition for the lush herbs. The males became much less sure of when to compete with each other, so intimidation of females no longer paid off as it does for male chimpanzees. As selection increasingly favored the less aggressive males as mates, the self-domestication syndrome emerged. Homosexual behavior emerged spontaneously and was then woven into the bonobo social system as a means to strengthen bonds and reduce tensions.

Genetic evidence indicates that the timing of this picture is a little more complicated than I have presented it so far. The first crossing of the Congo River by chimpanzee ancestors appears to have been followed by at least two other dry periods when ancestors of chimpanzees and ancestors of bonobos coexisted again and briefly bred together. This interbreeding did not have much genetic impact, however; less than 1 percent of genes in central African chimpanzees can be traced to bonobos.[42]

So a series of Pleistocene droughts allowed chimpanzee ancestors to cross the great river barrier and evolve into bonobos. Bonobos were left occupying a relatively small area compared with other great apes, and their numbers have been reduced by habitat loss and hunting so that now there are only between ten and fifty thousand in the wild. We are extraordinarily fortunate that they exist. They provide an illuminating contrast to chimpanzees, and they also attest to the power of Belyaev's rule. They give us the best indication yet found that the process of becoming peaceful can have effects in the wild similar to those in captivity.[43]

—

Bonobos open a window on a previously unseen world. If we peer through it, we should eventually expect to see the domestication syndrome in many places. A reduction in reactive aggression should prove to be a common evolutionary phenomenon.

The combination of two features of bonobos makes them a particularly intriguing species. Their social behavior includes a series of unusual patterns, including markedly less aggression than is found among chimpanzees; and they are one of our two closest relatives. The combination might be interpreted as suggesting that something about bonobos' close relationship to humans, such as high cognitive ability, predisposes them to self-domestication. Yet that suggestion is unwarranted. Self-domestication should depend solely on whether natural selection happens to favor a reduced tendency for reactive aggression, regardless of whether the species is closely or distantly related to humans. Sometimes a ready propensity for reactive aggression is an advantage. If the competitors who are most ready to fight tend to win the competition for status, food, and mates, to have more infants, and to survive better, reactive aggression is favored and self-domestication will not occur. But changed conditions of life can change the costs and benefits of a given behavior. Being too quick to lose your temper might then no longer pay dividends. The evolution of modern chimpanzees and bonobos from a common ancestor offers a compelling example of different environments favoring different levels of reactive aggression.

Animals living on islands suggest another example. Islands are like natural experiments that provide insights on evolutionary processes. Islands are almost always younger than the neighboring mainland. So species living on islands have normally evolved from the continental forms, rather than the other way round.

Diverse species living on islands have been compared with their close continental relatives. The comparison yields a pattern so strong that it is called the Island Rule. It applies to mice, lizards, sparrows, foxes, and many other species. The Island Rule starts with size. It says that large animals isolated on an island tend to get smaller. Bones of different species of miniature elephants are found on islands off the coast of California, in the Mediterranean, and in Southeast Asia. Being stuck on an island has produced the same consistent results: charmingly small elephantoids, some only one meter (three feet) high at the shoulder. As a parallel generalization, small animals, those weighing less than about one kilogram (2.2 pounds), tend to get larger on islands. For example on islands in the Indian Ocean, an ancient fruit pigeon became the dodo, to the delight of the sailors, feral pigs, and introduced monkeys, who ate it to extinction in the seventeenth century.

The Island Rule applies not just to size, but also to many aspects of species' growth and reproduction. Island animals tend to have delayed sexual maturity, to have fewer offspring in their litters, to live a longer time, and to have reduced sexual dimorphism—in other words, males are physically more similar to females than on the continent.[44]

The reason that island animals are interesting for thinking about self-domestication is that effects on behavior are equally pervasive. Island animals tend to be less reactively aggressive than their ancestral kin. Lizards, birds, and mammals all show the trend. Some animals abandon any efforts to defend territories, even though the continental relatives are fully territorial. If a territority is maintained on an island, it is relatively small, overlaps more with the neighbors' territories, and might be shared with subordinates. In experiments, two animals put into the same cage are less likely to fight each other if they come from an island than if they are from the mainland. All these behavioral changes may be influenced by evolved psychological differences—specifically, reduced reactive aggresssion in island animals.[45]

The reduction in reactive aggression is explained by an island's being too small to hold a full complement of predators, which means that the risk of being killed is less than on the continent. As a result, animals survive longer on islands and live at higher population densities. Island populations are therefore relatively crowded, which means that being too aggressive can be overly exhausting. For example, defending a territory may not be an effective strategy when, as soon as the territory holder has chased one invader out, three more appear. If aggression does not pay, it is better not to waste time and energy and incur high risks by fighting. Under these conditions selection favors the less aggressive.[46]

The generalization that island animals are relatively unaggressive to members of their own species leads to a simple prediction: island animals should tend to show the self-domestication syndrome. That hypothesis has not been investigated systematically, but some cases are strongly suggestive. Take the Zanzibar red colobus monkey, a species that lives only in Zanzibar.

Zanzibar is a political unit in the Indian Ocean, an archipelago including two main islands, Unguja and Pemba. The islands both lie twenty to thirty kilometers (twelve to nineteen miles) off the Tanzanian coast and are the only places in the world where Zanzibar red colobus monkeys live. Pemba has been separated from the African continent for a million years or more. Molecular data indicate that the Zanzibar red colobus monkey, as a species, is about 600,000 years old, suggesting that it evolved there not long after Pemba became an island. The species looks strikingly different from all the other red colobus monkeys on the African continent, of which some sixteen species have been described.[47]

Almost every feature that distinguishes the Zanzibar red colobus monkey from all the other red colobus monkey species fits the domestication syndrome. Compared to all but one of the continental forms, Zanzibar red colobus are smaller and more lightly built and their faces are relatively short. The reduction in body size is greater for males, so much so that some authorities believe that Zanzibar red colobus females may be larger than males. Zanzibar red colobus are also paedomorphic, meaning that adults retain the juvenile characteristics of related species. Among the Ugandan red colobus monkeys that I am familiar with, a pink outline around their mouths is seen only in infants less than a few months old; yet that same pink shape is retained throughout the lives of the Zanzibar monkeys. The shape and size of the whole skull are also paedomorphic, including large eyes, small face and a relatively small braincase. In all of these features, the Zanzibar red colobus monkey is to a continental red colobus as a dog is to a wolf.[48]

As research into self-domestication proceeds, islands should prove particularly rewarding, because they seem likely to provide repeated opportunities to test the key inference stemming from Belyaev's work. In my view, however, bonobos have already provided the crucial breakthrough. Bonobos support the simple prediction that comes from research on silver foxes: selection against reactive aggression will produce the domestication syndrome even in the wild.

6

Belyaev's Rule in Human Evolution

BELYAEV'S RULE, AS I called it in the previous chapter, is that, in captivity, selection against reactive aggression causes a domestication syndrome. Belyaev's rule now appears to apply to the wild too—almost certainly in bonobos, likely in Zanzibar red colobus monkeys, and perhaps in many other species as we begin to look for evidence. How the selection happens should not matter for Belyaev's rule to work. The species might have been deliberately domesticated by humans, as mink were. It might have self-domesticated in the presence of humans, as dogs likely did when wolves became increasingly drawn to the refuse of human camps. Or, like bonobos, it might have self-domesticated entirely in the absence of humans. Belyaev's rule apparently applies in all these circumstances. Select against reactive aggression, and the domestication syndrome emerges.

Belyaev's rule appears to be so strong that we can use it in reverse by inferring, from the presence of a domestication syndrome, that a species has undergone selection against reactive aggression. This reversed version of Belyaev's rule gave us an explanation of many behavioral and biological oddities of bonobos. We can now apply the idea to humans. Helen Leach has already identified humans as having a domestication syndrome, based on our skull and skeleton. According to Belyaev's rule, the implication is clear. In the course of evolution, humans experienced selection against reactive aggression.

We should be able to tell when the selection happened by discovering when the domestication syndrome started. All we need is to find a good fossil record. That takes luck. With some species we have no luck. Not a single fossil is known for bonobos, so we do not know when their domestication syndrome began. A reasonable guess is that it happened shortly after the bonobo line first separated from the ancestors of chimpanzees, at least 875,000 years ago. Maybe eventually fossils will test that idea.

Humans, by contrast, have left a rich fossil record that allows us to trace the ancestry of our genus for two million years and more. Given Belyaev's rule, the fossil record becomes highly instructive, because a domestication syndrome is indicated as having been present during just one phase and species of the genus *Homo.* That phase is the last three hundred thousand years, and the species is *Homo sapiens.*

To offer a précis, let's greatly simplify a rich and complex story. Two kinds of *Homo* dominated our evolution for at least the last 250,000 years. One is a series of robust, archaic types of *Homo;* and the other is us, the more lightly built, gracile *Homo sapiens.*

These two types were admittedly not the only players. The small-brained *Homo naledi* populated part of southern Africa, where bones found in a deep, dark cave were dated to about 300,000 years ago. On Flores Island in Indonesia, there was a diminutive species that also had small brains, the so-called hobbits or *Homo floresiensis.* Hobbits lived as late as 65,000 years ago and possibly as early as 700,000 years ago. *Homo naledi* and *Homo floresiensis* are fascinating and mysterious, but they represent side branches that show no signs of having contributed to our ancestry.[1]

The critical time and place for the origin of *Homo sapiens* was the Mid- to Late-Pleistocene in Africa. The 2.6-million-year Pleistocene era is the time during which our lineage changed from a chimpanzee-sized prehuman into a culturally sophisticated and psychologically modern *Homo sapiens.* When the Pleistocene began, our ancestors were habilines, variably called *Australopithecus habilis* or *Homo habilis.* The habilines' uncertain naming reflects their status as part ape (they had small bodies and large jaws) and part human (their brains were bigger than those of apes). Shortly after 2 million years ago, the habilines spawned the first undisputed members of our genus, *Homo erectus.*

By the time the Pleistocene had its final ice age and gave way to the warmer Holocene, 11,700 years ago, the only living descendants of *Homo erectus* were *Homo sapiens.*[2]

The various species of *Homo* all appear to have originated in Africa, but several also colonized other parts of the world. At least four times during the Pleistocene, Africa introduced populations of *Homo* into Europe and Asia. *Homo erectus* reached Indonesia and China by 1.8 million years or earlier. A subsequent expansion gave rise to a population that some call *Homo antecessor,* living in Spain by 800,000 years ago, which was very similar to another European group, *Homo heidelbergensis. Homo neanderthalensis,* or Neanderthals, entered Europe probably by way of the Middle East around five hundred thousand years ago. Each new species flourished for a time, only to be replaced by the next wave coming from Africa. *Homo erectus, Homo antecessor, Homo heidelbergensis,* Neanderthals and their ancestors were all members of the archaic type of *Homo,* whereas the last species to emerge from Africa was the less hefty, more gracile form, *Homo sapiens.*

When *Homo sapiens* first reached Europe and Asia is uncertain, but an expansion 100,000 to 60,000 years ago seems to have been the critical move that led them to diversify into most of the various familiar populations around the world, including spreading within the African continent. By the end of the Pleistocene, almost 12,000 years ago, *Homo sapiens* were hunting and gathering with sophisticated tools. Some populations were already occupying settled villages, living with dogs, decorating cave walls with multicolored paints, using pottery, and grinding grain. Shortly afterward, about 10,000 years ago, the agricultural revolution began.[3]

Too few fossils have been found for us to be sure when and where an archaic form first started differentiating into *Homo sapiens.* To be unambiguously *Homo sapiens,* skulls must be markedly round (globular) in profile with a clearly flexed base, and have such a small face that it is tucked mostly under the cranium. The earliest examples with these features come from the Omo River in Southern Ethiopia, dated to 195,000 years ago.[4] Shortly after that time *Homo sapiens* is found more broadly in Africa and later in the Middle East.

So where and when did definite *Homo sapiens,* such as those from the

Omo River, originate? Fossils that seem to be transitional toward being full *Homo sapiens,* and possibly the earliest of our species, come from a desert area on the west coast of Morocco, a site called Jebel Irhoud. Bones and teeth were unearthed there during mining operations in the early 1960s, and more material was excavated recently, leading to a total of at least five individuals, including three skulls. Compared to other Mid-Pleistocene *Homo,* features that hint at later developments include a less protruding face, slightly smaller chewing teeth, and less prominent brow ridges. In 2017 paleoanthropologist Jean-Jacques Hublin and his colleagues dated the bones to 315,000 years ago (± 34,000 years). Hublin's team argued that while the ancient Moroccans were certainly very different from contemporary humans (their faces are still relatively huge and their braincases show no trend toward being rounded), a few anatomical changes in face and teeth nevertheless mark them out as the forerunners of a new evolutionary direction. The Jebel Irhoud people, Hublin's team concluded, represent the first intimation of *Homo sapiens,* an early premodern version.[5]

To call the Jebel Irhoud material *"Homo sapiens"* is controversial because the name *Homo sapiens* is normally reserved for populations having the round skull and short face that are absent in Jebel Irhoud, and that are found only from about 200,000 years onwards.[6] Hublin's team called their Moroccan material *"Homo sapiens"* not because it conformed to the standard definition, however, but because it seems to have been launching a trend. Future discoveries might cause *Homo sapiens* to be given a different starting point. In this book, however, in line with Hublin's proposal, I will treat the Jebel Irhoud population as the earliest known version of *Homo sapiens.*

The date for the Jebel Irhoud type, which we can approximate as being 300,000 years ago, accords comfortably with recent finds from genetics and archaeology that suggest the same period for the deep origins of *Homo sapiens.* Based on genetic differences among living peoples, the ancestors that gave rise to everyone alive today are estimated to have lived between 350,000 and 260,000 years ago. That timing meshes, too, with archaeological finds indicating that cultural developments were starting to accelerate. The "Levallois" method for knapping stone tools is an important example of such a cultural development, because compared to earlier technologies it demands enhanced cognitive ability. The Levallois technique requires the user to shape a rock before trying to produce flakes from it. The technique leads to smaller, more elegant, and more efficient stone knives than before, and is first known from 320,000 years ago in Kenya's Olorgesailie Basin. Other recent discoveries in the Olorgesailie Basin show that by 320,000 years ago, humans were becoming more selective in their choice of raw materials for making stone tools. Instead of putting up with low-quality local sources, for example, they would obtain high-quality material such as obsidian from as far away as ninety kilometers (fifty-six miles). The Olorgesailie population is also the earliest that is known to have collected red ochre, which they presumably used as a pigment. Thus fossil, genetic, and archaeological evidence converge on 300,000 years ago as a time of change. Between a half and a quarter million years ago, the line uniquely leading to *Homo sapiens* was apparently beginning to emerge.[7]

The immediate ancestors of the Moroccan *Homo sapiens* are not well known and do not even have an agreed-upon name. In the past they have sometimes been called "archaic *Homo sapiens.*" However, to refer to predecessors of the Jebel Irhoud types as a type of *Homo sapiens* is confusing, because they also gave rise to additional species besides *Homo sapiens:* Neanderthals are the best-known example of their other descendants. Other labels sometimes given to the Mid-Pleistocene forerunners of *Homo sapiens* refer to fossils named *Homo heidelbergensis* from Europe or *Homo rhodesiensis* from Africa. But which if any of these names is a good fit is unknown: fossils that can clarify the relationships are simply missing for the moment. Paleontologist Chris Stringer prefers to call our pre-*sapiens* ancestors by the noncommittal term "Mid-Pleistocene *Homo.*" The Middle Pleistocene lasted from 780,000 to 130,000 years ago, so it clearly covers the period of interest. I will therefore follow Stringer by referring to the archaic *Homo* ancestors of *Homo sapiens* as Mid-Pleistocene *Homo.*[8]

Suppose you could travel backward in time to encounter those ancient people. The behavior of the Mid-Pleistocene *Homo* would in some ways seem familiar. If you saw a small group in the distance, images shimmering in a dry-season haze, you would instantly recognize them as human: similar size, same shape, same stride. As you came closer, you would start to see some less familiar features. The people are heavily muscled, both men and women, more like wrestlers than runners. Their faces are strikingly broad and strong, particularly those of the males. Heads are somewhat slanted, sloping forward from the peak down to a great big brow ridge, and lacking a prominent forehead. That brow ridge is wide and thick, giving a daunting look to the eyes. Large mouths surmount a heavy, chinless jaw.[9]

An informative picture of the kind of camp life they would have enjoyed comes from a well-documented open-air site in Israel next to an ancient lake. Around 780,000 years ago the site, which is now called Gesher Benot Ya'aqov, was used for about 100,000 years. Which species of *Homo* lived there is unknown, but to judge from the timing it may have been *Homo erectus,* the forerunner of Mid-Pleistocene *Homo.* Whatever name we give them, they indicate a sophisticated system of hunting and gathering. A team led by archaeologist Naama Goren-Inbar has studied the rich remains of nutshells, animal bones, and tools of wood and stone. Goren-Inbar's team found that these *Homo* ate dozens of different kinds of plants depending on the season, including seeds, fruits, nuts, vegetables, and water plants. They used fire so continuously throughout the period of habitation that they could apparently make it at will. Combined with the evidence of butchery, fire use suggests that the camp would have routinely smelled of delicious roasting meat, often from deer but including animals as large as elephants. People also knapped rocks for various uses. Among various tools were sharp-edged cleavers, scrapers, and small flint flakes that they likely hafted onto spears. They brought thin slabs of basalt to camp to use as stone "boards," apparently for cracking nuts or pounding meat. The preparation of some foods, such as prickly water-lily nuts, would have been demanding. Based on the way people harvest and prepare the same species today, the Mid-Pleistocene *Homo* would have had to dive under water to gather the nuts, dry them, roast them, and perhaps pop them. These were well-organized foragers.[10]

When enough fossils are found so that paleontologists can properly characterize different populations of Mid-Pleistocene *Homo,* it is possible that Africa in that era will prove to have held more than one species (in addition to the mysteriously small-bodied and small-brained *Homo naledi* of southern Africa, which was contemporaneous with the Jebel Irhoud population). The environment was as diverse then as it is today. At different times and in different places, there were closed forests, open valleys, and vast bush. There were millennia of drought, and others of high rainfall. Sometimes desert or water created barriers that kept populations apart for long enough so that evolutionary differences could have emerged. Depending on exactly when and where your time machine happened to drop you, you might find yourself encountering any one of a number of populations, each representing a specific period in a particular region. But the potential differences need not concern us. With the question of self-domestication in mind, the key point is that, prior to *Homo sapiens,* every population of large-bodied *Homo* had a relatively broad, heavy skull and thick-limbed skeleton. Even after the first *Homo sapiens* evolved, some typical Mid-Pleistocene populations persisted. Their archaic looks were of a species that differed from *Homo sapiens* rather as a chimpanzee does from a bonobo, or a wolf from a dog.[11]

Over time, the morphological transitions in the direction of *Homo sapiens* that are first detectable in the Jebel Irhoud people became elaborated and intensified. Fossils from elsewhere in Africa show that, sometime after 200,000 years ago, there were further reductions in the size of the face and brow ridge. Sex differences also diminished, as male faces became more feminine. Much later, during the Upper Paleolithic from 40,000 years ago onwards, the whole body also became lighter, to judge by a reduction in the diameter of the femur (the thighbone). The limbs became less robust in a further way too: they were less bony. The effect is seen in a cross-section of an arm- or legbone. The wall of bony cortex that surrounds the marrow-containing cavity became thinner. In the last 35,000 years, sex differences in height and tooth size have likewise been reduced. In all these ways, modern *Homo sapiens* is a less forcefully male species than our ancestors were 300,000 years ago. Our ancestors became feminized.[12]

The anatomical components of the domestication syndrome identified for humans by Helen Leach were smaller bodies, shorter faces, reduced sexual dimorphism, and smaller brains. As we saw, the first three of these are found during the greater part of *Homo sapiens*'s evolutionary history. Smaller bodies are indicated by thinner femurs from at least 200,000 years ago. Smaller faces have been used to characterize the Jebel Irhoud fossils as belonging to *Homo sapiens*. Whenever it has been possible to assess sex differences in anatomy, males have become more feminine.

For the greater part of *Homo sapiens*'s existence, brains did not reduce in size. Instead, they grew larger. There are not enough preserved skulls from the Middle Pleistocene to be sure how big the brains of the earliest *Homo sapiens* were, but by 300,000 years ago they were probably about 1,200–1,300 cc, a little less than the average of about 1,330 cc in living people. For the next quarter of a million years the brains of *Homo sapiens* continued to increase in size, up to an average of a little more than 1,500 cc.[13] Meanwhile, as the brains of *Homo sapiens* were slightly increasing in size, their shapes changed also. By 200,000 years ago their skulls had become increasingly round, or globular.[14]

While the continuing increase in brain size during the Middle and Late Pleistocene shows that humans did not completely conform to the domestication syndrome, in one respect the brain growth in *Homo sapiens* does echo a pattern found in dogs. Christopher Zollikofer has shown that the skulls of *Homo sapiens* are somewhat paedomorphic compared to those of Neanderthals, because by the time that *Homo sapiens* skulls stop growing, their shape resembles the shape of a Neanderthal skull that is in its penultimate phase of growth.[15] Essentially, Neanderthal skulls (and by inference, their brains) continued to grow beyond the final point reached by *Homo sapiens*. *Homo sapiens* did not evolve from Neanderthals, but with respect to this aspect of skull growth Neanderthals appear to be reasonable models for the population from which *Homo sapiens* did evolve.[16] The slower growth of the skull of *Homo sapiens* presumably reflects a slower growth of the brain also, suggesting that not only the skull but also the brain of *Homo sapiens* was paedomorphic with respect to their immediate ancestors.

Eventually, during the last 35,000 years, *Homo sapiens* experienced a reduction in brain size of around 10 to 15 percent to reach today's levels. As I noted earlier, there is debate about what the reduction in brain size means, since populations were becoming lighter-bodied at the same time. But some scientists regard the brain reduction as a further instance of the domestication syndrome. The drop in brain size is remarkable considering that for the last 2 million years the lineages leading to *Homo sapiens* have experienced a steady increase in brain size (from around 600 to 800 cc), and given that brain size continued to increase during most of the evolution of *Homo sapiens.*[17]

Our plunge into human paleontology has led to a prediction. We wanted to know when the domestication syndrome evolved, because that would indicate a likely time when reactive aggression was selected against. The domestication syndrome can be seen as beginning around 315,000 years ago, in the first glimmerings of the smaller face and reduced brow ridge that signal the evolution of *Homo sapiens.* Over time, the domestication syndrome became more exaggerated, reaching its greatest extremes in the most recent periods, when males have become more female-like than ever, faces have become shorter, and, possibly part of the domestication syndrome, brains have become smaller.

So the whole process of becoming *Homo sapiens* might be linked to self-domestication. If self-domestication was indeed responsible for the changes associated with the origin of *Homo sapiens,* then the selection pressures that caused it must already have started before 315,000 years ago. The process appears to have accelerated over time, suggesting that the selection pressure against reactive aggression became increasingly strong since then and up to the present time.

How long before 315,000 years ago is a matter of conjecture. Doubtless, the earliest *Homo sapiens* emerged before the Jebel Irhoud population: scientists are never lucky enough to catch the very beginning of a prehistoric process. The self-domestication process could have begun, very slowly, around 400,000 years ago. But, obviously, that is a guess. Another 100,000 or 200,000 years would be equally reasonable, taking us back to 500,000 to 600,000 years ago.

We cannot go much beyond that, however, because DNA analysis limits the possibilities. The Mid-Pleistocene *Homo* that gave rise to *Homo sapiens* had earlier spawned a line that left Africa. In Western and

Central Europe and the Middle East, they became the Neanderthals. In Siberia and beyond, they became the Denisovans. Neanderthals and Denisovans all died out as distinct populations, but interbreeding with *Homo sapiens* led to the survival of some of their genes in modern non-Africans. When the Denisovans went extinct is unknown. Neanderthals last existed in Europe, in Greece and Croatia. By 43,000 years ago, *Homo sapiens* had entered Europe and were living in the fertile river valleys and along the coasts. Neanderthal sites quickly disappeared. Some populations hung on in the mountain areas, but by about 40,000 years ago, Neanderthals were gone.[18]

The Denisovans are known only from three tooth fragments and a finger bone, whereas we have so many fossils of Neanderthals that paleoanthropologists can reconstruct even their growth rates. Study of the anatomy of Neanderthals is helpful for understanding the evolution of *Homo sapiens,* because, unlike our lineage, Neanderthals show no evidence of reduction in aggressive anatomy or a domestication syndrome.[19] Their skulls and faces remained robust in Europe and Asia. Neanderthals therefore provide a model for Mid-Pleistocene *Homo,* for which fewer specimens have been found.[20]

Because Neanderthals show no signs of the changes experienced by *Homo sapiens,* the process that gave rise to *Homo sapiens* likely started after the split between the two ancestral lines. Accordingly, the question is when the split between the lines leading to Neanderthals and *Homo sapiens* occurred. Based on a high-quality genome sequence from a Neanderthal woman who lived in Altai, Siberia, in the same cave as was previously occupied by Denisovans, the answer is between 275,000 and 765,000 years ago.[21] Even though the estimate includes a wide margin of error, it is helpful. The genetic data say the split did not begin before 765,000 years ago. Remember that the fossil data put the earliest evolution of *Homo sapiens* as being in process by 300,000 years ago. So the period between these extremes is when the distinctive evolution of *Homo sapiens* began.[22]

For the sake of simplicity, I will call this "around 500,000 years ago." Half a million years, in other words, is a generous estimate that should take us back to before the evolutionary beginnings of *Homo sapiens.* If self-domestication produced us, the process should have begun sometime between then and 200,000 years ago, which is the date for the earliest fossils that are undoubtedly *Homo sapiens.*

Considering the sheer cosmological fascination of understanding where we come from, the reasons for the very existence of *Homo sapiens* have been discussed surprisingly little. Studies of our species's origin have been focused mostly on when and where, rather than on how or why. In 2008, paleoanthropologist Dan Lieberman captured the state of ignorance: "A key unresolved question is: What were the selection pressures that favored the evolution of modern humans in Africa around 200,000 years ago?"[23] Even now few researchers have explored this problem that lies at the heart of our existence.

One ambitious proposal, by the archaeologist Curtis Marean, is a rare example that gives an ecological context for understanding the origin of *Homo sapiens.* Marean identified the "dominant adaptation" of *Homo sapiens* as our ability to accumulate cultural adaptations. We cannot live without the cultural knowledge that enables each new generation to re-create its society's way of life. Naïve animals dropped into a new environment can often work out for themselves how to find food and survive. By contrast, humans mostly have to learn from others how to make a living by digging for edible food, cooking, fashioning tools, building houses, making boats, irrigating farmland, taming horses, making clothes, and so on. Without the learned skills passed down to us by previous generations, we are in trouble. With them, we dominate the planet.[24]

According to Marean, three features of *Homo sapiens* enabled us to accumulate those kinds of cultural skill: we are highly intelligent, we are highly cooperative, and we excel at learning from others—so-called social learning. To judge from brain size, as indicated by the internal volumes of fossil skulls, the intelligence of other Mid-Pleistocene species of *Homo* was close to that of *Homo sapiens,* but *Homo sapiens* had the edge. For example, one analysis of fourteen fossils from the period of 200,000 to 76,000 years ago finds that the brain volumes of Neanderthals were relatively small: eight specimens of Neanderthals averaged 1,272 cc (78 cubic inches), compared with 1,535 cc (94 cubic inches) in six specimens of *Homo sapiens.* Between 75,000 and 27,000 years ago, by contrast, the internal volume of fossil skulls was indistinguishable, averaging 1,473 cc (90 cubic inches) in both species.[25] But even if levels of intelligence were somewhat similar across the *Homo* species, exceptional cooperation and social learning seem to have been unique to *Homo sapiens.* Marean speculates that this combination of abilities resulted from a crucial advance in food production.

Prior to *Homo sapiens,* Marean suggests, humans lived at low density in small societies, like chimpanzees. Then one population, which he thought might have lived on the southern African coast, developed an ability to gather and hunt so well that their food resources became far more productive. The population naturally grew to the point where there was competition over the food supply, and soon groups were fighting over the best territories. Success in war became imperative. Groups accordingly allied with one another, giving rise to large societies of the type that hunter-gatherers form today. Cooperation among warriors within groups was so vital for winning conflicts that it evolved to become the basis of humans' exceptional propensity for mutual aid. Sociality became more complex, learning became more vital, and culture became richer.

Marean's ideas are in the mainstream in linking the success of *Homo sapiens* to culture. The relationship is well supported by archaeological evidence. Pigments, innovative tools, and various kinds of symbolic artifacts (such as decorative seashells) were being used by 100,000 years ago. From then on, cultural diversity grew quickly. Marean's scenario also points to social features, not mere intelligence, as being especially critical for the origin of *Homo sapiens.* It identifies success in war as a plausible explanation for *Homo sapiens*'s outcompeting other species of *Homo.* It stresses that the process of producing *Homo sapiens* was not a single momentous event but rather a continual development, which fits the fact that our species has never stopped evolving, culturally or biologically. The paleontological evidence of brain size and cultural flourishing points are widely accepted, and the hypothesis that intergroup competition and warfare promoted sociality are constructively tied together in Marean's scenario.

However, two important problems concerning the evolution of

Homo sapiens were not addressed by Marean's theory—or any other theory of our origins, for that matter. First, none of the alternative theories accounts for the self-domestication syndrome evident in *Homo sapiens.* Marean's scenario highlights the importance of cooperation as an important ability, but ignores the fact that cooperation depends on a very low propensity for reactive aggression. Blumenbach, Darwin, and many subsequent thinkers would surely have seen this as a critical omission. Given how much less emotionally reactive humans are than chimpanzees, bonobos, or most group-living primates, a low propensity for reactive aggression cannot be taken for granted in our Mid-Pleistocene ancestors. Reduced reactive aggression must feature alongside intelligence, cooperation, and social learning as a key contributor to the emergence and success of our species.

Docility should be considered as foundational of humankind, not just because it is unusual, but because it seems likely to be a vital precondition for advanced cooperation and social learning. The importance of tolerance was indicated in chimpanzees by research led by the comparative psychologist Alicia Melis. In the wild, chimpanzees cooperate in territorial patrols and alliances against others, but in captivity, they often show little interest in working together. Melis's team wondered if the reason for this failure to cooperate was that their social relationships in captivity were too tense. To find out, the team assessed tolerance within pairs of chimpanzees by recording how inclined individuals were to share food with each other. They then tested how well the two chimpanzees cooperated. Sure enough, the pairs who shared most food with each other also cooperated best in pulling on a rope at the same time to bring in a reward.[26] Among captive chimpanzees, the less reactivity within a pair, the greater their ability to work together. A study of spotted hyenas focused specifically on aggression and showed that cooperation was better in pairs that were not merely more tolerant but also less aggressive. Similar links between tolerance and cooperation have been seen in a number of mammals and birds, including macaques, marmosets, ravens, and keas (a terrestrial parrot species).[27]

Most such studies have been conducted within a species, but the same idea seems to work between species. I joined Brian Hare, Alicia

Melis, and others in a study of cooperation among captive bonobos. Since bonobos are more tolerant and less aggressive than chimpanzees, we suspected that they would cooperate with one another more readily than chimpanzees do. We were right. On a rope-pulling task, they cooperated better than chimpanzee pairs did.[28] In an interesting extension of this idea, species of macaques that are more tolerant and egalitarian are more skillful at inhibiting themselves socially, and at using communicative cues, than less tolerant macaque species.[29] Overall, considerable evidence supports the idea that the evolution of cooperation depends on tolerance. The fact that most scenarios of human evolution, such as Marean's, have not considered how our species came to be so unaggressive seems to me an important shortcoming.

Equally, Marean did not ask why *Homo sapiens* experienced all of the characteristic anatomical changes by which the species is recognized in the fossil record. According to the paleontologist Chris Stringer (and much as Helen Leach described), the changes include: a tall skull, relatively rounded in profile; a small and less protruding face; small brow ridges that have a gap above the nose; a lengthened period of being a juvenile; a chin, even in infants; and a narrow trunk and pelvis.[30]

Two main kinds of explanations have been proposed to account for these features. Neither of them incorporates self-domestication. One kind is that these characteristically human anatomical features arose because the individuals who were most successful at surviving and reproduction happened, quite by chance, to have some unusual characteristics. For instance, a population that developed a particularly productive way of obtaining food might have happened to have heads that were unusually rounded. If such a population expanded over the whole continent, their chance traits would spread with them, and in this hypothetical example, the genes for globularized heads would become more common. Theoretical models suggest that this idea, which is called genetic drift, is a mathematically plausible route by which a species can gain new features. But genetic drift is an explanation of last resort because it works best for traits with little biological significance. By contrast with that expectation, having smaller teeth or being less masculine as a fighter are likely to be very important influences on an individual's ability to survive and reproduce.[31]

The other kind of explanation involves finding a series of adaptive reasons, one for each of the various anatomical changes in *Homo sapiens,* much like the parallel adaptation hypothesis that has been the traditional explanation of the domestication syndrome. As I noted in chapter 3, different features of *Homo sapiens* have been argued to be individually related to such factors as warm climates, more cooking, better hunting weaponry, or falling body weight. These and many other suggestions imply a variety of causes for a variety of effects. Some or all may be relevant, at least to some extent. For instance, a warmer climate might have fostered a more gracile skeleton; cooking may have led to decreased jaw and tooth size. But those one-by-one ideas do not solve the problem that Leach identified. Why are the characteristic features of *Homo sapiens* congruent with the domestication syndrome found in animals? Leach's persuasive solution was that the reason that human features are described by the domestication syndrome is simple: humans are a domesticated species.[32]

In Leach's hands, the idea that humans are a domesticated species ran into trouble, however, because she proposed that the process happened very late in our evolution, at a time after many people stopped being nomadic and became sedentary, mostly during the last ten thousand years. This concept not only ignored the evidence of a domestication syndrome reaching back to the origins of *Homo sapiens,* but also made Aristotle's mistake. It implied that some populations of humans (those that have always remained as nomadic foragers) had not undergone domestication.

A further difficulty of Leach's explanation is the mechanism that she proposed for self-domestication. She suggested that once people began building houses, the "artificial protective environment" would lead to "conscious or unconscious interference in breeding." She speculated that this would lead to the domestication of plants, animals, and humans, but she did not explain how this would happen, nor did she tie her thinking to Belyaev's insight that selection against reactive aggression is the key influence producing domestication.[33]

To understand why humans exhibit the domestication syndrome, we must identify a mechanism, a specific selective force that could have operated throughout the time of *Homo sapiens.* The factor or factors responsible for domestication must apply to our species as a whole, which means that it goes back at least 60,000 years and probably to the genetic roots of our species, around 300,000 years ago.[34] The selective force must also be unique to *Homo sapiens,* not occurring in Neanderthals or other species of *Homo.* Most importantly, the mechanism must explain how the propensity for reactive aggression is reduced. Precisely such a mechanism was hinted at by Charles Darwin, and has been elaborated by Christopher Boehm. Their proposal accounts for the emergence of the domestication syndrome in *Homo sapiens* by a reduction of reactive aggression through punishment of reactive aggression, imposed by humans on ourselves.

7

The Tyrant Problem

THE DOMESTICATION SYNDROME indicates that a less aggressive psychology began to emerge during the African Mid-Pleistocene by 300,000 years ago, and came to define *Homo sapiens.* With the passing of time the skull became increasingly feminized, the domestication syndrome became more prominent, and human neural-crest-cell genes experienced recent positive selection. While the trends suggest that our ancestors became more and more docile, they give no indication of how or why reactive aggression would have been selected against. Fortunately there is an explicit explanation, which we can call the execution hypothesis. The execution hypothesis is purely a scientific explanation, without any ethical implications: it is not intended to suggest that capital punishment nowadays is a social good. Its core claim is nevertheless somewhat unnerving. It proposes that selection against aggressiveness and in favor of greater docility came from execution of the most antisocial individuals.

The execution hypothesis can be traced to Darwin, which is surprising, given that Darwin considered that self-domestication had not occurred in humans. Recall that he had asked himself whether humans had gone through an evolutionary phase of becoming domesticated, and he had decided the answer was no. King Frederick William I of Prussia had failed in his attempt to select humans artificially. If an imperious sovereign could not breed humans, surely no one could. For that and other reasons, Darwin concluded that humans had not been domesticated.

Yet, in Darwin's 1871 discussion of human evolution, *The Descent of Man, and Selection in Relation to Sex,* he nevertheless sketched a simple version of the execution hypothesis as a way to explain the evolution of two important features—reduced aggressiveness and increased social tolerance—that nowadays we regard as central to domesticaton. The reason Darwin wanted to explain how aggressive tendencies had been reduced, despite his dismissal of the idea of self-domestication, was that he considered the evolutionary reduction of aggressiveness to be a problem of morality rather than of domestication. He was anxious to provide an evolutionary explanation for positive moral behavior.

The kind of moral behavior that Darwin was most concerned about was selfless helping. The conventional wisdom during Darwin's time regarded the moral sensibilities responsible for such self-sacrificial cooperation as a blessing provided by a beneficent God. But the idea that morality was God-given posed a challenge to Darwin's evolutionary theory, because Darwin proposed that all of life's features had evolved without the intervention of a deity. If evolutionary theory was to be as complete as Darwin hoped, he had to explain morality without invoking the influence of religious beings.[1]

Darwin focused on aggression, the opposite of moral virtue. He wanted to know why humans are in many ways so unaggressive. He asked himself what happens to hyperaggressive men. He appeared to take for granted the idea that men tend to be more violent than women, a sex difference that has been richly confirmed.[2]

Darwin had an answer to his question about the fate of exceptionally aggressive men. "In regard to moral qualities," he wrote, "some elimination of the worst dispositions is always progress even in the most civilised nations. Malefactors are executed, or imprisoned for long periods, so that they cannot freely transmit their bad qualities. . . . Violent and quarrelsome men often come to a bloody end."[3]

Darwin's observation came from contemporary society. Nowadays, he was saying, criminals and aggressive wrongdoers are punished by the law. If "they cannot freely transmit their bad qualities," their traits are less likely to be inherited by the next generation. If comparable kinds of punishment had been applicable throughout human evolution, genes promoting aggressive behavior would have been steadily selected against. Generation by generation, less aggressive, more positively moral behavior would tend to spread. At first blush, the idea seems irrelevant to the Pleistocene. Punishment of criminals in Darwin's time, the Victorian era of nineteenth-century Britain, was made possible by features of contemporary society not found among nomadic hunter-gatherers. Police, written laws, trials, and prisons all contributed to sanctioning the violent. Until recently, our ancestors had none of these institutions. But Darwin recognized that even if prehistoric human societies were different from today, they might still have found ways to deal harshly with "violent and quarrelsome men." If exceptionally aggressive men were always routinely punished in ways that reduced their reproductive success, there would have been eons of prehistory in which the culling of violent men could lead to evolutionary change. Darwin's conclusion was forthright. The morality problem could be solved by an ancient system of execution leading to the eradication of selfishly immoral individuals, which would lead to selection against selfish tendencies and in favor of social tolerance. Through this kind of natural selection, he wrote, "the fundamental social instincts were originally thus gained."[4]

Darwin's writings on this topic reveal a remarkable contradiction. In chapter IV of *The Descent of Man,* he denied that human self-domestication could have happened, because human "breeding has never long been controlled, either by methodical or unconscious selection."[5] But in chapter V of the same book, he proposed that what he called the evolution of social instincts, which is clearly closely equivalent to self-domestication, could have happened when the breeding of aggressive men was socially controlled—whether by imprisonment or by execution. Darwin's consideration of human self-domestication thus included both an explanation for its core feature of reduced aggressiveness (in chapter V) and rejection of it happening at all (in chapter IV). Apparently, the great evolutionist never noticed this inconsistency. He described morality as evolving as a result of reduced aggression, but he did not think of domestication in the same way.[6] His confusion is understandable. Almost a century was to pass before Belyaev's experiments showed that selection against reactive aggression is the key to producing a domesticated animal.

Still, even though Darwin did not recognize the significance of his own idea about execution, in thinking about morality he had produced a stark and simple notion that readily applies to the problem of self-domestication. Selection favored moral behavior thanks to the "bloody end" of "violent and quarrelsome men." Extreme violence had thus been selected against. The trend had continued long enough, he surmised, to have favored the "social instincts."

Thanks to Darwin, human docility—reduced reactive aggressiveness—had the start of an evolutionary explanation, the first version of the execution hypothesis.

Darwin's explanation for the evolution of the "social instincts" was provocative and reasonable, so you might expect that it would have generated a lot of interest. It did not, however. It was overshadowed by a second explanation for the evolution of morally positive behavior, now called the parochial altruism hypothesis.

Darwin was the first to propose the parochial altruism hypothesis, though he did not call it that and he later also suggested it was wrong. Nevertheless, the parochial altruism hypothesis has remained popular. It is complementary to the execution hypothesis, rather being than an alternative, since the two ideas are designed to explain slightly different components of moral behavior. The parochial altruism hypothesis argued why cooperation was favored, while the execution hypothesis proposed why aggression was reduced. I also believe the parochial altruism hypothesis is wrong, but it has been such a popular and appealing idea, and has been so important in distracting scholars from the execution hypothesis, that its value and problems are worth considering. Like the execution hypothesis, an early version of the parochial altruism hypothesis for the evolution of cooperation appeared in *The Descent of Man* alongside his execution hypothesis. It was based on the benefits of cooperation rather than the costs of aggression. Darwin observed that the success of competing societies is often affected by how well each side's warriors unselfishly support one another. In an account that neatly linked two peculiarities of human social nature, cooperation and war, he raised the possibility that exceptional cooperation was the result of exceptional war. "It must not be forgotten," he wrote in a famous passage,

> that although a high standard of morality gives but a slight or no advantage to each individual man and his children over the other men of the same tribe, yet that an advancement in the standard of morality and an increase in the number of well-endowed men will certainly give an immense advantage to one tribe over another. There can be no doubt that a tribe including many members who, from possessing in a high degree the spirit of patriotism, fidelity, obedience, courage and sympathy, were always ready to give aid to each another and to sacrifice themselves for the common good, would be victorious over most other tribes. . . . At all times throughout the world tribes have supplanted other tribes; and as morality is one element in their success, the standard of morality and the number of well-endowed men will thus everywhere tend to rise and increase.[7]

Darwin's notion was that groups that were in conflict against neighboring groups needed internal solidarity. His idea was received warmly. In 1883, the political philosopher and essayist Walter Bagehot applied it to modern life: "The compact tribes win, and the compact tribes are the tamest. Civilisation begins, because the beginning of civilisation is a military advantage."[8]

This type of explanation—that ingroup solidarity fosters success in between-group competition—continues to attract scholars up to the present day. One can see its appeal. As the historian Victor Davis Hanson has argued, armies whose soldiers cooperate better really do succeed better in war, such as when ten thousand united Athenians were able to defeat thirty thousand Persians at the Battle of Marathon in 490 BCE.[9] The benefits of unity in the face of an enemy explained the communal spirit in New York after the 9/11 attack, or the story of the Israelites battling from Egypt to Canaan, or the motto of the Three Musketeers: "One for all, and all for one!" International solidarity is often imagined as our earthly response to the arrival of space aliens. In fiction and reality, war can promote cooperation within groups.

In 2007, economists Jung-Kyoo Choi and Samuel Bowles defined parochial altruism as self-sacrificial behavior in war and quantified the conditions under which it would be expected to be positively selected. Falling on an exploding grenade to protect one's fellow soldiers would be an extreme example. Choi and Bowles argued, much as in Darwin's passage quoted above, that parochial altruism would have been evolutionarily favored when the value of a group's defeat of its opponents was greater than that of selfish behavior within groups. Bowles later supported the claim with data on hunter-gatherer death rates in war, and on genetic differences among hunter-gatherer groups. The argument invoked group selection, the contested idea that selection among groups can lead to the evolution of traits that are beneficial for groups, even when some individuals suffer as a result.[10]

Parochial altruism theory is an elegant idea but has several lines of evidence against it. Choi and Bowles's account is designed to explain a specifically human feature, the evolution of positive moral tendencies, and should therefore invoke specifically human selective forces. However, in intergroup conflicts chimpanzees have very similar rates of death as in hunter-gatherer war. By Bowles and colleagues' argument, chimpanzees should therefore show evidence of self-sacrifice in battle. Yet parochial altruism is unknown among chimpanzees. This difficulty for parochial altruism theory has not been resolved.[11]

Theoretical problems aside, a fundamental issue regarding parochial altruism is that it has not been shown to occur among human hunter-gatherers. Choi and Bowles admitted that most warfare among hunter-gatherers follows the safe style displayed by chimpanzees, avoiding conflict unless victory is all but assured, but they also claim evidence of self-sacrifice in humans' pitched battles (as opposed to raids). The evidence that they cited in support of this claim is of a single "battle" from Australia involving at least seven hundred warriors. Choi and Bowles perhaps assumed that the Australian aboriginal warriors would support each other at high personal risk, like medal-winning heroes in contemporary wars. Yet in the Australian example no deaths were reported, nor any altruistic risk-taking. The flavor of the contest can be judged from the way it ended, after a man got angry from being hit by three spears. According to an eye-witness: "With an infuriated yell and a string of native oaths, he bounded for shelter, from which he produced a hidden muzzle-loaded gun and, ramming a charge into it, he returned to the battleground to face the enemy, who dispersed in disorder."[12] No one, apparently, wanted to get hurt, much less killed. No mutual support was described. The episode was typical of hunter-gatherer battles. Political scientist Azar Gat reviewed warfare among Australian aboriginals. He found much evidence of heavy death tolls among victims of raids (in which attackers clearly aim to avoid being hurt, while killing by surprise). Battles, however, "mainly involved spear throwing at a distance," so that "little blood was shed."[13] Hunter-gatherers in face-to-face battles tended to stop fighting after the first injuries, just like the Dani farmers in New Guinea studied by Karl Heider.

Until evidence is produced that hunter-gatherers exhibit risky self-sacrifice in war, parochial altruism should be regarded as a culturally induced behavior rather than an evolutionary product of selection. Japanese kamikaze pilots who flew their planes into enemy ships during World War II, or Islamist suicide bombers, responded to intense cultural pressures more than innate inclinations. The evidence that parochial altruism is a generalized tendency found universally among humans as a species is currently inadequate.[14]

Darwin also decided that the effects of war for promoting solidarity within groups is cultural. The social instincts could not have evolved as a result of intergroup fighting, he said, because, even within the most cooperative and morally virtuous tribes, some people would be more selfish than others; and the more selfish people would have more babies than the moral. "He who was ready to sacrifice his life . . . would often leave no offspring to inherit his noble nature. . . . Therefore it seems scarcely possible . . . that the number of men gifted with such virtues, or that the standard of their excellence, could be increased through natural selection, that is, by the survival of the fittest." Parochial altruism, or self-sacrifice in war, might be explicable in specific societies by their military cultures or promotion of risk-taking ideals, but not by evolution.[15]

In sum, the parochial altruism hypothesis that warfare could have led to the emergence of self-sacrificial behavior seems to apply only in terms of its cultural effects, not as an evolutionary selective force. Nevertheless, the theory has been important, because it has dominated efforts to explain the exceptionally positive aspects of human sociality. The focus on self-sacrifice distracted from the question of why humans are so docile and eclipsed Darwin's speculation about the "bloody end" of "violent and quarrelsome men." For a century, the question of reduced aggression was forgotten, and the execution hypothesis ignored.

Explaining humans' extremely high cooperative tendencies has garnered more attention than explaining our strikingly low reactive aggression, but eventually the aggression problem was taken up again. Three decades before Bowles and colleagues proposed that self-sacrifice in war is the root of human goodness, evolutionary biologist Richard Alexander proposed that reputation is the key. Although Alexander did not raise any questions about humans' reduced reactive aggression, his scenario would lead back to Darwin's focus on violent and quarrelsome men. Alexander's question was how could natural selection favor the evolution of a kindly morality? The core problem was why, in a dog-eat-dog world, humans' sense of virtue has developed beyond the level seen in other animals. Resuscitating an idea that Darwin had sketched, Alexander focused on the biological value of a good reputation, meaning an evaluation of personal traits that is shared among two or more judging individuals. In his 1979 *Darwinism and Human Affairs,* Alexander argues that at some unknown point in our evolution, language skills developed to the point where gossip became possible. Once that happened, reputations would become important. Being known as a helpful individual would be expected to have a big effect on someone's success in life. Good behavior would be rewarded. Virtue would become adaptive.[16]

This explanation would help solve the problem of why cooperation is uniquely elaborated in humans by attributing it to a feature, language, found only in humans. If reputation depends on the judges sharing their evaluations with each other, chimpanzees are not expected to care about it. Chimpanzees can demonstrate negative feelings toward others, but they cannot explain why they feel that way. They cannot gossip about the guy who bit a rival, or slapped a female, or stole some food, or whether a particular chimpanzee is trustworthy, generous, or kind. Their communication skills are simply inadequate. Obviously, this implies that chimpanzees would not care about their reputation.

To assess whether chimpanzees care about their reputations, the cognitive psychologist Jan Engelmann found out whether chimpanzees change their behavior when they are observed by others. He and his colleagues arranged a setup whereby a chimpanzee could steal food from another chimpanzee. Sometimes the potential thief was watched by a third individual. If being observed matters to chimpanzees, they should be less likely to steal when a third individual is watching. As expected, the presence of a third individual made no difference. The same was true when the experiment was about helping rather than stealing. Chimpanzees behaved selfishly or helpfully according to their own proclivities, and did not modify their behavior according to whether they were observed. Reputation did not appear to matter.[17]

Chimpanzees do have individual personalities. Some are more timid, others more aggressive, still others more generous. One might tend to groom readily; another might be more selfish. Such differences are noticed. Individuals choose whom to interact with, depending on how they have been treated in the past. Good cooperators tend to be favored. Bad cooperators tend to be shunned. The same is true in many species.[18] So the absence of concern about reputations among chimpanzees does not stem from any lack of individual differences, nor from an inability to evaluate others. Chimpanzees know that partners vary in quality, but they have to use that information on their own. They cannot talk about it.

Humans have no such problem. Gossip surrounds us, so we care about our reputations. Our caring can be subconscious. People are more likely to give to charity, or clean up a mess, when they are being watched than when they are alone. The watchers do not even have to be real people. Two big blobs drawn on a mug to resemble a pair of eyes, when placed in a room containing a box for charitable donations, are enough to make us increase how much we give.[19] Our sensitivity to what others think starts when we are young. Engelmann tested five-year-old preschool children in a setup similar to the one he had used for chimpanzees. Unlike the chimpanzees, for the preschoolers the presence of an observer mattered. When they were being watched, they stole less and helped more.

Alexander's idea was that the social pressures wrought by the effects of a good or bad reputation underlie the evolution of morality. In the short term, individuals with bad reputations might mend their ways and become conforming members of society. But over the long term, the effects of a bad reputation would have genetic evolutionary consequences. Individuals who were too feisty, short-tempered, or selfish ever to adapt successfully to the criticisms of their peers would have poor chances of surviving and reproducing well. Ostracized by their group, these nonconformists would have passed on fewer genes than those with the good reputations. Selection would thereby have favored the kind, cooperative, tolerant types: the morally positive ones, less aggressive than their ancestors. Our ancestors would have evolved to be a nicer species. Language begat reputation, and reputation begat morality.

Alexander's scenario fits evidence from small-scale societies. The biological anthropologist Michael Gurven led a study among the Aché, a settled group of people in Paraguay who had recently been hunter-gatherers. Some individuals had a reputation for being generous, regardless of how much they were normally able to give. When those who were renowned for their generosity ran into difficulties, they were helped more than those with a reputation for stinginess. They were given more food, for instance. As predicted, reputations mattered.[20]

Supported by chimpanzee comparisons and data on humans, the reputation hypothesis has looked promising. The psychiatrist Randolph Nesse felt it could be extended from cooperation to docility. "It seems plausible," Nesse wrote in 2007, "that humans have been domesticated by the preferences and choices of other humans. Individuals who please others get resources and help that increase fitness. Aggressive or selfish individuals get no such benefits and are at risk of exclusion from the group, with dire effects on fitness. The result is thoroughly domesticated humans, some of whom can be enormously pleasing."[21] In this passage, Nesse inadvertently raises the problem that returns us to Darwin: Why are "aggressive or selfish individuals . . . at risk of exclusion from the group"?

Alexander's hypothesis assumes that individuals care about their reputations. Not everyone does care, however. Even among humans, some individuals ignore the complaints of others. What if someone with a poor reputation simply helped himself to food? We can probably all remember bullies from our school years who were so powerful that they did not care what some of the less well connected children thought of them. They were big and bold, and if others resented them, so what? The bullies got what they wanted without the approval of the mutterers. They were not stopped simply by gossip. They had to be stopped by someone fighting back, or by adults who whisked them off to detention.

Translate those types into the past and we are faced with a vital question unsolved by the reputation hypothesis. Why would a bad reputation matter to a male who was bold and strong enough to take matters into his own violent hands? How would his bad reputation have inhibited him if, like an alpha chimpanzee, he did not care what others thought? The apes have not evolved to be sensitive to social disapproval. Reactive aggression is much more readily provoked among chimpanzees than among humans. Until we evolved our calmer dispositions, we would have behaved more like those apes. There would have been a lot of fighting, and the strongest, toughest, most persistent fighters would have won.

A female might have preferred a kinder, gentler male as a mate. But how was she to stop a domineering male from coercing her? More tolerant men might provide more meat. But what would prevent a powerful individual who was willing to throw his weight around from refusing to be denied, aggressively seizing more than his share? Tyrants who did not care about their reputations could bully their way to getting more than others—whether more food, more matings, the best sleeping-spot, or more social support. That is what happens among chimpanzees. Among Mid-Pleistocene *Homo,* who was to stop the despots?

Shunning would be insufficient to affect an individual who is able to intimidate or defeat all others in a fight. Subordinates' resentment could be translated to effective resistance only by coalitionary force. Cooperation among the weaker individuals is needed.

One use of social power is to teach the aggressor to accept defeat. Among bonobos, when a male is too pushy, a collection of females chases him; he presumably learns not to annoy them. Aggressive tendencies in male bonobos could have been reduced evolutionarily partly by the ability of female power to reduce the advantages to males of being violent. Could early *Homo* females have similarly ganged up on the alpha male to stop him when he asserted himself too selfishly?

Despite its logical appeal, the hypothesis is improbable. Among hunter-gatherers, women are not known to support one another in physical fights against violent men. In the Pleistocene, males were stronger and more robust than they are today, so it would have been even riskier for females to confront them in a fight. Whether females could count on one another, as bonobos can, also seems dubious. Among hunter-gatherers, men are important as providers of food and as protectors, and women compete with one another to become the wives of the best husbands. There are no such dividing forces among female bonobos.

Among hunter-gatherers, as we will see, aggressors are stopped not by repeated coalitionary chases, and not by females acting on their own. When teasing and pleading and ostracizing and moving camp all fail to change a man's violent behavior, the last resort of the coalition is execution, as Darwin foresaw.

If we think ourselves back to the origin of *Homo sapiens,* probably both sexes were feistier than they are today. But, to judge from facial anatomy, males were the sex in which the behavioral difference from *Homo sapiens* would have been particularly acute. Recall the macho power of our Mid-Pleistocene ancestors. Male faces were big and imposing, wide and long, with thick, protruding brow ridges above their eyes.[22] The facial traits of exaggerated maleness in Pleistocene *Homo* and early *Homo sapiens* are typically associated with increased aggressiveness. Male skulls are relatively feminized in bonobos and relatively masculinized in chimpanzees; and male chimpanzees are more aggressive. Male skulls became feminized in Belyaev's selected foxes; males in the unselected lines have the more masculine skulls, and are more aggressive. In general, domesticated animals have reduced sex differences in the skull compared with their wild ancestors; and domesticated males are less aggressive. The effect may be due partly to differences in the level of testosterone production at puberty. Men with wider faces, for example tend to produce more testosterone.[23]

Even among today's men (but not women), since 2008 facial breadth has been discovered to be correlated with a propensity for reactive aggression. The time when male faces become relatively broad compared with female faces is in puberty, apparently under the influence of testosterone. In professional hockey games the number of minutes spent in the penalty box tends to be greater for broader-faced men than for men with narrower faces. In general, among European whites, wider-faced men have been found to have a higher propensity not only for aggression and retaliation but also for more self-centered and deceptive behavior, less cooperative negotiation, a higher score on the psychopathic trait "fearless dominance," and a higher score for self-centered impulsivity. Wider-faced men are also better fighters, which may explain why in one study of more than one thousand United States skeletons, broader-faced men were less likely to die in a fight than men with narrower faces. These statistical effects have been found repeatedly even in samples of less than a hundred men, but they are too weak to predict a man's aggressiveness from his facial ratio. Nevertheless, in experiments, subjects who are unaware of these findings tend to treat broader-faced men warily, as if recognizing that a relatively broad face is a signal of aggression. This unconscious sensitivity to a man's facial breadth suggests that during human evolution, broader-faced men would have demonstrated more socially undesirable behavior, and that our broad-faced Pleistocene male ancestors would have been relatively impulsive, fearless, uncooperative partners who were swift to act aggressively in defense of their selfish needs.[24]

The difficulty with the reputation hypothesis is that it does not explain why such tendencies would have been reduced. Physical aggressors could have succeeded in bullying their way to the top. The same challenge faces all ideas for explaining the evolution of human goodness that do not address the control of aggression. The evolutionary anthropologist Sarah Hrdy proposed that cooperative tendencies became heightened when our ancestors began looking after one another's young, enabling them to rear more offspring than before. The primatologist Carel van Schaik argued that cooperation would have been so important for hunting that hunting would have favored the development of tolerant bonds among males. The psychologist Michael Tomasello suggested that, as males learned to recognize their offspring, they would benefit by reducing their level of aggression and becoming more involved in parenting. He also thought that, in the face of predators, a need for teamwork could select in favor of a cooperative personality. There is much to be said for these surmises; all of them may contribute to understanding cooperative tendencies. But none of them solves the problem of domineering aggressors. Even if a man's hunting success was feeble because no one would hunt with him, if he was sufficiently aggressive he could commandeer the prey killed by others.[25]

The only proposal sufficient to explain how our ancestors overcame the problem of determined bullies, which was surely a vital first step to the elaboration of cooperative behavior, is the elaboration of Darwin's idea that "violent and quarrelsome men" came to a "bloody end." The execution hypothesis claims that, during the Pleistocene, a new kind of ability crystallized. For the first time, coalitions of males became effective at deliberately killing any member of their social group who was prepared to use violence on his own behalf and simply did not care what others thought about him. In the end, execution was the only way to stop such a male from being a tyrant.

8

Capital Punishment

On a summer night in 1820, sixteen-year-old Stephen Merrill Clark of Salem, Massachusetts, was arrested for having set fire to a stable. He was a member of a respectable family, and because no one was hurt by his action, he might have expected some leniency. But he had already acquired a bad reputation for petty crime, and in this case the fire spread, destroying three homes and five other buildings. Arson of a dwelling was a capital crime in Massachusetts, and Salem had old-fashioned standards. Clark was tried, found guilty, and sentenced to death. Despite impassioned appeals, the governor stood firm. On the appointed day, the boy climbed a scaffold and gazed over a crowd of hundreds. So feeble in his terror that he had to be held up on either side, he listened to his own words being read by a minister.

> "May the youth who are present take warning by my sad fate, not to forsake the wholesome discipline of a Parent's house. . . . May you all pray to God to give you timely repentance, open your eyes, enlighten your understandings, that you may shun the paths of vice and follow God's commandments all the rest of your days. And may God have mercy on you all. To the world at large I bid Farewell!"[1]

To the sound of sighs and groans in the anxious crowd, he was hanged.

Poor Clark. The modern debate over the merits of capital punishment began in the eighteenth century. Only a year or two later, he would have been sent to prison instead of the gallows. His execution was one of the last in America for a crime that involved no personal harm. It now seems barbaric that anyone, let alone an adolescent, should have been put to death for setting a fire that accidentally burned a house. But if our recent past seems harsh compared with a generally more enlightened present, it is we of the modern world who are odd, rather than the New Englanders of two hundred years ago. For Clark's death connects us to a way of life that has prevailed throughout history and long before, possibly to the beginnings of *Homo sapiens*.[2]

In the seventeenth century, hundreds of felonies in America had been capital crimes. In New England, you could have been executed for witchcraft, idolatry, blasphemy, rape, adultery, bestiality, sodomy, and, in New Haven, masturbation. You could be put to death for being "a child of sixteen or older who was a 'stubborn' or 'rebellious' son, or who 'smote' or 'cursed' a parent." These penalties were far from theoretical. From 1622 to 1692, Essex and Suffolk Counties of Massachusetts recorded the executions of eleven murderers, twenty-three witches, six pirates, four rapists, four Quakers, two adulterers, two arsonists, and two each charged with bestiality and treason.[3]

The principle of execution was popular. It was not unusual for criminals to be hunted down by citizens, tried, convicted, sentenced, and executed in less time than the four days that were supposed to elapse after the pronouncement of a death penalty. Communities would sometimes stretch the rules to achieve a desired verdict. The historian David Hackett Fischer tells of a one-eyed servant named George Spencer living in New Haven. He "had often been on the wrong side of the law, and was suspected of many depravities by his neighbors. When a sow gave birth to a deformed pig which also had one eye, the unfortunate man was accused of bestiality. Under great pressure, he confessed, recanted, confessed again, and recanted once more. The laws of New England made conviction difficult: as a capital crime bestiality required two witnesses for a verdict. But so relentless were the magistrates that the deformed piglet was admitted as one witness, and the recanted confession was accepted as the other." Spencer was executed.[4]

Spontaneous mob killings reflected the community's enthusiasm for savage justice. Fischer describes how women in Marblehead whose husbands had been taken by Indians "seized two Indian captives and literally tore them limb from limb." The countrywide toll of such events was substantial. From the first recorded capital sentence in America in 1622 (for theft) to 1900, there were perhaps eleven to thirteen thousand legally determined executions; mob violence or lynchings were thought to have killed an additional ten thousand during the same period. Occasional lynchings and unofficial executions still bedevil the powerless.[5]

In short, a combination of tough laws, determined citizens, and enthusiasm for eliminating outcasts has traditionally made America a lethally dangerous place for those who offended society's norms. It was not until the late eighteenth century that capital punishment started to fall in popularity. Before then, if you challenged the rules, you risked death. The counterpoint was impressive. Trouble was rare. Home owners could sleep with their doors open. They did not have to lock up their valuables. As long as you followed the rules, New England was a place of peace.

According to the execution hypothesis, a similar dynamic has applied to our species in general. Capital punishment has been the ultimate source of law and order.

Executions and the reasons for them have been described in every society with written records. Capital punishment was present in all the earliest civilizations, from Egyptian, Babylonian, Assyrian, Persian, Greek, and Roman to Indian, Chinese, Inca, and Aztec. It happened not only for violent crimes but also for nonconformism (as in Socrates's case), for minor felonies, and even some heartbreakingly trivial matters such as malpractice in selling beer (according to the Code of Hammurabi), or stealing the keys to one's husband's wine cellar (according to the laws of the early Roman Republic). Executions were an accepted part of life, often drawing great public crowds to watch spectacles of unequivocal cruelty. They continued in all historically known societies until the 1764 publication of *On Crimes and Punishment,* by the Italian jurist Cesare Beccaria. Beccaria's arguments against the death penalty helped launch the change that continues today in community attitudes toward capital punishment. Prisons then increasingly took over the responsibility for social control.[6]

Although agrarian societies have a universal record of prevalent capital punishment, however, they tell us little about our deep past. Until the 1980s, capital punishment had not been systematically studied among hunter-gatherers. The omission is easily understood. Executions were mostly secret. Any single group was too small to perform many executions. Missionaries and governments try to stop all kinds of killing, and Western intellectuals tend to frown on the idea of the death penalty. Cultural change has been rapid, especially with respect to the control of violence. As soon as executions become illegal in the eyes of the state, they are likely to disappear from small-scale societies.

The death of Michael Rockefeller in New Guinea in November 1961 illustrates the difficulty of documenting killings. Rockefeller, a twenty-three-year-old American adventurer, was exploring remote areas in search of exotic artwork. He disappeared near a village called Otsjanep, occupied by Asmat hunter-gatherers. In December 1961, four men from Otsjanep told a local missionary their story. They had found Rockefeller alive, lifted him into their canoe, speared him, killed him, and eaten him. The killing was a response to deaths their village had suffered four years earlier, when a Dutch officer had killed four men from Otsjanep. The men included credible details. The missionary, who knew the men personally and understood the cultural context, was convinced by their story. But years of investigation never found any confirming material evidence, such as Rockefeller's clothes or eyeglasses. The missionary's church, the government, and the Rockefeller family did not want to believe a tale of murder and cannibalism.

When the writer Carl Hoffman revisited the case in 2012, he met a wall of silence. A video that he took in 2012 shows why. In the video, an Otsjanep man speaks in Asmat to a group of his peers. Hoffman was the only outsider. The speaker was unaware that his words might later be translated. He could not have been more explicit.

> "Don't you tell this story to any other man or any other village because this story is only for us. Don't speak. Don't speak and tell the story. . . . Don't talk to anyone, forever, to other people or to another village. If people question you, don't answer. Don't talk to them, because this story is only for you. If you tell it to them, you'll die. . . . You keep this story in your house, to yourself, I hope, forever. Forever. I hope and I hope. If any man comes and has questions for you, don't you talk, don't talk. Today. Tomorrow and for every day, you must keep this story. Even for a stone ax or a necklace of dogs' teeth, do not ever share this story."[7]

In 1986, capital punishment in small-scale subsistence societies finally became a topic of scientific research. The anthropologist Keith Otterbein was inspired by *Elizabeth R,* a television series in which the head of Mary, Queen of Scots, was last seen rolling on the ground, to investigate the hypothesis that executions would be found only in state societies. His premise was that only powerful political leaders would be in a position to dispose of those who threatened them. Otterbein defined capital punishment as "appropriate killing within a political community of a person who has committed a crime." To his surprise, he concluded that capital punishment is a human universal.[8]

Hunter-gatherers offer the opportunity for understanding whether capital punishment might have stretched back into prehistoric times. The ethnographic record is not perfect. Otterbein found some hunter-gatherer cultures in which it was unclear whether executions occurred, such as the Andaman Islanders, east of India. But he judged that all societies for which there was adequate information conformed to the wider trend. Christopher Boehm later amplified Otterbein's work by surveying hundreds of ethnographies to identify the best-studied cultures of hunter-gatherers. He reported capital punishment in every inhabited continent, notably among Inuit, North American Indians, Australian Aborigines, and African foragers.[9]

Much like the combination of legal executions and mob lynchings in America, among hunter-gatherers communal support for killing was achieved in various ways. Sometimes executions were preapproved.

Everyone might then want to be involved. The anthropologist Richard Lee studied the Ju/'hoansi (or !Kung San), the famously peaceful hunter-gatherers of Botswana's Kalahari desert. Lee reported how a community came together to solve the problem of a member of their group who had killed three men. The murderer's name was /Twi. "In a rare move of unanimity," wrote Lee, "the community . . . ambushed and fatally wounded /Twi in full daylight. As he lay dying, all the men fired at him with poisoned arrows until, in the words of one informant, 'he looked like a porcupine.' Then, after he was dead, all the women as well as all the men approached his body and stabbed him with spears, symbolically sharing the responsibility for his death. . . . It is as if for one brief moment, this egalitarian society constituted itself a state and took upon itself the powers of life and death."[10] /Twi's death was like Julius Caesar's, whose body was said to be stabbed thirty-five times by the twenty or more senators who had conspired against him.[11] They all wanted to show they were part of the coalition.

Alternatively, preapproved decisions could be left to a single executioner. In his 1888 book, *The Central Eskimo,* Franz Boas described a case in which a man named Padlu had eloped with another man's wife. When the husband came in search of her, Padlu killed him. On subsequent occasions, the husband's brother and a friend of the husband tried separately to rescue her, and they, too, were killed. After this, a headman went to Padlu's camp and asked every man whether Padlu should be killed. "All agreed; so he went with [Padlu] deer hunting . . . and near the head of the fjord he shot Padlu in the back." What might have looked like a one-on-one murder was actually the completion of a community plan.[12]

/Twi and Padlu were executed for being murderers. Many transgressions other than violence justified execution, making capital punishment a serious threat to anyone. One important source of guilt was disregard for cultural rules. The rules tended to favor the men. The anthropologist Lloyd Warner spent three years with Australia's Murngin people (now called the Yolngu) in the 1920s. As in many societies, women were required not to intrude on the men's secret goings-on on pain of death.

> Some years ago the Liagomir clan was holding a totemic ceremony, using its carpet snake totemic emblems (painted wooden trumpets). Two women stole up to the ceremonial ground and watched the men blowing the trumpet, went back to the women's camp and told them what they had seen. When the men came back to camp and heard of their behavior, Yanindja, the leader, said, "When will we kill them?" Everyone replied, "Immediately." The two women were instantly put to death by members of their own clan with the help of [men from another group].[13]

Rules were so important that an individual could kill for the sake of them regardless of his own personal preference. The amateur ethnographer Daisy Bates wrote about Western Australians at the beginning of the twentieth century. Men were required to behave themselves in their sexual liaisons. They were killed for having sex with a betrothed girl, or for taking a woman away when she was supposedly secluded during her menstruation, or for having intercourse before he was initiated into manhood. While Bates was there, a young woman fell in love with a youth during the segregation period of his initiation. The adoring woman followed him to express her feelings. He knew the penalty. Rather than face being killed himself, the young man killed her. "He was able to prove to her kinsmen that his action was justified and he escaped punishment."[14]

A romantic view holds that life in a small-scale society is delightful. In many ways, this is true. Unlike centralized tyrannies, the leaderless bands of nomadic hunter-gatherers or villages of slash-and-burn farmers are genuinely pluralistic societies. Disputes are solved communally. Anyone's voice can be heard. No one is allowed to be sad. The sense of social support is enormous, and, as we have seen, the tenor of daily life is very peaceful. To add to the pleasures, the bands or camps are not subjects of any polity. They are members of a larger network of groups in which everyone speaks the same dialect or language and shares the same culture. Neighboring groups may sometimes have disputes and can even turn to violence against one another, but they all operate at the same political level, without hierarchy among groups, or group subjugation.

So, in many ways, individuals in these small-scale societies are freer than those in larger, agrarian groups, where individuals are subject, in the words of the social anthropologist Ernest Gellner, to the "tyranny of kings." But liberty has its limits. In the absence of domineering leaders, a social cage of tradition demands claustrophobic adherence to group norms. Gellner called it a "tyranny of the cousins." The cultural rules are paramount. Individuals have limited personal freedom; they live or die by their willingness to conform. Gellner's "cousins" did not have to be literally relatives. "Cousins," in the context of a small-scale society, is a metaphor for the group of adults whose decisions held sway. Their power was absolute. If you did not conform to their dictates, you were in danger.[15]

Lucas Bridges experienced such a threat directly. In 1874 he became the first European born on the island of Tierra del Fuego, the son of the earliest missionary to settle there. The island lies on the southern tip of South America. It was occupied by two hunter-gatherer societies, the coastal Yamana (also known as Yaghan) and inland Selk'nam (or Ona). Bridges had as intimate an experience of hunter-gatherer life as any European has had. He grew up speaking the Yamana language and also participated extensively in Selk'nam life. His in-depth experience made him vulnerable. When he joined a men's society, he was insructed that if he revealed their secrets to women or uninitiated males he would be killed. A tattletale could be killed by his father or brother. The male coalition was more important than kinship. Bridges would wait more than half a century to publish his remarkable account of living with hunter-gatherers, by which time the societies that he had known as a boy had disappeared.[16]

The cultural anthropologist Bruce Knauft had an opportunity to learn how gossip among men in a small egalitarian community could lead to someone's being pushed out of the group. In the early 1980s, he spent almost two years with the Gebusi people in a remote lowland rain forest of New Guinea. The Gebusi were hunter-gardeners (or slash-and-burn farmers) who had first been briefly contacted in 1940, then brought under government control in the 1960s. Their society was small, a total of about 450 people. Like mobile hunter-gatherers, they lived in small villages, averaging fewer than thirty people, with no leaders holding authority within the group. Social relations within the society were in the classically gentle mold: serene conversations, rich humor, no boasting. Their political system was egalitarian, but, as always, not everyone was equal. Among the Gebusi, sorcerers, like witches, could be accused of bringing misfortune to others. But whereas witches were born, sorcerers were supposedly evil by their own will. Perhaps that made them harder to forgive. Certainly they were frequently killed, on grounds that we would often call trivial. The killings served to solve conflicts and maintain social cohesion.[17]

Knauft described how consensus could build toward capital punishment. Based on transcripts of actual conversations, he portrayed a time when villagers convened in the hope of curing a seriously ill member of the community. As is often the case in animist societies, the patient's sickness was assumed to be caused by evil. Such a meeting would typically take place at night, in a longhouse, in a heady atmosphere. Men's faces are lit only by a few glowing coals in the fire. Soft voices provide a background stream of song. One man is the medium to the spirit world. Mostly, he is in a trance, silent and still.

Every now and again, the medium wakes up. He lurches and shouts. He smacks at people and hits at flaming logs. He throws anything he can find. People whoop and holler and roll around to escape the chaos. Then all becomes calm again.

Trouble starts when the patient seems to be failing. An accuser floats the idea that the guilty party must be a sorcerer. He makes his case gently.

> "Well, I don't know who could have sent this sickness. I really don't. . . . But we had this séance, and, well, the medium said that it was a man from this settlement and, well, he pulled the leaves here [supposed to be the sorcerer's magic leaves]. And . . . there's no one else but [names someone]."

The named sorcerer is present. He is now in peril. If he gets angry and denies any responsibility for the illness, he may be seen as unrepentant. His best chance is to admit to a minor crime—in other words, confess to being a sorcerer, and agree to stop making the patient sick. That is what he does. Straining to maintain composure, he pleads for his life:

> "I don't know either. I don't know anything about it. He's my own relative, too, I couldn't make him sick. I don't know. . . . I was so sorry for him when I heard he was sick. I only heard a few days ago. We had been out in the bush and I didn't even know. When I finally heard, I said to my wife right then, 'We'll have to go and make sure he's all right. I wonder, what sickness could he have?' . . . I don't know about it, but I'm sure that now the leaves are pulled . . . if you throw them away he'll get better. I might have been a little angry from not eating enough fish lately, but I certainly wouldn't make my own relative sick like that."[18]

The accused man would likely leave the longhouse, right to be nervous about what might happen.

Knauft reported that the fate of the accused would depend on divinations following a death. Unless the outcome was clearly helpful to his case, he would almost certainly be executed within a few months. The accuser would quietly ensure the support of the community. He and his trusted friends might convene a séance at a time when they knew the guilty party had few male kin at home. During a sleepless night of joking and shouting in the longhouse, the men would grow more and more enthusiastic about the idea that the alleged sorcerer was responsible for the death. Consensus is reached. All decide the accused is guilty.

At dawn, they stage an ambush. They kill with clubs or arrows. Sometimes there is torture first. Then they butcher and cook him. In precontact times, cannibalism was rife, but only sorcerers were eaten.

Knauft's account provides an unusually intimate understanding of how dangerous the social dynamics of a small anarchic group can be. His experience was with a particular group of recently contacted horticulturalists. However, the essential deadliness created by accusation and consensus building seems typical of the political networking that leads to executions in small-scale societies.

To most people growing up in comfortable state societies, the idea that a group of closely connected people who know one another well can agree to kill one of their own is disturbingly strange. Yet, during times of crisis, the same practice has sprung up in the richest nations. It happened in the Second World War concentration camps, where prisoners were desperate for food. Food sharing was common despite the suffering, a mark of humanity in dark times. But stealing was common, too, and it was reviled. Prisoners worked out how to stop it. Rudolf Vrba wrote about the "bread law" in Auschwitz: "If a man stole your food, you killed him. If you were not strong enough to carry out the sentence yourself, there were other executioners; it was rough justice, but it was fair because to deprive a man of food was to murder."[19] The bread law was also enforced in Soviet and other Nazi camps, according to Terrence Des Pres. It was not a trivial matter. Des Pres described it as the "one law and one law only which all prisoners knew and accepted . . . And in a definite and clear-cut sense this particular 'law' was the foundation and focal point of moral order in the concentration camps."[20] The bread law sustained the community of sufferers. Eugene Weinstock was a survivor from Buchenwald who described its merits: "If hunger so demoralized a man that he stole another's bread, no one reported him to the SS or even to the Block Leader. The room attendants themselves took care of him. . . . If he did not die of the beating, they so incapacitated him that he was fit only for the crematorium. . . . We approved of this rule because it actually helped us maintain a certain standard of morale and mutual trust."[21]

The purposes of capital punishment are in some ways different in state and small-scale societies. In state societies, capital punishment has often served to remove individuals who challenge the leader. In keeping with those aims, states often execute ostentatiously in public, especially during the early years of a state's existence, when power structures are still unstable. Kim Jong Un, the North Korean president, inherited his position from his father in 2011. In the first four years of his reign, he reportedly executed at least seventy of his subjects, including his vice-premier, his minister of defense, and his uncle. He supposedly was theatrical in his deeds, killing his uncle with an anti-aircraft gun in front of an assembled crowd of senior politicians. Today's Westerners are horrified by such behavior, but cavalier disregard for the welfare of even close colleagues has been typical of despots who can get away with it—which, until recently, most have been able to do. The constrained behavior of contemporary European royals bears no resemblance to the unfettered cruelties of their ancestors.[22]

By protecting the "tyranny of kings" in state societies, capital punishment achieves a goal that does not exist in small-scale society. Small-scale societies have not protected the power base of a single leader simply because there has normally been no leader, no "king." What capital punishment has done instead among hunter-gatherers and other egalitarian groups is to protect the "tyranny of cousins," both from challenges to social norms, and from selfish aggressors.

The decisive form of social control represented by the killing of aggressive males could clearly have had far-reaching significance in human evolution. With regard to the idea that *Homo sapiens* self-domesticated, the critical question is whether individuals with a particularly high propensity for reactive aggression tended to be killed. The characteristic fact of egalitarian relationships indicates that the execution of would-be despots was indeed systematic. Even now, although married hunter-gatherer men normally respect one another's autonomy, occasional individuals have been reported trying to control others. Like alpha chimpanzees, these would-be despots defend their top status by reacting ferociously to challengers. In a world without prisons or police, those bullies whose reactive aggression was particularly egregious could be stopped only by execution. Thus the egalitarianism found among all mobile hunter-gatherers indicates that the most aggressive individuals were eliminated. The ironic and disturbing conclusion is that egalitarianism, a system that appeals because of its lack of domineering behavior, is made possible by the most domineering behavior in the human arsenal.

To understand what it means to be egalitarian, consider the structure of hunter-gatherer society. A typical society averages close to a thousand people, sharing the same unique language (or dialect) and cultural practices, such as the rituals that they practice at funeral ceremonies. Societies are too large for everyone to live together, because the environment does not afford enough resources to support hundreds of people in a small area. People typically live in bands averaging fewer than fifty people. Each band occupies its own subregion within the society's territory and tends to stay in one place for a few weeks at a time. When foraging becomes too arduous because nearby resources are exhausted, the band moves on, normally reoccupying a previous campsite. Those regular shifts explain why hunter-gatherers are called "mobile" or "nomadic." A band typically comprises only ten to twenty married adults; roughly half the population is children, and some adults are unmarried, either young or widowed. Band membership can be fluid. Families might move to a different band to join relatives or escape troublesome relationships. Band size varies but is always small.[23]

The egalitarianism that is such a special feature of relationships among hunter-gatherer men is centered on the five to ten married men within a band. Those few husbands are the "elders," or Gellner's "cousins." The system found in Tierra del Fuego was typical of other hunter-gatherers. Lucas Bridges captured it with a celebrated anecdote about an Ona (Selk'nam) man called Kankoat.

> A certain scientist visited our part of the world and, in answer to his enquiries on this matter, I told him that the Ona had no chieftains, as we understand the word. Seeing that he did not believe me, I summoned Kankoat, who by that time spoke some Spanish. When the visitor repeated his question Kankoat, too polite to answer in the negative, said: "Yes, Señor, we, the Ona, have many chiefs. The men are all captains and all the women are sailors."[24]

The story applies to mobile hunter-gatherers in general. Bands might have a headman, and some men are respected more than others, but the men are all equal in the sense that everyone must work to provide his own food, and no married man has authority over any other.[25] When a group decision is needed, the situation determines who has most influence. The elders behave like a boardroom without a chairman. Everyone has a voice, but they all show considerable reluctance to use it. Men are so averse to grandiosity that self-deprecation is a highly regarded part of public behavior. The anthropologist Kenneth

Liberman recorded the effect among Australian Aboriginals. It was important to show shame and embarrassment, he wrote, because they "demonstrate to others that one does not have a conceited view of oneself."[26]

As Kankoat indicated, egalitarianism can be less strict with regard to women and children. The degree of patriarchy varies. Equality among the Ju/'hoansi has been claimed to apply to all adults, but if a woman is beaten by a man, his punishment is minimal at most.[27] Tanzania's Hadza hunter-gatherers are said to be egalitarian, but if a hot area has few shade trees, the men get the shade while the women sit in the sun.[28] Still, even if some individuals' voices hold little sway, there is no such thing as a group leader who can demand obedience.

The egalitarian mobile hunter-gatherer system stands in contrast to the larger, typically hierarchical farming groups, in which individuals such as chiefs, monarchs, dictators, or presidents hold positions that give them authority over others. Of course, even settled societies headed by an individual may incorporate strongly egalitarian components. The Declaration of Independence may have seemed revolutionary when it stated, "All men are created equal," but within small groups, people have always tended to adopt norms of equality. Those who try to better themselves or take command in small groups can easily become targets of resentment.

Most group-living primates, by contrast to humans, have a clear dominance hierarchy enforced by brute fighting ability. Typically, the alpha is a male, but, whatever the sex, the alpha is the one that has physically defeated all challengers.[29] Hunter-gatherer men are entirely different from primates such as chimpanzees or gorillas, because ordinarily men's status does not depend on violence. Individual apes and other primates become alpha by soundly defeating the prior alpha in a physical fight. Among mobile hunter-gatherers who follow social norms, by contrast, there are no fights and no equivalent of being an alpha male.

To the extent that there is leadership in hunter-gatherer bands, such as in taking initiative for group decisions, prestige is the important criterion. People compete for influence mostly by producing good arguments, creating good plans, being the best mediators, telling the best stories, or seeing the future most convincingly. A person who is skilled in these ways might be recognized as a leader or headman, but that role would be earned by his or her being wise and persuasive rather than assertive, pushy, or a good wrestler. Although leaders can be admired and respected, they cannot enforce their ideas, nor can they use their position to take anything from other members of the band. An inability to dictate to others means that, among mobile hunter-gatherers, there is no alpha position.[30]

What explains the absence of alphas among hunter-gatherers? Given the view of hunter-gatherers as harmless, peaceful people, a theoretical possibility would be that they are born that way. A few species of primate show that psychology can indeed evolve to be strikingly uncompetitive. Male tamarins, for example, breed together as co-consorts of a single female without ever fighting over her. Such unconcerned sharing is rarely found in humans, however. Rather, nomadic hunter-gatherers appear to be psychologically just like us. Whether born in the Kalahari or New York, infants show the same mixture of self-centered pushiness and helpfulness. They have the same potential for competitive psychology everywhere, and, as we have seen, sometimes hunter-gatherer men do grow up to challenge the social norms by trying to throw their weight around.[31]

In 1982, the anthropologist James Woodburn, then the principal ethnographer of the Hadza hunter-gatherers of Tanzania, reported cases of Hadza men trying to dominate others by ordering them about, or taking their wives or possessions. He also noted the danger inherent in conflicts among men armed with deadly weapons. Hadza men, he wrote, recognize "the hazard of being shot when asleep in camp at night or being ambushed when out hunting alone in the bush." He suggested that the absence of alphas among nomadic hunter-gatherers was caused by killing: "The means to kill secretly anyone perceived as a threat . . . acts directly as a powerful leveling mechanism. Inequalities of wealth, power and prestige . . . can be dangerous for holders where means of effective protection are lacking."[32]

In 1993, Christopher Boehm surveyed dozens of the best-known hunter-gatherer societies to find out whether Woodburn's idea that egalitarianism was enabled by the killing of alphas applied more broadly than just to the Hadza. On the one hand, he found, there was a universal social norm: men were expected not to attempt to dominate others. Surprisingly, however, Boehm found many accounts of self-aggrandizing men who tried to intimidate or deceive one another. His conclusion supported Woodburn's: despite the social norms, "a potential bully always seems to be waiting in the wings." For all their famous egalitarianism, hunter-gatherer men can be unpleasantly competitive.[33]

Boehm even found occasional records of men achieving an alpha position, at least for a time. A newly assembled group of Inuit consisted of a few men who had come together after being exiled from their original bands. One, an especially combative individual, took the wife of another man, declared her his own, and taunted the defeated husband. Later, he took a wife from a second man, and commanded him never to speak to his lost wife again.[34]

The question is why tyrannical behavior such as was exhibited by the Inuit wife-stealer is not more common. The answer is that would-be tyrants are constrained by whatever social pressure is necessary.

The process of social control normally starts at a very low level of conflict, perhaps prompted by a small display of pride. The anthropologist Richard Lee described his own experience of being humbled. A young man at the time, he wanted to give a big Christmas present to the band of Ju/'hoansi hunter-gatherers that he was studying, so he surprised them by buying a magnificent ox. The Ju/'hoansi had no money and had to hunt for days to get a large animal. Their response to the splendid gift of meat shocked Lee. The men were insulting. They said the ox had no meat on it. They called it a bag of bones. They said the ox was so thin they would have to eat the horns.

Eventually, an elder called Tomazo explained what was going on: the gift had made Lee look arrogant. "When a young man kills much meat," said Tomazo, "he comes to think of himself as a chief or a big man, and he thinks of the rest of us as his servants or inferiors. We can't accept this. We refuse one who boasts, for someday his pride will make him kill somebody. So we always speak of his meat as worthless. This way we cool his heart and make him gentle."[35]

The insults directed at Lee were a typical leveling mechanism. The Ju/'hoansi wanted Lee's meat, but they did not want Lee to feel superior. The same attitude explains why, in Woodburn's words, "A Hadza returning to camp having shot a large animal is expected to exercise restraint. He sits down quietly with the other men and allows the blood on his arrow shaft to speak for him."[36] Often a few words are all that is needed. Everyone knows how he or she is supposed to behave. The anthropologist Elizabeth Cashdan captured the ethos: "The proper behavior of a !Kung [Ju/'hoansi] hunter who has made a big kill is to speak of it in passing and in a deprecating manner; if an individual does not minimize or speak lightly of his own accomplishments, his friends and relatives will not hesitate to do it for him."[37]

"The slightest gesture or facial expression," wrote Ernest Burch about Inuit camps, "could serve to communicate a complex idea or sentiment." If an offender did not respond to subtle disapproval, his or her resistance was clearly deliberate. The protests then became more obvious. Shaming, ridicule, and cutting references occur. A song of derision might be sung in the offender's face while he stood before the assembled community.[38]

If more drastic measures are needed, shunning or ostracism is often effective because being avoided and excluded is deeply painful to most people living in small groups, and will normally do the trick. When the anthropologist Jean Briggs was living with a remote Arctic group, the Utku, she made the mistake of losing her temper. Slushy ice had dropped from the roof of the igloo onto her typewriter. She threw a knife into a pile of fish and railed against the endless fish diet. The igloo emptied fast, and over the ensuing weeks, Briggs found herself left alone in her tent, with no one calling on her anymore. She found the experience desperately painful until she finally found a way to explain herself.[39]

The notion that within human communities people control one another by shaming, ridicule, and ostracism was proposed by sociologist Émile Durkheim in 1902 and has been a core principle of anthropology ever since. Woodburn's hypothesis that such control extends to execution among hunter-gatherers, where it accounts for the absence of alpha males, now seems compelling. An ultimate sanction is needed to control those who are immune to verbal and social pressures, when nothing other than capital punishment can work. By bringing down even the most domineering of men, capital punishment underlies the unique human phenomenon of an egalitarian hierarchy in which men normally succeed in suppressing a desire for dominance.[40]

If the offender disregards even the highest level of social disapproval, however, "people are prepared by their ethos to deal with such a threat summarily."[41] That was what happened to the Inuit man from Greenland who stole the wives of his companions, laughed at the husbands, and commanded their acceptance. He was killed by two of the men that he had offended.[42]

One way in which the practice of capital punishment can reduce societal aggressiveness is by encouraging conformity. Punishment is important everywhere. In the 1960s the rural Bang Chan people of central Thailand, a largely Buddhist population of rice growers, became renowned as one of the world's most peaceful societies, a place where fighting, domestic violence, and child abuse were virtually unknown. Psychological anthropologist Herbert Phillips, whose studies brought the Bang Chan's gentleness to light, was clear about where their serenity came from: "What is more impressive about aggression in Bang Chan is not its absence but the amount and kinds of control that are exercised over it."[43] A similar dynamic is found among mobile hunter-gatherers, whose societies are, like the Bang Chan's, marked by gentleness, restraint, and strikingly respectful relationships, yet whose delightful nature again owes much to parenting and social control. Ju/'hoansi children and adults are regularly reminded not to be too pushy. Anthropologist Polly Wiessner found that in conversations among Ju/'hoansi hunter-gatherers criticism was eight times more frequent than praise.[44] Every culture seems to have learned the trick of taming the next generation through socialization, using social control more than reward. The fear of capital punishment can doubtless contribute to encouraging a spirit of conformity and restraint.

Important as such sociocultural effects of the threat of execution may be, however, they are not the focus of the execution hypothesis, which is concerned with long-term genetic consequences. The idea of the execution hypothesis is that during thousands of prehistorical generations, the victims of capital punishment were disproportionately those with a high propensity for reactive aggression. Killing or repression of such individuals is supposed to have happened so often that our species evolved a calmer, less aggressive temperament. Unfortunately to quantify the past rates of execution, or to calculate the Pleistocene selection pressures, is impossible. The concept that the domestication syndrome results from capital punishment is supported, however, by the human system of male egalitarianism, because the primate-style alpha males that are missing from human society are characteristically reactive aggressors.

In the millennia before groups found a way to control the bullies, reactive aggression would have dominated social life in the same way as it does in most social primates such as chimpanzees, gorillas, and baboons. In those species, alpha males achieve their position at the top of their group's dominance hierarchy by defeating each rival in turn in physical and often bloody fights. The process can take years. Having cowed every challenger the alpha can afford to be a tolerant dominant, but underneath his cloak of calm lies a potentially violent reaction. If another male fails to give the appropriate signals of subordinacy, the alpha exhibits an explosive reaction of aggressive chasing that ends, if necessary, in a ferocious beating of the subordinate. The alpha's bullying is strongly correlated with having high levels of testosterone, which appear to support his motivation to dominate others. To judge from the ubiquity of such behavior in the social primates, our ancestors once followed the same brute fashion; and in view of the massive faces of the ancestors of *Homo sapiens,* they most likely continued to physically fight, one on one, until at least the Mid-Pleistocene. Human competitiveness still has elements of the primate system of achieving status by individual combat. High-testosterone men are not particularly aggressive unless challenged, but when confronted they are more likely than low-testosterone men to respond with aggression. Primate-style alpha males, in sum, are individuals with a high propensity for reactive aggression. Execution of alpha males selected against reactive aggression.[45]

Even though capital punishment has been widespread, the idea that it would have been sufficiently frequent to have evolutionary effects on aggression may seem surprising at first. But considering the length of time, and the number of generations within that period, the rate of evolution for self-domestication appears relatively slow. As we have seen, the process of self-domestication could have begun at least 300,000 years ago. Three hundred thousand years is equivalent to about twelve thousand generations; the domestication syndrome apparently intensified in the later times. The number of generations is considerably greater than the number for any mammals to become domesticated, such as the roughly 15,000-year evolution of dogs from wolves. The average generation time for wolves is between 4 and 5 years, suggesting that dogs have been separated from wolves by fewer than four thousand generations. Even faster is the 80-year evolution of domesticated mink from wild mink; in this case, humans were deliberately selecting each generation for docility, greatly accelerating the speed of domestication. Thus, humans probably self-domesticated slowly, in accord with relatively gentle social pressure, with occasional elimination of individuals who displayed especially high levels of reactive aggression being only one of many factors affecting human evolutionary change. The question is: was the rate of killing group members who had a high tendency for reactive aggression sufficient to be a significant evolutionary force?[46]

The rate of deaths caused by killing within groups can be high. The largest recorded rate in an egalitarian small-scale society appears not in hunter-gatherers but in the Gebusi horticulturalists studied by Bruce Knauft. Within-group killings among the Gebusi served to solve conflicts and maintain social cohesion. The victims were said to be people full of "lethal anger."[47]

Knauft collected data on 394 deaths during a forty-two-year span, from 1940 to 1982. He found that the killing of one in four men (24.4 percent) and of almost one in six women (15.4 percent) resulted from accusations of sorcery. There was no difference in the execution rate for young (unmarried) and married men.[48]

Such a high death rate could clearly have rapid selective effects, but I am not suggesting that the Gebusi kill rate is representative of human evolution. We cannot even assume it is typical of the Gebusi. Since 1982, they have become a changed people, dancing to disco and adopting Christianity. Their high rate of execution during the years that

Knauft analyzed might have been due partly to a failure to organize their marriages properly, a result of demographic pressure from other Gebusi clans. Their exceptionally high rate of killing was probably a specific phenomenon peculiar to a particular culture, time, and place.[49]

Still, it opens our eyes to the possibility that within-group killing was sufficiently frequent to be a very significant selective force. At about the same time, in highland New Guinea, Ray Kelly was documenting witchcraft killings among the Etoro; the victims were typically people known for their "archetypical selfishness and lack of responsiveness to others." Kelly found that 9 percent of fifty-five adult deaths were by execution. Rumor, fear, competition, and ignorance can combine into a surprisingly deadly brew.[50]

Nevertheless, how often reactive aggressors were victims of capital punishment is a matter of guesswork, and how long capital punishment has been practiced is equally unknown. Even the oldest direct hint is tentative and evolutionarily very recent. Paintings on rocks in eastern Spain made by hunter-gatherers around 8,500 years ago include representations that have been interpreted as execution scenes. In one, ten figures are lined up together, waving their bows, seemingly looking at a figure lying on the ground who has been hit by five arrows. Others show dead bodies hit by arrows. But whether such portrayals mean that executions actually happened is unknown.[51] Despite these uncertainties regarding the importance of execution for producing the domestication syndrome in humans, the evidence that it did so seems sufficiently persuasive to justify thinking about when capital punishment might have started. Traits found in all humans, such as language or hunting-and-gathering, are normally credited with a minimal date of 60,000 to 100,000 years ago, because that was when *Homo sapiens* extended their range from Africa, carrying those "cradle traits" with them into the rest of the world.[52] Capital punishment is regarded as one such cradle trait, an original feature of humanity that has "become fundamental in all human thinking," in Otterbein's words.[53]

How much further back capital punishment was practiced can be guessed from considering the evolution of language. Chimpanzees sometimes kill adults of their own communities, but there is no evidence that they can plan to do so, as I discuss in chapter 11. Language seems necessary for planned killings of specific individuals. The criticisms and ridicule given in response to an aggressor depend on gossip for their conspiratorial power. When those gentler methods of social control do not work, people start floating the idea of killing the offender. For that, linguistic ability is vital, and considerable skill is needed. In Boehm's words, "The tactical problem is obvious: whoever speaks up first may be putting his life in danger. The danger may involve a physical attack, or in bands that have shamans it may also incorporate sorcery." One can see why the ebb and flow of suspicion that Knauft illustrated among the Gebusi is handled so carefully.[54]

Gossip solves the coordination problem by allowing individuals to test their feelings cautiously and to generate shared plans. Julius Caesar's more than twenty assassins talked in small secret groups for weeks to gain confidence in one another. After individuals sound one another out sufficiently, a consensus builds to the point where, if the troublemaker does not respond appropriately, he or she can be killed by collaborative decision. Even then it can be dangerous for the initiators. Caesar's first assailant, Casca, was left shouting to his brother to help him before his co-conspirators all joined in.[55]

While we cannot be certain how rapidly or over how long language had emerged before 60,000 years ago, we can put some bounds on it. One view is that language developed to the level of contemporary sophistication between 100,000 and 60,000 years ago—or in other words, no earlier than can account for the universal abilities of humans today. More radical thinkers suggest that modern language originated substantially earlier. A helpful touchstone is a comparison with Neanderthals. Recall that Neanderthals' occupation of Europe and West Asia is recorded from about 200,000 to 40,000 years ago. Genetic data indicate a Neanderthal divergence from our lineage between 275,000 and 765,000 years ago, and in their skulls and skeletons Neanderthals do not show the signs of self-domestication that *Homo sapiens* do. Because Neanderthals did not achieve the same degree of cultural complexity as *Homo sapiens,* and particularly because their symbolic culture was much more limited, most experts think that their linguistic ability was less than ours. There are also some tentative biological indications supporting this claim. For instance, the temporal lobes of the brain, which are involved in language functions (as well as memory and social behavior), are relatively larger in *Homo sapiens* than in Neanderthals. The evidence of skull and skeleton, cultural complexity, and brain all suggest that Neanderthals did not have language as we know it. On this basis we can conclude that before the split between *Homo sapiens* and Neanderthal lineages, between 275,000 and 765,000 years ago, our Mid-Pleistocene ancestors' language was significantly less sophisticated than it is today.[56] This all implies that language became increasingly effective in our direct ancestors subsequent to 765,000 years ago, though at a pace that cannot be directly assessed.

Most likely, therefore, linguistic ability improved substantially in the *Homo sapiens* lineage compared to all other *Homo*. With that improvement came the ability for individuals within a group to form coalitions that excluded or ostracized a member of the group who had become a domineering aggressor. Those coalitions enabled human selection against excessively aggressive men. The result was continuing change in the direction of a more cranially gracile, paedomorphic, and tolerant species—or self-domestication.

In short, the ability to murmur about how much we resent some other individual, and to float the idea of doing something drastic about it, has certainly been part of the human legacy for at least a few thousand generations. If competent language began to develop five hundred thousand years ago, its increasing social significance would help explain how our ancestors started controlling the alpha males and thereby bred a new kind of *Homo*. In this model of self-domestication, language was the key feature of *Homo sapiens* that allowed many tools of social control, from gossip to killing.

But language is not the only feature that has been proposed as the key to unlocking the domestication syndrome. Various scholars have focused on the use of weapons as a way to explain the transition from a primate system of alpha males to a human system of egalitarianism and cooperation. Weapons would have been useful in organizing and launching the move against an alpha male, because thrown rocks or spears could make a proactive attack more overwhelming and easier to carry out safely. Several points seem to me to undermine the importance of weapons, however, compared with the development of language.

Weapons are not needed for capital punishment. Animals such as wolves, lions, and chimpanzees kill using collaboration, not weapons. Humans can also kill without weapons. In his worldwide survey of small-scale human societies, Otterbein admittedly found that stoning, spearing, and shooting were frequent execution techniques, but also noted hanging, burning, drowning, beating, throwing off a precipice, enforcing suicide by requiring a guilty party to jump off a tall tree, and (for rape) forcing a briar into the penis. Another method was to give the victim to a vengeance party from a hostile neighboring group and let them deal with the problem, as happened in Australia.[57]

Weapons may also increase the danger for killers, as illustrated by an execution reported among Yanomamö hunter-gardeners. Some men had been repeatedly annoyed by the arrogance and bullying of one of their compatriots, but they never had the courage to stop him. One day, a group of them encouraged the bully to climb high into a tree to extract honey. Unsuspecting, he laid down his weapons before climbing up. His assassins were now safe: they gathered the victim's weapons and merely had to wait for him to come down before easily killing him. The episode recalls a way that some organized crime kills supposedly have occurred—planned for weeks until the victim sat in a car next to the driver, with his killer poised behind him to strike. The importance of ensuring that the victim is helpless at the time of the attack dramatizes the importance of attackers' sharing their intentions with one another—making plans by using language.[58] Language, whether gestural or spoken, makes the sharing of intentions sufficiently sophisticated for scared, frustrated, and resentful men to devise a safe plot against a tyrant.

The time assigned to the development of weaponry sufficiently efficient to kill is also inappropriate to explain the emergence of self-domestication. Suppose that *Homo sapiens* indeed originated as early as 300,000 years ago, and that this represents the time when leveling mechanisms started the trend toward the egalitarianism that makes us unique as a species. Even that early date was long after *Homo* species had been using weapons sufficiently well to hunt or to chase lions off kills, a practice that could well have gone back to 2 million years ago. Another argument for the key role of language comes from comparisons across species. We have no evidence that chimpanzees or any other nonhuman can decide in advance how, when, and where to kill a member of their own social community. Without language, the task seems impossible. I therefore infer that the vital human novelty that propelled the origin of a new kind of political system was plotting. The ability to plot together, rather than the ability to make weapons, surely determined when the change occurred in the balance of power between the classic alpha-male type and the new coalition of subordinates.

The ability to plot together is an example of what psychologist Michael Tomasello calls "shared intentionality," defined as "collaborative interactions in which participants share psychological states with one another."[59] Humans excel at shared intentionality, which emerges in children at about the age of one year, whereas chimpanzees show barely any evidence for it. Tomasello considers that the uniquely human development of shared intentionality explains why humans can do many special human things, from using math and building skyscrapers to playing symphonies and forming governments. If the hypothesis that selection against reactive aggression led to the domestication syndrome is correct, however, none of those human abilities was as special as the one that enabled conspirators to trust one another sufficiently to collaborate to kill a bully. That ability could both have domesticated us and, as chapters 10 and 11 show, made possible many kinds of human cooperation.[60]

Shared intentionality and language are both critical parts of the story. How these vital advances in cognitive ability began is a matter of speculation. We do not even know whether or not they arose by chance. In theory, language and shared intentionality could have evolved arbitrarily, as has been argued for other cognitive features. For example the ability to recognize one's own image in a mirror—shared by humans, great apes, bottle-nosed dolphins, Asian elephants, and a bird, the magpie. Because the only mirrors in the wild are merely occasional still pools of water, no adaptive explanation for self-recognition in mirrored reflections makes sense. Instead, the ability for self-recognition happened to result from a rise in mental power that was favored for other reasons. The same could well be true of the initial emergence of language.[61]

Even if reasons for the origins of language remain mysterious, the magnitude of its impact points to the origin of *Homo sapiens* as the time when linguistic ability took a major step forward. As the paleoanthropologist Ian Tattersall noted, the algorithmic basis of language appears to be rather simple, which suggests that its invention was "more or less instantaneous." Tied as language is to symbolic thought, and being responsible for social consequences of great importance for human politics and social behavior, language most likely announced its reaching a sophisticated level by having visible effects. In the whole 2 million years of our Pleistocene ancestry, the changes in our neural and skeletal anatomy that began 300,000 years ago are the most dramatic.[62]

The development of increasingly skilled language thus provides the best basis for the ultimate explanation of human domestication. As we will see in the next chapter, the consequences went far beyond a mere loss of reactive aggression.

9

What Domestication Did

IN OUR SEARCH for the origins of humanity's peaceful virtues, we started with Blumenbach. In 1811, on the basis of humanity's exceptional docility, that genial anthropologist had insisted that we are a domesticated species. Now a mechanism has emerged to support Blumenbach's claim. The idea that we owe our current placidity to 300,000 years of selection against reactive aggression seems to work well, but it is a new way of thinking. In this chapter, I want to go deeper into the argument for self-domestication to see how far we can distinguish it from alternative hypotheses for our low aggressiveness. Two have been proposed that are worth considering in detail. One is that low reactive aggression was a feature of our ancestors much further into the deep past than 300,000 years ago; the other is that low aggression has not been directly selected, but is instead a by-product of selection for another trait such as increased self-control.

To explain anatomical changes in the evolution of *Homo sapiens,* conventional scholarly wisdom presents an idea different from self-domestication. As I previously noted, paleoanthropologists have traditionally interpreted the special anatomical features of *Homo sapiens* as a series of parallel adaptations, rather than as incidental by-products. To explain the lighter build, shorter faces, and feminization of our lineage, they have invoked forces such as changed climate, better diets, or increasing sophistication in tool use. Those proposals are very reasonable if one assumes, as scientists often do, that biological features always evolve by the direct action of natural selection.

Suppose the traditionalists are right, and the unique anatomy of *Homo sapiens* is not a mark of our domestication. In that case, what alternative theories might explain our being so relatively pacific?

An extreme kind of answer might suggest that our ancestors were forever docile, that none of our ancestors was ever as aggressive as, say, a chimpanzee or a typical monkey. For every one of our animal ancestors to have been particularly docile would be remarkable. About 542 million years have elapsed since the Cambrian explosion, when our warm-ocean ancestors first became bilaterally symmetric animals. For the last 360 million years, our lineage has been on land, first as salamander-like amphibians, eventually as mammals. The probability that a high propensity for reactive aggression was never favored in any of the hundreds of species that took their turn in evolving down the path toward us, from the Cambrian to the present day, is clearly low. Nevertheless, conceivably all those ancestors could have been amazingly unaggressive.[1]

More likely, a reduced tendency for reactive aggression evolved at least once but too long ago to be detectable. Suppose that our relative lack of aggressiveness emerged about the time when primates evolved from a shrewlike mammal, around 80–90 million years ago. Or imagine that our docility goes back to our great-ape origins, perhaps 25 million years ago. It would be a difficult task to reconstruct the biology of the more aggressive immediate ancestor from which we must have descended. But that is what we would need to do to infer self-domestication.[2]

Indeed, reductions in reactive aggression may well have occurred in the distant past. A large increase in docility could have happened, for example, in the evolution of the pre-*Homo* australopithecines, signaled by a shortening of the face and reduction in length of the weaponlike canine teeth. The paleoanthropologist Owen Lovejoy has long advocated this idea, but telling evidence is elusive. As I noted earlier, canine reduction was likely adapted to a diet that required more intense chewing, as it is in living mammals; and a large role for aggression is indicated by evidence that male australopithecines were much bigger than females, and had bigger faces than females, suggesting that selection favored aggression in males. There are arguments both ways. The past is murky. Thus one alternative to Mid-Pleistocene self-domestication is that our ancestral lineage adapted to docility much earlier than *Homo sapiens,* too long ago for us to be able to confidently reconstruct the process.[3]

An entirely different source for our Rousseauian tendencies has also been proposed. If our docility was an incidental by-product rather than an adaptation, it might have evolved without ever being positively selected. The evolutionary anthropologist Brian Hare raised this possibility based on the finding that species with larger brains are especially good at inhibiting their responses. Lower reactive aggression would thus be a by-product of a more general tendency toward lower emotional reactivity.[4]

The evidence came from a new kind of experiment that gave identical challenges to dozens of species. A team led by Hare and the comparative psychologist Evan MacLean presented the same two problems to thirty-six species of birds and mammals ranging in size from sparrows to elephants. For both problems, the solution demanded that an initial response be inhibited, allowing a second response to be expressed instead. The aim was to find out if species differences in the cognitive ability of inhibition are related to other known species differences.

One challenge that was used to test behavioral inhibition was how to get at a piece of food placed inside a short length of transparent plastic tube. In MacLean and Hare's experiment, individual animals were presented with a tube that was open at both ends and wide enough so that the animals could easily reach in and get the reward.

The single slightly difficult aspect was that the tube containing the food was presented so that the food inside was readily visible. So, although the animal could see the food, accessing it required shifting attention from the food itself and moving to the end of the tube to get at it. Some animals were so intent on their reward that they never moved away from their starting position. Clawing or pecking at the curved wall of clear plastic, they remained fixated on the food. They could not inhibit their first reaction to get directly at the food they saw, and they never got their reward. But others succeeded, thanks to their ability to abandon their initial attempts to get at the food directly—that is, inhibiting their first response. That enabled them to express a second response, the response that allowed them to achieve their goal. These animals moved away from the food item to one of the two open ends of the tube. There they could reach in effortlessly and take the reward.

Looking across the thirty-six species, what feature is associated with success at inhibiting initial reactions? The MacLean-Hare team found that the probability of success varied systematically with brain size. Species with bigger brains did better.[5]

The reason that brain size leads to better inhibition, or self-control, is most likely that bigger brains contain more cortical neurons. The cortex is the source of willpower and voluntary control over emotions. In species with bigger brains, a higher proportion of the brain is cortex, and more cortex means more neurons. Humans have more neurons in the cortex than any other species, about sixteen billion. The great apes and elephants follow, with a mere six billion or so. More neurons in the cortex—especially the prefrontal cortex, at the front of the brain—allow an animal to inhibit its emotional reactions better.[6]

The inhibition experiment raises the possibility that our low propensity for reactive aggression results from our large neural networks, which enable us to think before we act. But in practice, this explanation seems unlikely to contribute very significantly to our low aggressiveness, because our emotional responses tend to be too quick to be controlled completely by the cortex. Still, the idea offers a second alternative theory for Pleistocene self-domestication, that our species' increasingly strong control over reactive aggression came from having a big brain, rather than from selection against aggressive tendencies.

So there are at least two other explanations for the reduction of our emotional reactivity. It happened too early for us to study it; or it was an incidental result of selection for other cognitive abilities. Either way, the more completely we can probe the self-domestication hypothesis, the better. We want a test that distinguishes the hypothesis of recent self-domestication from other propositions.

The biology of living humans is helpful in this respect. I noted previously that animals can have a domestication syndrome in physiology and behavior, as well as in the anatomical traits that we have reviewed. So, according to the self-domestication hypothesis, humans should also have a domestication syndrome not only in anatomy, but in physiology and behavior, too. Other explanations for human docility do not make this prediction.

The ideal research strategy for testing the idea would be to compare *Homo sapiens* with our Mid-Pleistocene *Homo* ancestor, or, failing that, with Neanderthals. But, of course, both are extinct. Long gone, they are frustratingly uninformative. Fortunately, contemporary domesticated animals have something else to offer instead.

Compared with their wild ancestors, of course, all domesticated mammals show less reactive aggression. That is why we call them domesticated.

Aggression, however, is just one of many types of behavior modified by domestication. Among other things, domestication leads to changes in the age at which fear responses become stronger, in playfulness, in sexual behavior, in the speed and effectiveness of learning, and in the ability to understand human signals there are also physiological changes in the production of hormones and neurotransmitters, and in the size of the brain and its component regions. These features make up a domestication syndrome of physiology and behavior.

At first glance these various modifications seem to have nothing in common. But a single principle unites them: in one way or another they are all paedomorphic, or "child-shaped," referring to a juvenilized trait. For a trait to be recognized as paedomorphic the species in question has to be compared with its ancestor. A trait is paedomorphic if it is found in juveniles of an ancestral species and is retained into a later stage (i.e., adolescence or adulthood) of a descendant species.

Take the anatomy of dogs and wolves. Many adult dogs have delightfully floppy ears, such as Labradors and Weimaraners. Adult wolves never have floppy ears, but juvenile wolves do. Labradors, Weimaraners and other floppy-eared dogs have extended the juvenile wolf feature of ear-floppiness into adulthood. Because dogs evolved from wolves, and adult dogs maintain the floppy ears of juvenile wolves, floppy ears are paedomorphic.[7]

Dog breeds vary enormously, from short-faced breeds such as Pekingese to long-faced Afghan hounds or German shepherds. This variation in the head shape of adult dogs might seem to make the question of whether dog skulls are paedomorphic impossible to answer. However, the solution proves elegant if the shape of the skull is analyzed in a sufficiently sophisticated way.

In 2017, the paleontologist Madeleine Geiger led a team that compared how the skulls of dogs and wolves change in shape during growth. By the time dogs are adult, and even when they are juveniles, their skull shapes are in some ways specific to their breed. The evolutionary mechanism that has produced this change varies among breeds. In some cases, the breed-specific components are paedomorphic with respect to wolves, such as the short face of the Australian dingo. But in other cases, they are not paedomorphic. Afghan hounds, for example, have long faces, the opposite of paedomorphic. This feature of Afghans is novel, meaning that the facial shape results from a new kind of growth pattern not found in wolves. Breed-specific features, in short, have evolved by various different mechanisms. They are not consistently paedomorphic. Selective breeding has pulled dog crania in all sorts of different directions.[8]

But in addition to breed-specific features, some uniform species-wide aspects are found in the skulls of all dogs regardless of breed. These features of the shapes of dog skulls are what make a dog a dog. With respect to these "breed-general" components, the skulls of dogs and wolves are always similar, and the similarity is paedomorphic. At every age, the wolf skull that is most similar to a dog's always comes from a wolf that is younger than the dog. This is true starting at birth, even though dogs and wolves have the same gestation period (about nine weeks). The paedomorphic quality of these breed-general aspects of skull shape is found, early in life at least, even in breeds like Afghans, whose long faces make them look not at all like wolf cubs.

Paedomorphism appears to have played a role in the evolution of many skeletal features of domesticated animals, including the reduced maleness of their skulls, their short faces, and their smaller brains. In the case of foxes, the evidence of paedomorphic evolution is as clear as it is in dogs, because the process has been observed. Looking back over fifty years of the silver-fox study, Lyudmila Trut was impressed by the number of anatomical traits that were paedomorphic in the domesticated lines. She listed widened skulls, shortened snouts, floppy ears, and curly tails, all of which are also found in dogs but not wolves. But in some species, such as pigs, the skulls of domesticated animals take new shapes, including faces that are short without being paedomorphic. As research continues, it will be fascinating to learn how broadly the findings of paedomorphism apply to other aspects of the skull and skeleton of domesticated foxes. But we already know that anatomical paedomorphism is a regular feature of the domestication syndrome.[9]

Paedomorphism can evolve by various mechanisms with exotic names, such as neoteny, postdisplacement, and progenesis, but we will not be concerned with these differences. For our purposes, it is enough to note whether the descendant species has become, in a particular respect, juvenilized, regardless of precisely how it did so.[10]

The paedomorphism found in anatomical components of the domestication syndrome is satisfyingly complemented by behavioral traits. In her review of the fox study, Lyudmila Trut drew attention to the "emotional expression of positive responses to humans" as a juvenile trait that some domesticated individuals retained into adulthood. Her observation heralds an explanation of the high frequency of paedomorphic traits in domesticated animals. The suggestion is that tameness is basically a juvenile characteristic. Selection for docility means selecting for juvenility.[11]

An experiment with mice tested this idea directly. Inspired by Belyaev, the developmental psychologists Jean-Louis Gariépy, Daniel Bauer, and Robert Cairns asked if selection against aggression leads to behavioral juvenilization. They developed two lines of laboratory mice, characterized by low and high propensities for reactive aggression. The researchers assessed aggression by noting the response of mice to being touched. They bred from less or more aggressive mice for thirteen generations and repeatedly measured two male behaviors, attacking and freezing. An attack was a vigorous lunge toward the opponent with biting or slashing directed at any part of the body. Freezing was what it sounds like—remaining rigidly immobile. Freezing avoided reactive aggression.[12]

As expected, the mice selected for low aggression and high aggression quickly diverged. After thirteen generations, the low-aggression males attacked others at about one-tenth of the frequency of the high-aggression males, and even when they did attack, they waited five times longer to do so. They were also quicker to freeze, and they froze more often.

The changes are all paedomorphic. Unselected juveniles rarely attacked, and took longer to do so than did adults. Juveniles often froze, and were quicker to freeze than adults. As the lines diverged, adult male mice of the line selected for low aggression retained behaviors of unselected juveniles longer into adulthood, including low rates and speed of attack and high rates and speed of freezing.

Remarkably, this mouse study represents one of only a handful of experimental domestication programs other than the studies of foxes, rats, and mink launched by Belyaev. We are still in the early stages of learning about this aspect of domestication.[13]

It is easy to understand, however, why domestication generally produces behavioral paedomorphism. In every mammal, juveniles tend to be friendly. Compared with adults, they are strikingly unafraid and inquisitive. That is why petting zoos, where children can meet animals close up, use animals that are not merely domesticated but also juvenile.

The evolutionary reason for juveniles' friendliness is that they must learn whom to trust when they are older and on their own. Youth is the perfect time for them to learn, because their mothers protect them from the wrong kind of social interaction. As long as they are helpless wards of the mother, the juveniles can trust her judgment and do not need to be fearful. They can let their guard down and be open to developing trusting relationships.

But in any mammal's life, there is a predictable change as it grows older, more mobile, and less likely to be protected by its mother. It becomes more easily frightened and then, as a response, more aggressive. At an age that varies across species, a fear response kicks in.

Unselected silver foxes become fearful at about 45 days old (six and a half weeks). From that time onward, the fox puppies show fear and aggression toward unfamiliar individuals, whether other foxes or humans. Their socialization window, as it is called, has closed. The bonds they have formed up to that time can last a lifetime, but youthful innocence has gone. Henceforward, they will find it hard to trust unfamiliar individuals. After the socialization window closes, even dogs can be difficult to train.[14]

Domestication extends the socialization window. Trut and her colleagues found that the fear response of silver foxes was delayed from 45 days in the unselected line to a full 120 days (four months) in the line selected for low emotional reactivity. The result of this more than doubling of the socialization window is, of course, that foxes in the selected line have a greater chance to interact with humans without fear. As Trut's group noted, the effect is paedomorphic: in the selected foxes, a juvenile trait (low fear) ends later in life than in the unselected foxes.[15]

The extension of the socialization window is only one of many paedomorphic features in domesticated animals. A general principle is that domesticated adults adopt the physiological and behavioral reactions of a wild juvenile.

The closing of the socialization window is due to maturation of the physiological stress systems. One of these systems is the HPA axis (or hypothalamic-pituitary-adrenal axis), which mobilizes the body's response to stress by producing cortisol. The HPA axis links the hypothalamus (in the brain) with the pituitary (linking brain and blood) and the adrenal cortex (part of the adrenal gland, lying atop the kidney). The rise of the fear response is partly due to the increased adrenal production of the steroid hormone cortisol, ultimately caused by activity in the hypothalamus.

Cortisol, which is easily measured in blood, is produced at impressively lower levels in the selected tame foxes than in the unselected foxes. After twelve generations of selective breeding of foxes for docility, the average blood level of cortisol in adults had more than halved. After twenty-eight to thirty generations, it had halved again. These remarkable drops in the basal level of cortisol produced by an unstressed adult are paedomorphic. Cortisol production in a selected adult fox is at the low levels of an unselected juvenile.[16]

Cortisol does rise sharply in response to an emotional stressor such as the approach of a scary stranger, whether in a selected or an unselected fox. But the stress-induced rise in production is markedly less in the foxes selected for docility. The stress system of the selected foxes is more like the stress system of juvenile unselected foxes. Once again, it is paedomorphic.[17]

Dogs echo the foxes in all these effects of domestication as they relate to the socialization window and the HPA axis. The socialization window closes later in dog puppies (eight to twelve weeks) than in wolf puppies (six weeks). The development of the fear response is likewise less acute in dogs than in wolves. At the end of the socialization window, fear crystallizes sharply in wolves, whereas in dog puppies fear of new individuals rises more slowly and is only fully expressed by around three months of age. Similarly, dogs have a reduced stress response compared with wolves, including smaller adrenal glands.[18]

A second major stress system, the SAM axis (or sympathetic-adrenal-medullary axis), has been studied less well in domesticated animals. However, research in guinea pigs suggests it is also paedomorphic, because its basal activity level is high in wild adults and low in wild juveniles and domesticates. The serotonin system conforms to the same picture. Serotonin, as discussed briefly earlier, is a neurotransmitter in the brain associated with the suppression of reactive aggression. In silver foxes, serotonin tends to be present at higher levels in juveniles than in adults and, as therefore expected, at higher levels in selected silver foxes than in the unselected lines.[19] Retaining or exaggerating juvenile characteristics, both a muted fear response and reduced reactive aggressiveness, thus seems to be a typical effect of domestication in mammals.

In addition to the stress response system, a second paedomorphic suite of features is also widespread among domesticated animals: the retention of juvenile body form as a result of selection for a smaller, less intimidating body.

In the initial stages of selection, domesticated species tend to be smaller than their wild ancestors. As mentioned earlier, boys who are large and strong for their age tend to be more aggressive, both when young and in later years. The standard explanation is that larger individuals win fights more often than smaller individuals. So selection against aggressiveness would favor boys who are smaller and accordingly slower and more paedomorphic in their growth rate.[20]

Foxes, dogs, mice, guinea pigs, and some other animals afford superb opportunities for studying the effects of domestication because, in each case, the ancestor is known and can be compared with the domesticated descendant. In wild animals with low tendencies for reactive aggression, the ancestor-descendant relationship is less well understood, but, consistent with the evidence from domestication, paedomorphic features are pervasive.

Remember how bonobos were first discovered. As adults, they have a skull that resembles the skull of a juvenile chimpanzee. Bonobo skulls include features that look strongly paedomorphic.[21]

Other aspects of bonobo biology echo the same trend. Bonobo development is delayed in several ways compared with that of chimpanzees, including growth of the skeleton and increase in body weight, decline of juvenile tolerance, increase of social inhibition, and production of thyroid hormones.[22]

Mammals native to islands tend to have reduced reactive aggression, and frequently exhibit strikingly paedomorphic traits, as I noted in chapter 5. A few years ago, I visited the island of Zanzibar in order to assess the possibilities for studying the Zanzibar red colobus monkey. Recall that the Zanzibar red colobus has strikingly paedomorphic coloration on its face, retaining the pink lips that in Ugandan red colobus are seen only in the youngest infants. Its skull is also paedomorphic. I wondered if I would see evidence of paedomorphic behavior.

At the entrance to the national park was a poster providing information about the Zanzibar red colobus. These monkeys are different from all other primates, the poster announced, because suckling can be greatly prolonged. Even adult males, it reported, sometimes suckle from adult females. How much more paedomorphic can you get! The primatologist Thomas Struhsaker later showed me videos of adolescent males exhibiting this remarkable behavior.

Zanzibar red colobus have reduced sexual dimorphism and smaller male canine teeth than their continental counterparts. But whether or not they show less reactive aggression is unstudied. They illustrate what rich opportunities await investigators.

Much remains to be discovered. However, juvenilization has already emerged as a major pathway for reducing emotional reactivity in silver foxes, dogs, and possibly all domesticates. The implication for the self-domestication hypothesis is straightforward. If *Homo sapiens* have been selected for reduced reactive aggression for the last 300,000 years, our species would be expected to show evidence of paedomorphism in our behavioral biology.

Popular culture has sometimes embraced a misleading concept of human paedomorphism. In his 1939 novel *After Many a Summer Dies the Swan,* Aldous Huxley imagined what would happen if people could greatly extend their life span. Huxley's story had the Fifth Earl of Gonister eating raw entrails of carp, hoping to postpone death. The old man's gamble worked, but it proved to be a Faustian pact. By the time he was two hundred years old, he had transformed into an adult ape—a gibbering, hairy, uncontrollable animal who had lost his understanding of language. He was kept captive in an underground catacomb together with his same-aged housekeeper, whom he would slap, scream at, and chase in response to the most trivial provocation.[23]

Huxley's fantasy came from seeing humans as juvenilized apes. If people could stay alive long enough, he speculated, they would no longer be juvenilized; and as a result, they would no longer be human. This assumption was derived from science. In 1926, the German anthropologist Albert Naef had noted that human adults look more like infant than adult chimpanzees. The implication was appealing. Evolution produced us by paedomorphism.[24]

In 1976, the paleontologist Stephen Jay Gould built on Naef's observation to suggest an important role for paedomorphism in human evolution. He proposed that our large heads, large eyes, small teeth, sparse hair, and upright stance had all evolved paedomorphically from apes. In 2003, the zoologist Clive Bromhall took the idea to the limit. In a book called *The Eternal Child: An Explosive New Theory of Human Origins and Behaviour,* he argued that, just as dogs are infantilized versions of wolves (which, as we have seen, is an exaggeration), humans are infantilized versions of chimpanzees. These expansive evolutionary concepts have been complemented by a burst of pathbreaking recent work comparing development in humans and great apes. Results often support the idea that brain development is slower in humans than in chimpanzees.[25]

For example, some genes that are functional in the prefrontal cortex are identical in chimpanzees and humans, but the time when these genes actively make proteins and other products (in other words, the time when the genes are expressed) is several years later in humans than in chimpanzees. The geneticists Svante Pääbo and Philipp Khaitovich led a team that identified the most prominent differences in gene expression between the two species. The largest differences are found in genes that help form junctions between nerve cells, called synapses. In chimpanzees, the peak expression of these synapse-forming genes occur when the youngster is less than one year old. In humans, by contrast, peak expression is extended to five years old. Human brain development is thus greatly delayed.[26]

Nerve myelination is similarly delayed in humans. Myelination is a process that makes nerve impulses travel faster by coating neurons in a protective layer of fatty myelin (and thereby making them into the familiar "white matter" of the brain). The downside of myelination is that myelinated neurons lose their ability to grow and form new synaptic connections. In chimpanzees, myelination ends at around ten years of age, whereas in humans it continues until individuals are as much as thirty years old.[27]

Exciting as these findings are, however, the notion of the human as a juvenilized ape is flawed. Paedomorphism refers to specific features, not to the organism as a whole. In relation to apes, some human features are paedomorphic, but others (such as brain growth) are peramorphic.

And, most important in the context of this book, comparisons with apes are irrelevant to the hypothesis of Pleistocene self-domestication.[28]

The comparison that we need is between Mid-Pleistocene *Homo* and *Homo sapiens.* In chapter 6, I noted Christopher Zollikofer's finding that *Homo sapiens* has a somewhat paedomorphic skull morphology compared with Neanderthals. His conclusion refers to the fact that the skulls of Neanderthals and *Homo sapiens* change in parallel as they grow, but Neanderthal skulls continue to grow after reaching the equivalent of the end-stage for *Homo sapiens.* In that final period, which occurs in Neanderthals but not in *Homo sapiens,* Neanderthal faces grow bigger compared with their braincases.[29]

Neanderthal skulls are not identical to those of the less well-known Mid-Pleistocene *Homo,* which spawned our lineage, but they are sufficiently similar to offer a helpful model of our ancestor.[30] The comparison between Neanderthals and *Homo sapiens* suggests that *Homo sapiens* evolved partly by paedomorphism of the braincase and face. A faster rate of tooth growth[31] and their relatively accelerated juvenile growth period[32] in Neanderthals also suggest paedomorphism in *Homo sapiens.* So, compared with that of Neanderthals and our ancestral *Homo, Homo sapiens*'s behavior should be paedomorphic, too. More generally, human behavior should have been changed in directions that occur in domesticated animals.

In many ways, this prediction looks promising. Some genetic evidence supports it with exactly the comparison that is needed—namely, between *Homo sapiens* and Neanderthals. The Pääbo-Khaitovich team found evidence that a delay in brain development found in humans compared with chimpanzees occurred after *Homo sapiens* separated from Neanderthals. If their result is confirmed, it will suggest that human behavioral development was more delayed in *Homo sapiens* than in Neanderthals.[33]

Developmental delays have been found in the behavior of many domesticated animals compared with their wild ancestors, including in social relations, play, learning, sexual activity, and vocalization. Guinea pigs illustrate the changes compared with cavies, their wild South American ancestors.

As juveniles, guinea pigs spend longer times in contact with their parents, and they nurse to a later age than cavies. Although the amount of play does not differ between guinea pigs and cavies when they are young, adult male guinea pigs play more. Experiments show that guinea pigs also learn more rapidly than cavies. And they follow a general trend among domesticated animals of intensified sexual behavior, by displaying more overt courtship. Finally, they call more than cavies.[34]

These changes from cavies to guinea pigs all mirror ways in which humans are unusual compared with most species, suggesting that humans too have been through a domestication-like process. Humans have an especially long juvenile period. Weaning (at around two or three years) is two to three years earlier in hunter-gatherers than in apes, but after weaning, children continue to rely on food supplied by their parents for longer than is seen in any other animal. Humans are strikingly playful as young adults, and even later in life. Humans, of course, are the ultimate learners, both as children and as adults. Human sexual behavior is frequent and prolonged, and strikingly emancipated from its purely reproductive function. Language makes *Homo sapiens* a complete outlier with regard to the elaboration of communication.

These behaviors all fit the supposition of recent human self-domestication. But we know too little about Neanderthals to determine when human paedomorphic behaviors evolved. The test is merely suggestive.

In one area of behavior, however, the archaeological data offer an opportunity to infer a more interesting difference between Neanderthals and *Homo sapiens.* The behavior of interest is cooperation.

Europe is full of well-studied sites that give clues to certain aspects of Neanderthal behavior. The sites suggest that an important difference between Neanderthals and *Homo sapiens* could be our greater capacity for learning and cooperation.

Neanderthals occupied Europe for perhaps half a million years. Then, about 43,000 years ago, after thousands of years of hovering along the Middle Eastern fringe, *Homo sapiens* entered Europe from the south and the east, coming mostly along river valleys and the coast. The result was dramatic. By about 40,000 years ago, almost all the Neanderthals were gone. In a few mountain areas, they hung on until as late as 35,000 years ago. They interbred with the invaders, and nowadays they survive only in the 1 to 4 percent of their genes that are found in the DNA of non-African humans worldwide.[35]

A hot debate surrounds the reason why *Homo sapiens* replaced Neanderthals. The argument often centers on intelligence. The first fossils of Neanderthals were discovered in 1856, and in the immediate aftermath, Neanderthals were mostly assumed to have been brutish and mentally simple, or possibly even degenerate *Homo sapiens.* Nowadays the pendulum has swung back. Some archaeologists argue that there was no difference in behavioral capacity between Neanderthals and *Homo sapiens.* In that case, the reason Neanderthals lost out could have been bad luck. The invaders might simply have brought a new disease with them, as the smallpox and measles brought by Europeans ravaged North America even before the first Thanksgiving.

Supporting the notion that Neanderthals were merely unlucky is the fact that, for some 60,000 years, they held their own against the first onslaught of the aliens: populations of *Homo sapiens* from Africa had begun to collide with Neanderthals in the Middle East by 100,000 years ago.[36] Furthermore, their material culture is now known to resemble that of *Homo sapiens* closely. Neanderthals apparently lived much like recent mobile hunter-gatherers. They controlled fire, cooked their meals of hunted meat, and gathered plant foods or shellfish. Their prey ranged from doves to woolly rhinoceroses. They had cozy home bases, where they slept on furs and likely treated themselves with medicinal herbs.[37] They had status markers such as bird-wing cloaks, they sometimes buried their dead, and they used deep caves for activities presumed to be rituals. After 300,000 years ago they used the sophisticated Levallois method for making stone tools. They could make fine blades, pigments, ornamental beads, and engraved artwork. As early as 200,000 years ago, they synthesized pitch from birch bark "through a multi-step process that relied on strict control of temperature and required a dry distillation excluding oxygen."[38]

Yet, despite these definite similarities in behavioral capacity, there are differences that consistently credit *Homo sapiens* with having created a culture that was slightly more elaborate. Neanderthals expressed their creativity at a lower rate than *Homo sapiens.* Fewer than ten ornamental beads worked by Neanderthals have been found from 200,000 years of rich archaeology, compared with thousands wrought by *Homo sapiens* before and after their swift takeover in Europe. Similar comparisons apply to stone blades, stylized figurines, ritualized burials, and engraved symbols. In each case, Neanderthals had the skills, but they used them more rarely and later than *Homo sapiens.*[39]

There were also some abilities that Neanderthals appeared not to have had at all. They seem to have lacked facilities for the long-term storage of food. There is no evidence that they built sleds, despite living through harsh winters. Nor do they appear to have produced or used boats. *Homo sapiens* colonized Australia about 60,000 years ago, in a journey that requires multiple legs of sea travel, so they clearly negotiated water crossings somehow. *Homo sapiens* in South Africa made bows and arrows (evidenced by arrowheads), spear-throwers, and fine bone points by 71,000 years ago, but none of these important tools have yet been proved for Neanderthals. Although Neanderthals used fire, they did not use it to make better stone tools, as South African *Homo sapiens* did, or to heat water by using hot rocks.[40]

A common explanation for all this is that they were less intelligent than *Homo sapiens.*[41] However, the idea has no direct support, and is unconvincing, based on their brains, which were roughly the same size as those of *Homo sapiens,* and on the fact that occasionally they did show great ability. Because Neanderthals had big eyes, which must have been serviced by a relatively large visual cortex, it has been conjectured that the amount of processing power they had available for thought was marginally less than in *Homo sapiens.*[42] Nevertheless, their cognitive abilities are widely considered to have been roughly equivalent to those of modern humans by about 200,000 years ago. Even if they rarely produced the kind of objects that reveal high mental ability, they did sometimes. Their sheer problem-solving ability was likely just as good as that of *Homo sapiens.* Whatever the extant differences, they were subtle. In the words of the paleoanthropologist Ian Tattersall, "What the Neanderthal record shows most eloquently is that it is possible to be very smart indeed without displaying either the particular kind of intelligence that modern human beings have, or the odd behaviors that result from it."[43]

Differences in intelligence between Neanderthals and *Homo sapiens* do not seem sufficient to explain the differences in evolutionary success. Another kind of explanation for the greater cultural success of *Homo sapiens* seems more promising. Based on the idea that *Homo sapiens* self-domesticated and Neanderthals did not, *Homo sapiens* may have been better at cooperating.

Although group sizes have not been confidently reconstructed, much evidence points to Neanderthals living in groups that were small by comparison with *Homo sapiens.* The archaeologist Brian Hayden inferred that Neanderthal bands averaged twelve to twenty-four people living in nuclear families with a sexual division of labor, that they formed alliances with ten to twelve other bands, and even that they may sometimes have aggregated into groups up to three-hundred strong for mounting drives.[44] However, evidence of regular inbreeding in Neanderthals points to their social networks' being small. The genome of a Neanderthal woman from Siberia showed that her parents were close relatives, such as half-siblings or an uncle and niece, and that similarly inbred matings had occurred regularly among her recent ancestors.[45] In small-scale societies of humans, groups break up into smaller units when social tensions erupt. Groups so small that matings were often inbred could have resulted from a tendency among Neanderthals to react to one another overly quickly with aggression, and consequently breaking up into ever-smaller groups.

Another feature that may have distinguished Neanderthal and *Homo sapiens* is communication ability, and therefore the capacity to learn from each other. Curtis Marean suggested that the technologies created only by *Homo sapiens,* such as making tiny stone arrowheads and other components of projectile weapon systems, depended on high-fidelity transmission of expertise, which the Neanderthals could not manage.[46] Similar skills cited by Marean include the use of poisons, heat to modify stones for better knapping, and the production of glues. The construction of food stores, sleds, and boats would also be challenging for a species poor at learning from one another. Taking that idea one step further, Brian Hayden has argued that Neanderthals differed from

Homo sapiens in the construction of items that required coordinated labor.[47] The more elaborate the cultural skill, the more cooperation is needed. Marean's and Hayden's ideas thus suggest that the reason for the poorer manifestation of Neanderthal culture was that they were less adept socially, rather than intellectually.

The more globular brain of *Homo sapiens* indicates that their cognition was somehow different from that of Neanderthals, but the difference need not have been a matter of intelligence. From the admittedly limited evidence we have, it seems that domesticated animals are not consistently better or worse at solving problems than their wild ancestors, but when they congregate socially they have fewer aggressive interactions. We should expect a similar balance between *Homo sapiens* and Neanderthals. There is an additional reason in support of the idea that Neanderthals were less effective at cooperating: domesticated animals are consistently better at certain kinds of cooperation than their wild ancestors.

That domesticated animals cooperate better than their wild ancestors is shown by research on "cooperative communication," which refers to understanding the deliberate sharing of useful information. An interesting take on cooperative communication is to look beyond communication within a species to communication between species, which has been studied by finding out whether animals understand helpful signals given by humans. Is this useful cognitive skill purely a matter of intelligence, or might it be linked to domestication? Starting in 2002, Brian Hare began testing several species. Dogs were the easiest subjects.[48]

A typical test, the so-called object-choice test, begins with admitting a dog into a room where an experimenter is standing at an equal distance from two upside-down bowls on the floor. The bowls are identical, but under one of them a food treat lies hidden. The person points to one of the upside-down bowls. A successful dog understands that the person is trying to help and goes to the indicated bowl. Hare and a small team of researchers found that most dogs passed this test.

For dog lovers, this was not a big surprise. What was interesting was that wolves failed the test.

Even chimpanzees cannot normally pass the object-choice test. Chimpanzees are generally better at solving problems than dogs, but they did not solve this one. The communicative-cooperative ability was special to dogs.

Dogs, by contrast to chimpanzees, passed the test not only as adults but even as puppies a few weeks old. This was evidence that the communicative-cooperative understanding of dogs is genetically grounded, not a learned capacity.

There are two possible explanations for dogs' doing better than wolves and chimpanzees, depending on what domestication did.

One possibility is that, during the process of domestication, dogs had been selected specifically for their ability to read human social cues. That is the tempting inference, because it is easy to imagine that dog breeders would favor the dogs that were most cooperative.

The alternative idea is that the cooperative-communicative ability was not selected for, but, rather, that an improved ability to read human signals emerged simply as an unselected by-product of the process of domestication. In this case, cooperative-communicative ability would be part of the domestication syndrome.

Belyaev's work with foxes offered a way to test between the two explanations. The domesticated line of foxes had been selected for reduced emotional reactivity, but they had never been selected for their ability to communicate with humans. If the domesticated foxes could read human signals even though they had never been selected to do so, then the cooperative-communicative ability must be a by-product of the domestication process.[49]

By 2003, Lyudmila Trut was the director of Belyaev's research station at Novosibirsk, in southwestern Siberia. With her help, Brian Hare tested the foxes that Belyaev and Trut had experimentally domesticated. He performed the same tests with a control line of unselected foxes that had been reared and housed in the same way as the domesticated line.

The result was lovely. The domesticated foxes behaved like dogs. They tended to follow human signals. Even the puppies (or kits, as young foxes are properly known) did so. But the unselected control foxes behaved like wolves. They did not use human cues to find food.[50]

The implication is clear: Domestication can indeed lead to unselected changes in a species's cognition. The cooperative-communicative ability can emerge as a by-product of domestication.[51]

Hare's work indicates that the increased cooperative-communicative ability is a feature of the domestication syndrome. One reason that selection against reactive aggression might cause animals to be able to read a human's signal is that reactive aggression results from a fear response. Selection against emotional reactivity reduces fear; and reduced fear allows a dog to take a longer, more careful look at a human than a wolf normally would. Reduced fear is paedomorphic.

Subsequent studies of wolves support the notion that reduced fear, more than intelligence, is what enables animals to understand human signals. Wolves that have been sufficiently socialized by living with humans when young prove able to pass the upside-down-bowl test. So the increase in social understanding is due to a change in the emotional system, not to superior intelligence.[52]

The evidence from bones suggests that, with regard to reactive aggression, the difference between *Homo sapiens* and our Neanderthal-like ancestors resembles the difference between a dog and a wolf. Neanderthals were certainly highly social, just as wolves are. Neanderthals had living sites with multiple fireplaces, and sleeping locations so close together that they could roll over and touch one another in the night. But the dog-wolf comparison suggests that a reduction in reactive aggression in *Homo sapiens* would have led to fewer frayed tempers, less self-centered domination, and more paying of attention to one another. Less fear, more mutual gazing, and greater ability to work together might have given Neanderthals enough patience and mutual tolerance to be able to build a boat, store food, make more complex weapons, and better coordinate the activities of warriors.

If Neanderthals had been better cooperators, in other words, they might have been able to hold their own against the *Homo sapiens* invaders. They, instead of us, could have ended up as the one species of *Homo* left on our planet. As it was, a paedomorphic capacity for reduced aggression, more tolerance, and greater cooperation seem to have given our ancestors the edge, reducing the Neanderthal legacy to a few bits and pieces in our DNA.

In domesticated animals, increased cooperative ability emerges as an incidental consequence of selection against reactive aggression. Self-domestication is therefore likely to have enhanced cooperation among *Homo sapiens.* But even assuming that an elevated cooperative ability began as a nonadaptive effect, clearly it became quickly advantageous. Cooperation was a key to *Homo sapiens*'s domination of the earth.

By contrast, another exceptional human behavior that may have arisen from the domestication syndrome has no known adaptive functions in humans. Homosexuality is a characteristic feature of our species that, from an evolutionary perspective, remains an intriguing unsolved puzzle. A novel possibility therefore seems worth considering. If homosexual behavior is not adaptive, perhaps it evolved in humans as a paedomorphic by-product of selection against reactive aggression.

The hypothesis that human homosexuality is adaptive, meaning genetically advantageous, has not been rejected lightly. Homosexual behavior can be frequently found among wild animals, and traits that are widespread are likely to be adaptive. In animals, homosexual behavior is often adaptive. So when evolutionary biologists began to study human homosexual behavior, they tended to take an adaptationist perspective by searching for ways to explain how a same-sex preference might have been favored by natural selection. Homosexual behavior among animals offered some ideas.

Among primates, homosexual interactions between adults have been recorded in at least thirty-three species. In most of them, it occurs among both sexes. In some primates, it is more frequent among females; in others, more frequent among males. It tends to be well integrated into ordinary social life. Homosexual interactions occur in all kinds of social systems, from monogamous pairs to large groups of many breeding males.[53]

Homosexual behavior is particularly prominent among bigger-brained species in which sexual behavior is emancipated from hormonal control. In primates, it occurs among apes and monkeys, but not in the smaller-brained lemurs and lorises. It is found among many whales and dolphins. Male gray whales romp together in an obvious state of sexual excitement. Male river dolphins use one another's blowholes for sex. Similarly exotic accounts from diverse species fill the pages of biologist Bruce Bagemihl's review of animal homosexuality, *Biological Exuberance.*[54]

Close study reveals how homosexual behavior can be adaptive. Among Laysan albatrosses in Hawaii, two parents are needed for chicks to be reared successfully. When there are not enough males, females pair together. Their sexual behavior includes courtship and pseudo-copulation. Females in same-sex pairs are fertilized by an already mated male, who then ignores the resulting eggs and chicks. The female pair bring them up without male help. Female-female pairs produce fewer surviving chicks than female-male pairs do, but more than unpaired individuals. Their pairing thus allows them to spread their genes better than any other strategy available to them. Among females in homosexual relationships, those who breed successfully are likely to transition in subsequent years to pairing with males.[55]

In animals whose choice of sexual partner is not a response to a shortage of opposite-sex partners, homosexual behavior sometimes appears to be adaptive by promoting useful social relationships. In troops of Japanese monkeys, females form temporary homosexual mating partnerships (or consortships) with other females even when males are available. Among savanna baboons, males form alliances that they use in fights against others. Allies reciprocally fondle one another's genitals, apparently to demonstrate their commitment to the bond.[56]

As described earlier, homosexual interactions are particularly prominent among female bonobos. As a female bonobo enters adolescence and becomes old enough to start having sex, she abandons her mother and leaves the social community in which she was born. She enters a neighboring community, where everyone is likely to be a stranger to her. The males welcome her, but the females are initially less friendly. After a few weeks, she is invited to have sex with a resident mother. From then on, she has regular sex with all the adult females, and so joins their social network. Sexual interactions among females are nonexclusive, shared with many other females rather than special relationships with a preferred partner. These encounters, which female bonobos show every evidence of enjoying, appear to contribute a very important aspect of bonobo social life. If a female ever has a conflict with a male, a scream from her will bring other females in quick support to chase the male away. The aid is undiscriminating: all females help one another. Females also use homosexual interactions to diffuse tensions among themselves. For bonobos, therefore, a propensity for homosexual behavior probably emerged as a by-product of self-domestication early in their evolution from a chimpanzee-like ancestor, and has now been co-opted and evolved into a useful behavior.[57]

Researchers have sought evidence that the kinds of reproductive or social benefits that animals gain from same-sex sexual interactions might be found in humans. In theory, humans could be like albatrosses, forming same-sex partnership in response to a short supply of members of the opposite sex. Certainly, partner availability influences us. Women and men in single-sex institutions such as prisons, schools, monasteries, and ships often temporarily shift their sexual activity toward their own sex. Nevertheless, of course, many individuals feel an exclusive attraction to members of their own sex, regardless of the availability of the opposite sex. Frequency estimates for various homosexual preferences vary. In the United States in the 1940s, the sexologist Alfred Kinsey estimated that 5 percent of women and 10 percent of men are homosexual or bisexual. Subsequent studies have tended to find that a predominant or exclusive orientation toward the same sex occurs at lower rates, in around 1 to 2 percent of women and 2 to 5 percent of men. Difficulties of definition and privacy make any precise figures arguable, but exclusively homosexual orientation is clearly a regular feature of humanity. Same-sex sexual orientation is often stable over a lifetime, and there is good evidence that it is partly heritable. These features make human homosexuality different from most animal homosexuality. Human homosexual attraction is especially strongly developed in a small proportion of the population, and is not primarily an effort to make up for restricted opportunities.[58]

A couple of possible explanations for how homosexuality might be adaptive for humans have been explored. One hypothesis is that same-sex relationships confer advantages in social competition. Homosexual men might be more intensely supportive of each other, for example, like female bonobos and Spartan warriors. Although such social bonds could presumably indeed be beneficial, the reproductive benefits seem too low to explain the occurrence of exclusive homosexuality. Among 600,000 domestic pairs of same-sex individuals in 2000 in the United States, for example, 34 percent of female couples and 22 percent of male couples were raising children, compared to 39 percent of 16 million opposite-sex couples.[59]

If homosexuals tend to have few children of their own, their sexual orientation could in theory be adaptive if it leads them to give exceptional help to their genetic kin. In some cultures, such as Samoa, homosexual men do show a stronger-than-usual interest in helping their brothers and sisters. But even in Samoa, the kin effect is too weak to explain the evolution of the homosexual tendency; and in other cultures, such as Japan, there is no evidence that homosexuals show more interest in kin than heterosexuals do. The numbers do not add up.[60]

The small family sizes of exclusive homosexuals, together with the lack of evidence that they provide large benefits to kin, suggest that homosexual behavior in humans is not biologically adaptive. This leaves the fascinating question of why homosexual attraction is as prevalent and persistent as it is in our species.

Unfortunately, the conclusion that same-sex behavior is not adaptive has sometimes been associated with a negative view of homosexuality, as if a trait should be viewed positively only if it evolved because it was adaptive. However, whether a behavior evolved because it was directly selected for, or because it was a by-product of another adaptive feature, or, indeed, whether it evolved at all, should not color our moral judgments. Many tendencies that we regard as morally reprehensible clearly evolved, including numerous kinds of sexual coercion, lethal violence, and social domination. Equally, many morally delightful tendencies did not evolve, such as charity to strangers and kindness to animals. Our decisions about which behaviors we like or dislike should never be attributed to our understanding of their evolutionary history or adaptive value. There is no implied moral prejudice or value judgment in exploring adaptive or nonadaptive explanations for a trait.[61]

The evidence that exclusive homosexual preference is common but not adaptive makes it a prime candidate for being an evolutionary by-product. An association with selection against reactive aggression is indicated on several grounds.

First, the only nonhuman animal in which exclusive homosexual preference is known is a domesticated species—namely, sheep. Male lambs that are reared in the exclusive company of other males fall into two groups. The division depends on how, once they become adult, they respond to a ewe who is introduced to them when ready to mate. One group is the heterosexuals. On meeting an estrous ewe, these rams experience a rise in testosterone and are fully sexually interested in her. The other group are the homosexuals, who interact with the estrous ewes without showing any hormonal increase or sexual interest. Females are just not their thing; they like males instead. About 8 percent of rams brought up in single-sex groups adopt this homosexual orientation.[62]

No adaptive explanations have been proposed for homosexual orientation in domestic rams. In wild sheep, dominant rams have been seen to mount subordinate rams in apparent displays of dominance, but the behavior is rare (4 percent of all social interactions in all-male groups). The obvious implication is that homosexual preference is an incidental consequence of domestication.[63]

In support of that idea, the animal physiologist Charles Roselli found that domestic rams who experience relatively low levels of testosterone before they are born are more likely to become homosexual. The effect is mediated in a part of the brain that responds to androgens in the fetal stage. This brain region, the ovine sexually dimorphic nucleus of the preoptic area (oSDN), is larger in heterosexual rams than it is in ewes. In homosexual males, it is smaller, more similar in size to that of females. These differences appear to be critical. An experimentally induced reduction of the size of an adult ram's oSDN tends to change his sexual-partner preference from female to male.[64]

The sheep studies thus indicate that homosexual preference is more likely in males who receive low testosterone exposure before birth. Since reduced testosterone is a common effect of domestication, homosexual orientation in this species appears to be explicable ultimately as an incidental consequence of selection against reactive aggression.

Intriguingly, humans also have a sexually dimorphic part of the brain that (like the oSDN) appears to be related to sexual behavior. It is called the third interstitial nucleus of the anterior hypothalamus (INAH3). The INAH3 is larger in heterosexual men than in women, and has been found to be intermediate-sized in homosexual men.

As with rams, evidence in humans suggests that variation in exposure to testosterone before birth might influence homosexual tendencies. A standard method for assaying prenatal testosterone exposure is to measure the length of the ring finger (the fourth finger) compared with the length of the second finger: increased prenatal exposure to testosterone tends to be associated with relatively long ring fingers. The largest surveys of homosexual men in the United States, China, and Japan have found a tendency for homosexual women to have relatively long ring fingers, whereas homosexual men have relatively short ring fingers. Homosexual men also have somewhat feminized face shapes and shorter, lighter bodies than heterosexual men, most likely from relatively low exposure to testosterone in the womb. These conclusions are still somewhat tentative, since the results are not always consistent, particularly if the sample size is small. Overall, however, they support the idea that levels of prenatal androgen hormones, especially testosterone, may affect sexual orientation. In general, females who have been exposed to higher-than-usual levels of androgens, and males who have been exposed to lower-than-usual levels, appear to have a higher likelihood of being homosexual—consistent with the idea that homosexual preferences are a consequence of self-domestication through a change in exposure to steroid hormones.[65]

Unfortunately, whether domesticated animals in general show more interest than their wild ancestors in same-sex sexual interactions does not appear to have been studied. Bruce Bagemihl listed nineteen species of domesticated mammals and birds known to show homosexual behavior, though it occurs in their wild relatives as well.[66]

A second avenue for exploring whether homosexual behavior is associated with domestication is by looking at the human ape relatives, chimpanzees and bonobos. Homosexual behavior is so comparatively rare in chimpanzees that it has not been suggested to have any consistent social function. The elaboration of homosexual behavior in bonobos, which occurs in both sexes, clearly fits the hypothesis that selection against reactive aggression favors homosexual behavior. A potential explanation is again that, in bonobos, prenatal exposure to testosterone is relatively low. The ring finger is longer relative to the second finger in males than in females in both bonobos and chimpanzees, suggesting that, as in humans, relative ring-finger length provides an index of prenatal testosterone exposure. As expected from the domestication hypothesis, relative ring-finger length is shorter in bonobos than in chimpanzees. Interestingly, relative ring-finger length in humans is more similar to the ratio in bonobos than in chimpanzees.[67]

However, as I noted earlier, great apes are not the appropriate species for testing human self-domestication. Too much has happened in the more than 2 million years since our ancestors were ape-like australopithecines, or in the 6 to 9 million years since our ancestors were chimpanzee-like forest dwellers. The ideal comparison is between contemporary *Homo sapiens* and Mid-Pleistocene *Homo,* or, failing that, Neanderthals or early *Homo sapiens.* Remarkably, a few data are available from five Neanderthal individuals, and even from a 100,000-year-old *Homo sapiens* from a site called Qafzeh, in Israel. The number of individuals is too small to provide much confidence, but the Neanderthals had significantly longer ring fingers (relative to second fingers) than modern people while the finger-length ratio of the 100,000-year-old *Homo sapiens* at Qafzeh is in-between the ratios for living humans and the five Neanderthals. These fascinating hints suggest that living *Homo sapiens* have indeed been subject to lower levels of prenatal testosterone than Neanderthals, in keeping with the self-domestication hypothesis.

The extensive homosexual behavior of bonobos appears to be a juvenile characteristic, retained into adulthood as an incidental consequence of selection against reactive aggression. Juvenile male primates often have penile erections, and they readily mount any available partners, in sexual interactions that look, to human observers, largely playful. In a study of rhesus monkeys, the psychologist Kim Wallen showed that juvenile males were equally likely to mount males and females. After the males entered puberty, their rate of mounting increased five times, almost entirely because they were mounting females more, and as they became adult, they almost exclusively mounted females. The shift in preference toward female partners appeared to result both from an increase in testosterone at puberty, and from the rewarding experience of mating females.[68]

The evidence from sheep, nonhuman primates, and humans thus suggests that domestication may have regularly led to the evolution of increased homosexual behavior. Certainly other important factors influence homosexual behavior in humans, and there are topics that I have not considered here. One major complexity for humans is that, among both females and males, individuals can take on relatively female or relatively male roles. Homosexual men who take a strongly male sexual role, for example, seem less likely to have had low exposure to prenatal testosterone than those taking a more female role. The question of why selection has not eliminated homosexuality, if it is a by-product of self-domestication without biological advantages, is also fascinating. A possible answer is that self-domestication has been continuing too recently, up to historical times, for selection to operate strongly against incidental consequences.

A full account of the biological and cultural evolution of homosexual orientation thus lies in the future. Nevertheless, the self-domestication hypothesis seems to contribute a helpful new component for understanding a special feature of humanity. The evolutionary retention of a more youthful sexual physiology and cognition could in theory have led *Homo sapiens* toward more same-sex mating preferences. And we can make a prediction that will probably, alas, never be testable: Neanderthals would have engaged in less homosexual behavior than we do.

The aim of this chapter has been to see if we can distinguish the idea of Mid-Pleistocene self-domestication from other explanations of human docility. The hypothesis of recent selection against reactive aggression predicts that humans exhibit a behavioral domestication syndrome, and that the syndrome should be largely paedomorphic.

Scientists have long claimed that humans exhibit a series of behaviors that are paedomorphic. Unfortunately, the claim has traditionally been made by comparing humans with apes, which is not the comparison we need. Ideally, our comparison would have been with the behavior of our ancestors, the Mid-Pleistocene *Homo,* about which little is known. Given the near absence of information about them, Neanderthals are a reasonable stand-in. Their culture was more limited than that of our ancestors. On the basis of the evidence in this chapter, an intriguing possibility is that Neanderthals' ability to learn socially and to cooperate was limited by their reacting to tension too quickly with aggression. Differences between *Homo sapiens* and Neanderthals may have been due more to emotion than to intellect.[69]

The evidence that *Homo sapiens* have been self-domesticating for three hundred thousand years, and how it happened, suggests that we are a thoroughly unusual primate. But, in accounting for our docility and capacity for cooperation, the self-domestication hypothesis goes only so far. To say that, behaviorally, *Homo sapiens* is to Mid-Pleistocene *Homo* as a dog is to a wolf, or as a guinea pig is to a cavy, or as a bonobo is to a chimpanzee, entirely underestimates modern human achievements. Dogs, guinea pigs, and bonobos are delightfully tractable species. Humans are more than merely tractable.

Higher intelligence and the ability to learn culturally are two familiar reasons humans have achieved more than other domesticates. Another reason, which, like self-domestication, has been argued to come from capital punishment, is that, in addition to a reduced propensity for reactive aggression and an increased ability to cooperate, capital punishment appears to have given us a new kind of moral system.

10

The Evolution of Right and Wrong

In the late nineteenth century, an Inuit widow named Kullabak lived in a traditional community on the northwestern coast of Greenland. Her son was a bachelor, a big man with an arrogant attitude and an annoying sense of humor. He cared so little about others' opinions that he played tricks such as asking people to come and help him, then pelting them with rotten eggs. Being hit with stinking slime must be especially unamusing when you have only one change of clothes, doing laundry is a nightmare, and you share a tiny living space.

To make matters worse, he threatened other men's pride. In Inuit culture, a husband could legitimately share his wife with another man. The prankster took advantage of the sexual norm. He lied to a woman that her husband had invited him to have sex with her. She innocently agreed. The deception emerged, and her husband was outraged.

The strapping offender laughed it off, but Kullabak was ashamed. Feeling obliged to save the honor of her house, she made a sealkskin noose and, one night, while her son was asleep, slipped it round his neck and strangled him. It was better that a family member be responsible.

Was Kullabak criticized and punished for her proactive homicide? Not at all. Her grim deed brought respect. She remarried and lived for many years as a renowned figure in the community, "her great, booming voice always an asset at parties."[1]

Many in the West would condemn Kullabak for following her moral sense at the expense of her son's life. But, however much humans debate the solution to a particular moral dilemma, there is one thing we all take for granted: from hunter-gatherers to the Pope, we all live by a moral compass.

The commitment to living beyond a narrow view of our own interests makes us different from animals and creates a series of biological puzzles concerning our lack of selfishness and our willingness to condemn others. The moral sense was once explained purely by religion. Now an evolutionary account is needed. As we have seen, Darwin launched the hunt. After a century and a half of interesting ideas, questions about how and why morality evolved have reached considerable agreement.

There is a wide consensus that moral psychology has two components. On the one hand it includes a strong tendency to serve "enlightened self-interest," as psychologist Jonathan Haidt put it.[2] Our intuitive responses often benefit ourselves. Those responses are easily explained. Selfish behavior leads to evolutionary success.

On the other hand, humans are remarkably group-minded. We care about loyalty, justice, fairness, and heroism. At times we might even experience what sociologist Émile Durkheim called "collective effervescence," a shared sense of awe that can have us losing our sense of individuality and feeling as if we were becoming embodied into a larger whole. Such tendencies contribute to our highly cooperative nature, making us behaviorally more like honeybees than like the fundamentally selfish chimpanzees. As Haidt put it, "Human beings are 90 percent chimp and 10 percent bee." The groupish components of human moral behavior tend to benefit a larger group.

Human groupishness begets a challenging question. Natural selection is expected to favor behaviors that are strictly selfish from a genetic perspective. The evolution of emotions that benefit a wider group, even at the expense of the individual's short-term interests, is therefore a puzzle. It has been explained in two main ways.

One approach suggests that group-directed moral responses evolved because they were good for the group. Those kinds of moral behaviors might have made groups better at warfare, as scholars from Charles Darwin and Jonathan Haidt to Christopher Boehm, Samuel Bowles, and the primatologist Frans de Waal have suggested. It could have enabled survival for individuals who could not obtain food on their own, as the developmental psychologist Michael Tomasello proposed. It could have helped groups use culture, as the evolutionary psychologist Joseph Henrich argued. Or maybe it evolved because it facilitates many kinds of cooperation regardless of the specific context, as the philosopher Elliot Sober and biologist David Sloan Wilson advocate.[3]

However, group benefits are not the only reason that morality might have evolved. Even when moral behavior leads to group benefits at an apparent expense to the individual, the individual's behavior might actually serve his or her own selfish ends. There are various forms of this idea. A benign version is that group-directed moral responses lead to the individual's being able to form useful alliances for cooperation, as the philosopher Nicolas Baumard and others propose. A darker form is that morality is all about self-protection. I have suggested that capital punishment emerged with language in the Mid-Pleistocene. Afterward, an individual who challenged the dominant culture could be in mortal danger. Sensitivity to social disapproval would have been favored as never before. As a result, individuals might behave in morally correct ways in order to promote their own survival. Group benefits would then be incidental.

According to this idea, which Christopher Boehm advanced in his 2012 book *Moral Origins,* the evolution of human morality was largely a response to the same power plays that self-domesticated us: we evolved to fear the killing power of the men in the group. The idea explains why group-directed moral emotions have been elaborated far more strongly in humans than in any other species.[4]

The biblical parable of the Good Samaritan tells how a man helped a stranger from a different religion. The story illustrates an attractive feature of morality: moral principles can promote altruism. In line with that idea, evolutionary theorists from Charles Darwin onward have commonly treated morality as being concerned only with altruism and fairness. However, being moral can include not only acts of kindness but also deeds of conformity and violence.[5]

Being morally good might mean restraining yourself. You can be good, according to some societies' moral prescriptions, if you refrain from various individual activities that are deemed "wrong," such as committing suicide, masturbating, or burning your national flag.

More consequentially, whether an act is felt to be good depends on whom it is done to and why. Kullabak killed her own son, but in Inuit eyes, her act was good because her son was bad. Whether an act is felt to be good also depends on the critical distinction between "us" and "them." In his autobiography, Robert Graves recalled his school days: "We considered it no shame to cheat, lie or deceive where a [teacher] was concerned, though the same treatment of a school-fellow would have been immoral."[6] It was the same in concentration camps of World War II.[7] Stealing from fellow prisoners was "theft"; stealing from prison guards was "organizing."

Just as cheating, lying, and stealing are morally good if done to "them," so, too, is killing. Australian aborigines provided one of many examples worldwide cited by the anthropologist Maurice Davie in 1929.

> The Australian has two sets of mores, one for his group-comrades or friends, the other for outsiders or foes. "Between the males of a tribe there always exists a strong feeling of brotherhood, so that . . . a man can always calculate on the aid, in danger, of every member of his tribe," but towards strangers there reigns inveterate hatred, and any means are justified in dealing with them. [A]mong the natives of Torres Straits, "it was a meritorious deed to kill foreigners either in fair fight or by treachery, and honour and glory was attached to the bringing home of skulls of the inhabitants of other islands slain in battle."[8]

The killers who committed genocide in World War II, Cambodia, and Rwanda were caught up in societies where moral boundaries became excessively crystallized. Yet most were not sadistic monsters or ideological fanatics. They were unremarkable individuals who loved their families and countrymen in conventional moral ways. When the anthropologist Alexander Hinton investigated the Cambodian genocide of 1975–79, he met a man called Lor who had admitted to having killed many men, women, and children. "I imagined Lor as a heinous person who exuded evil from head to toe. . . . I saw before me a poor farmer in his late thirties, who greeted me with the broad smile and polite manner that one so often encounters in Cambodia."[9] The combination of horror and ordinariness is routine. According to the anthropologists Alan Fiske and Tage Rai, "When people hurt or kill someone, they usually do so because they feel . . . that it is morally right or even obligatory to be violent."[10] Fiske and Rai considered every type of violence they could think of, including genocide, witch killings, lynchings, gang rapes, war rape, war killings, homicides, revenge, hazing, and suicide. Their conclusion was clear. Most violence is motivated by moral emotions. So the definition of morality that I will follow here is not limited to altruism or cooperation. I take moral behavior to be behavior guided by a sense of right and wrong.[11]

To explore the idea that the moral feelings that so vitally influence our behavior are derived from a prehistory of capital punishment, I consider three questions about the ways in which humans have evolved to behave morally.

First is the Good Samaritan problem. Why have we evolved to be nicer to one another than other mammals are? We have already seen why we are less reactively aggressive than our ancestors were, but a reduction in aggression does not explain why we can be so positively helpful.

Second is a question about how we reach moral decisions. Our emotions help us determine what, in our judgment, is right or wrong. What selective pressures caused our emotions to evolve as moral guides?

The last is about the interfering aspect of morality. Why did we evolve to monitor not only our own behavior, but others' behavior, too?

Answers to these questions are sometimes framed in terms of advantages for the group. The thought that in prehistorical times dissenting individuals courted mortal danger, as the threat of capital punishment indicates, leads to different ideas.

The Good Samaritan problem concerns altruism, cooperation, and fairness, categories of helpful behavior that are jointly called "prosocial." According to the standard theory of animal behavior, the emotions that elicit prosocial behavior have evolved because they lead individuals to promote the spread of their genes. Prosocial behavior spreads our genes in two main ways. We help our genetic relatives, who share our genes; or we invest in partners who help us in return, or are at least expected to do so. Those complementary ideas are called kin selection and mutualism respectively, and they tidily handle most cases of nonhuman prosocial behavior. Individuals in species such as baboons, wolves, and bottlenose dolphins tend to act kindly toward their kin, and they cooperate in self-interested ways with familiar nonkin. Humans also practice kin selection and mutualism.

But we also do something else: we often act morally toward people who cannot be expected to return the favor. Moral rules such as "It is wrong to steal" or "It is wrong to lie" are not meant to be limited to family and friends. In theory, they apply to all our interactions—even the person whose wallet, full of money, we may find when we are out for a walk on our own and no one can see us. The universal nature of the rule is the problem. When strangers are the recipients and our behavior is guided by conscience, ordinary biological theory does not explain why we should be prosocial.

The greater the self-sacrifice, the stronger the puzzle. Moral expectations might lead individuals to kill their own offspring, as Kullabak did. They can lead people to give their lives for the sake of others, as the "very gallant" Captain "Titus" Oates did in 1912, when he walked off into an Antarctic blizzard and was never seen again. He apparently thought his death might leave enough food for the three remaining members of Robert Falcon Scott's South Pole expedition to survive.

Perhaps you are saying to yourself that human intelligence solves the problem. In theory, moral behavior could come simply from socially useful rules that humans have invented and that have been taught down through the generations. In favor of this idea, people certainly get a lot of moral guidance from their upbringing. Kullabak's solution made sense in Greenland but would be anathema to the average New Yorker. Captain Oates had been brought up to regard honor as the highest form of virtue. Moral behavior, according to this idea, could result entirely from cultural indoctrination. Frans de Waal names this the "Veneer Theory" of morality, the notion that human morality is a purely rule-bound system lying atop a foundation of ancient animal-derived behavior that lacks morality, like a veneer of fine lacquer on a wooden box.[12]

The Veneer Theory is a nonstarter, however, because moral actions are produced partly by evolved moral emotions. Untutored children have prosocial inclinations that are not explained by kin selection or mutualism. The developmental psychologist Felix Warneken showed that eighteen-month-old toddlers will help a random adult who asks for help. The infants will pick up accidentally dropped objects, for instance, or hold a door open to allow the grown-up to put toys away. Significantly, experiments show that these kinds of helpful behavior are not explicable by the infants' merely trying to interact, or being intimidated, or wanting to be stimulated. They simply want to help, even when it costs them. They will give up their own food to an experimenter when the experimenter's bowl is empty, or sacrifice their own toys for the sake of others.[13]

Such lessons are not confined to prosociality. Children do not need to be taught to understand a simple version of good and bad. The moral attitudes of babies have been tested using puppet shows. Eight-month-old babies watched an antisocial puppet being mean to others. Then a "good" puppet hurts the antisocial puppet. Amazingly, the babies tend to prefer looking at the good puppet. Before we can talk or walk, we are programmed to recognize norm violators—those whose antisocial behavior classifies them as "bad."[14]

Education is important, of course. It can work both ways, to encourage both prosocial and antisocial behavior. People often expect religious believers to act in especially prosocial ways, but actually religiosity is not always a predictor of moral kindness.[15] A study of sharing among 1,170 children in six countries and four continents found a consistent tendency for those brought up in more religious families to be less altruistic than those from nonreligious families.[16] The effects are not necessarily due to anything special about religion. The psychologist Paul Bloom found that powerful effects on both good and bad behavior come from a variety of social inputs, such as those concerned with personal identity and group affiliation.[17]

Society influences what we care about, but evolution has produced the fact that we care. Nature sometimes beats nurture in this interaction. Innate emotions, such as sympathy in response to suffering, produce moral intuitions that are so strong that children sometimes trust their own feelings over the instructions of supposed authorities such as parents or teachers. Even at the age of three, children will disobey commands that seem likely to lead to harmful outcomes.[18]

On the face of it, adults seem more rational and less dependent on their feelings than intuitive infants, because adults can consciously think about moral problems. Adults are certainly more articulate when they explain an action by saying something like "I thought it was the right thing to do because . . ." However, in making moral choices we tend to act first and think later. As Jonathan Haidt has shown, moral reasoning is usually "a post-hoc process in which we search for evidence to support our initial intuitive reaction." Haidt compares the process to the actions of a press secretary working for a secretive administration, "constantly generating the most persuasive arguments it can muster for policies whose true origins and goals are unknown." Again, our biologically given emotions exert critical influences on our moral decisions.[19]

Since 1982, the Ultimatum Game has provided a standardized context for studying moral choices. The game allows investigators to study people's choices about sharing a resource with a stranger. Conventional economic theory predicted that decisions would depend on self-interest. However, in worldwide tests in more than thirty countries, from hunter-gatherers to the Harvard Business School, both adults and children are spontaneously and routinely more generous than expected by theories of economic maximization. This result makes humans very different from chimpanzees—and probably any other nonhumans.[20]

The Ultimatum Game has two players, Donor and Decider, instructed in the rules by a researcher. Donor and Decider are told that if they play right they will be allowed to share a sum of money that the researcher will give to Donor. All that Donor and Decider have to do is to agree on how to split it. Suppose the pot is ten dollars. The game begins when Donor offers Decider any amount, from nothing to the whole ten dollars. Decider then chooses whether or not to accept the offer. If Decider accepts the offer, the deal proceeds. Decider receives the offered amount, and Donor keeps the rest. But—and this is the key—if Decider rejects Donor's offer, neither player gets anything. Either way, that is the end of the game. It is played only once, and the two players never meet or learn each other's identity.

Self-interest theory would predict that Donors give the minimum (say, one dollar). Decider's best interest would then be to accept the paltry amount, since nothing that Decider can do would produce a bigger reward.

In fact, however, Deciders almost always reject small offers such as one dollar, or, indeed, anything less than about a quarter of the pot. When they do so, both Donors and Deciders get nothing, as Deciders are well aware. In other words, Deciders knowingly pay a price in order to punish Donors for being too tightfisted. When interviewed afterward, Deciders who reject a low offer describe having been angry at being treated unfairly by Donors. Their behavior is guided by a sense of what is morally right or wrong.

In practice, Donors normally behave as if they anticipate Decider's rejection of a low offer. On average, they offer around half the pot. This is big enough for Decider to accept, which leaves both Donor and Decider content. They both get something.

Deciders' rejection of small offers, thus not maximizing their self-interest, is virtually universal. Regardless of whether they will ever meet Donors, Deciders act by a different principle than economic maximization.

Chimpanzees do not play the Ultimatum Game in the same way as humans. Ingenious modifications of the Ultimatum Game, using food rewards instead of money, show that captive chimpanzees behave like the imagined *Homo economicus,* a species in which individuals always try to maximize their personal economic gains. Decider chimpanzees accept even the smallest rewards offered by Donor chimpanzees; unlike humans, they never reject "unfair" offers. The stark difference draws attention to the uniqueness of the human moral senses.[21]

So the first big problem about our evolved moral psychology is why, when we give or receive, we do not behave like *Homo economicus* or chimpanzees. Those imagined and real species are rational maximizers.

We are not. We give more than a theoretical economist would predict that we should, and we reject donations that we deem to be unfair. Why have we evolved such apparently self-sacrificial tendencies?

As I noted earlier, a popular kind of solution is group selection. Group selection theory suggests that self-sacrifice by an individual can be favored over evolutionary time if it provides sufficiently large benefits to the individual's group, which normally means a social breeding unit such as a hunter-gatherer band. Very often, however, the group that benefits from an individual's generosity is not a social breeding unit. As Robert Graves's recollection of his school days reminds us, the beneficiaries might be only a subgroup of a given social network. In the group as a whole, moral behavior might benefit some individuals at the expense of others.

Hunter-gatherers offer a chilling example. If there was a conflict of interest between men and women, moral rules would typically favor the men at the expense of the women. Male hunter-gatherers throughout Australia used their women as political pawns. Wives could be required to have sex with multiple men at special ceremonies. They could also be lent to a visiting man, or given sexually by a husband to a man with whom the husband had quarreled, in order to erase a debt or make peace. Women could be sent on sexual missions into dangerous situations. When potential attackers were seen approaching a group, one response was to send women out to greet them. If the strange men were willing to forgo their attack, they signaled their intent by having sexual intercourse with the female emissaries. If not, they sent the women back and then attacked. The final stage of peacemaking between two tribes almost always involved an exchange of wives. Apparently, women did not enjoy these coercive encounters. In 1938, the anthropologist A. P. Elkin reported that Australian Aboriginal women would live in terror of the use that was made of them at ceremonial times. In those cultures, all this was the moral practice of ordinary men. The men behaved prosocially toward one another while exploiting their wives and female kin. To say that these behaviors were good for the group implies a very restricted definition of "the group." The behaviors were designed to be good for the group of married men who made the rules, but not for the women.[22]

Coercive practices and unequal expectations of self-sacrifice cast doubt on the idea that moral practices are necessarily "good for the group." We need other ways to explain how the moral emotions underlying self-sacrificial behavior have evolved.[23]

The second major problem is how we come to classify some actions as "right" and others as "wrong." Scholars looking for the consistent application of moral rules have traditionally considered two main ideas. They are the "utilitarian" and the "deontological" principles. Both work sometimes, but neither is followed all the time, which means they fail as general explanations.[24]

The utilitarian principle states that people should act to maximize the general good. Sometimes experimental subjects presented with moral problems conform to this idea. A popular dilemma for philosophers imagines a train hurtling down a track. An observer sees that, if she does nothing, five people are going to be killed. But she can pull a lever that diverts the train onto a side track, where it will kill only one person. Should she pull the lever? Ninety percent of those asked say yes. Pulling the lever saves more lives than not pulling the lever, so it maximizes the general good. It is utilitarian.

The deontological principle, by contrast, states that right and wrong are absolutes. You cannot argue about them. Sometimes people follow this principle instead. In one experiment, subjects are told that a doctor has five patients who will die unless they receive donated organs. She also has a patient whose organs could be used to save the five. Should she kill the one to save the five? Ninety-eight percent say no. When asked why, they simply say it is wrong to kill.

The two cases show that we follow different principles in different situations. In the train problem, most people follow the utilitarian principle, not the deontological principle, which states that killing is always wrong. In the doctor problem, most people follow the deontological principle, not the utilitarian principle, which would state that the more survivors there are the better. Many real-life cases are similarly inconsistent. Some anti-abortionists believe that, although killing in general is wrong, killing an abortionist is legitimate.

People do not follow any general moral principle. Instead, moral decisions are influenced by a series of unconscious and unexplained biases. Three are particularly well studied.[25]

The "Inaction Bias" pushes us to do nothing rather than something. Suppose you are caring for a terminally ill patient. Most of us would rather deliberately withhold life-prolonging treatment than give the patient a lethal injection. We prefer an act of omission to an act of commission.

The "Side Effect Bias" tells us not to have our main goal be to create harm. Imagine that you are in command of a bombing raid that will kill a certain number of civilians regardless of the target. You have to choose the goal. Would you prefer to order your bombers to kill civilians so as to break the enemy's will, or to attack an army base so as to reduce the enemy's military capacity? Given that the number of civilian deaths is expected to be the same either way, most people would prefer to bomb a military target, letting the civilian deaths be an inevitable incidental outcome. The bias is against doing intentional harm.

The "Noncontact Bias" concerns physical touch. Other things being equal, most people would prefer an action that allows them to avoid having to touch someone who is being harmed.

These strong moral biases are well recognized by psychologists, but their rationale is debated. The psychologists Fiery Cushman and Liane Young suggest that they are derived from general cognitive biases unconnected with morality.[26] Against that idea, however, it would be surprising if such important influences on behavior are a chance result of pre-existing biases, rather than having their own adaptive logic. Others, such as Moshe Hoffman, Erez Yoeli, and Carlos Navarrete, explain the biases as being useful to the individual. As will become clear, that approach fits well with the self-domestication hypothesis.[27]

The third major puzzle about our moral psychology is why, as a species, we have evolved to be so sensitive to the abstract notions of right and wrong that we monitor one another's behavior and even sometimes intervene to punish someone whose behavior we disapprove of.

Whether other animals have a sense of right and wrong, a primitive version of what humans experience, is not known for certain. Conceivably, chimpanzees might have a mild version of social norms—in other words, expectations about how others should behave. Captive chimpanzees in Switzerland were shown video footage of wild chimpanzees hunting monkeys, being aggressive to adults, or being aggressive to infants, including killing one. The movie-watching apes looked longest at the infanticide scene, which suggested to the researchers that the chimpanzees were particularly fascinated by this unusual behavior. Interestingly, the chimpanzees showed no increase in arousal when watching the infanticide, which suggests that they experienced something other than sheer emotional disgust. The researchers speculated that the chimpanzees were responding to a social norm that disapproved of infanticide. Chimpanzees, they suggested, might be "sensitive to the appropriateness of behaviors that do not affect themselves."[28]

The possibility that chimpanzees have social norms is intriguing. But even if they do, the significance of such sensitivity is very limited compared with that of humans. Consider the response of chimpanzees to real infanticide rather than to videos.

In August 1975, Passion and Pom, a mother and daughter living in Tanzania's Gombe National Park, were members of the sixty-strong Kasekela community studied by primatologist Jane Goodall. Passion was around twenty-four years old, and Pom, her only daughter, was a ten-year-old adolescent who was already starting to mate and would soon be settling down with her own offspring. Pom behaved like a typical female of her age. She followed her mother everywhere, and often played with her little brother, four-year-old Prof.

The youngest among the dozen mothers who also shared the Kasekela territory was fifteen-year-old Gilka. Gilka had been orphaned since she was nine years old, and had had two pregnancies but no surviving offspring. It was a happy day for Goodall when Gilka overcame a difficult childhood by producing her first daughter, Otta.

Three weeks later, the delight turned to sadness. Passion and her family found Gilka and Otta sitting together on their own. For no obvious reason, Passion suddenly charged at Gilka. Gilka ran away screaming, clutching Otta to her belly. After sixty meters (two hundred feet), Passion caught them and attacked. Pom quickly joined in. Gilka defended powerfully, but her aggressors collaborated too well. Passion grabbed Otta and chased Gilka away. The kidnapped baby clung to Passion, who calmly dispatched her by deliberately biting into her skull. While Gilka watched, Passion, Pom, and Prof all cannibalized the victim.

The proactive attack proved part of a pattern. Over the next three years, Passion and Pom killed at least three more very young infants and possibly as many as six. Other females have since been observed carrying out similar attacks. Eerily, the killers and their victims had often relaxed together with no hint of animosity, apparently unconcerned about any possibility of violence. Tiny babies are vulnerable, however. A particularly helpless infant in the arms of a competitor appears to stir something dark in the mind of a female chimpanzee. In Goodall's words, it is as if a switch is thrown. Out of nowhere, a familiar companion turns into an enemy, without any seeming provocation.[29]

The grisly behavior does more than provide meat. The mothers whose infants Passion and Pom killed had spent much of their time in the same area as their tormentors, competing for access to the choicest fruit trees. Fear of lethal attack would likely keep their competitors away. Over many months, the attacks could be expected to lead to extra food for the killer mothers, which meant that the infanticides were selfish acts conducted at the expense of everyone except the killers' family. They were not the novel acts of broken minds. They were adaptive behaviors to which others might be expected to respond.[30]

Yet life in Gombe went on much as before. The mothers of the victims tended to avoid the killing pair. Sometimes when Passion and Pom attacked, alert males intervened. Males tend to protect the weaker females. Males likewise support new female immigrants against longtime residents, apparently to encourage them not to leave the community: male policing of conflicts among females seems to be a selfish behavior. Immediate protection was the most anyone did. Too often, Passion and Pom prevailed. The suffering was widespread: tensions rose, infants died, mothers were bereaved, adult males lost offspring. In the long run, the community was weakened by loss of numbers and reduced cooperation among mothers.

If adult males had acted in concert, they could certainly have stopped Passion and Pom, because the collaborative power of males is enormous: they kill prime adult males without themselves getting a scratch. But though the males had the means to punish or kill Passion and Pom, they did not have the mind-set.

The contrast with a human community is obvious. The equivalents of Passion and Pom would never have got away with it. One would expect human infanticidal murderers to rapidly suffer some combination of being gossiped about, hunted down, arrested, tried, and imprisoned or executed.

Humans punish norm breakers more than chimpanzees do, and humans are more spontaneously generous than chimpanzees are. In 1871, Darwin wrote: "A moral being is one who is capable of comparing his past and future actions or motives, and of approving or disapproving of them. We have no reason to suppose that any of the lower animals have this capacity."[31]

Subsequent evidence has supported Darwin's claim of the difference between humans and other animals. Even the most impressively prosocial primates, such as chimpanzees and capuchins, only go so far. They have abilities like empathy, perspective-taking, concern, and self-inhibiting, all of which humans can use while making moral decisions. But those abilities are mere starting points. They provide the psychological basis for being able to make moral decisions, but they are not sufficient to create moral beings. In de Waal's words, "We have moral systems and apes do not."[32]

Only humans have community standards that decide the crucial difference between right and wrong. So the third question involves understanding not only why humans are sensitive to what is right or wrong, but also why humans punish those who do wrong, whereas chimpanzees hardly do so at all.

The three moral questions that I have outlined concern why humans are exceptionally prosocial, why we are guided as we are in deciding right from wrong, and why we care so much that we are prepared to intervene when we see wrong done. Christopher Boehm suggests that a solution to these problems lies in the claustrophobic setting of small groups in which capital punishment was a realistic threat for troublemakers.

"Whenever it was that human groups became militant about their egalitarianism," Boehm wrote in 2012, "logically it became highly adaptive for a band's alpha types to very carefully hold their dominance tendencies in check. . . . Over time, the apelike, fear-based, ancestral version of personal self-control would have been augmented, as there appeared some kind of a protoconscience that no other animal was likely to evolve."[33]

Let us suppose, as Boehm does, that, in the first stage of using coalitionary proactive aggression, the power of the subordinates was used for nothing else than acts of counter-dominance, to control domineering males. Females would be largely unaffected. Among males, there would be selection against the ambitious or bad-tempered physical aggressors. How readily subordinates joined the coalition would not matter much so long as the deed could be done. The counter-dominant coalitions would produce selection against high propensities for reactive aggression, as I discussed in chapter 9. Increasingly, males would be born with milder tempers; fewer would try to bully others physically. Self-domestication had begun.

In this initial phase of generating a more peaceful species, the moral emotions would be little affected. The target of the newfound coalitions was only the hyperaggressive males.

The next stage would be critical for the evolution of moral sensibilities. In developing the ability to kill the physically imposing alpha, the subordinate males had discovered an irresistible coalitionary power. They could now coordinate to kill anyone. So all kinds of troublemakers became at risk. Any kind of noncompliance with the interests of the killing coalition could in theory provoke an intimidating threat. Women and young men were as vulnerable as domineering bullies to the power of the male elders.

In nomadic hunter-gatherers, as in small-scale societies in general, aspiring alphas are not the only victims of the tyranny of cousins. Young men can be executed for messing with the old men's wives. Women can be executed for breaking a seemingly trivial cultural norm, such as seeing the magic trumpets, or treading on the men's secret path, or for having sex with the wrong men. Anyone who flouts the rules that the men impose can be executed.

The result is a society in which the male coalition not only holds the power but uses it. The anthropologist Adamson Hoebel documented the legal systems of small-scale societies. He found that belief systems were typically founded on a religious statement such as "Man is subordinate to supernatural forces and spirit beings, which are benevolent in nature."[34] That kind of idea legitimizes the belief system by referring it to forces beyond human control. A series of postulates would follow. For the Inuit, Postulate VII was "Women are socially inferior to men but essential in economic production and childrearing."[35] No society inferred the reverse system, by which men might be socially inferior to women.

The anthropologist Les Hiatt summarized the effects in Australian Aboriginal society. Women often had strong traditions of independence and cultural autonomy. In some places, they had secret societies. They could have the largest role in choosing whom their daughters married. But, even though women were not subservient, the genders were not equal. Sanctions against women for discovering male secrets included rape and death. By contrast, there were no physical reprisals for men who intruded into a women's ceremony. Men could arrange for gatherings with neighboring societies; there was no equivalent for women. Men could insist that women provide food for the all-male secret ceremonies, or provide sexual services to whomever the men required. Religious knowledge, controlled by men, justified their dominance. The gods were kind to them.[36]

The elders decided what was a crime against society, which explains why among hunter-gatherers those who are executed are not just the excessively aggressive and violent. Among the Inuit: "Threats and abuse may lead to the same end. The obnoxious person is first ostracized, then liquidated if he continues his bothersome behavior." The execution of liars was reported across the span of Inuit territory, from Greenland to Alaska. Everywhere the same was true. A coalition of men controlled life and death according to rules that they set.[37]

Of course, most conflicts are settled before they escalate to the point of capital punishment. Once men dominate the society through their control of death, their word becomes law. Everyone knows the importance of conforming. People accept the inequities. Men get the best food, have the most freedom, and are the ultimate arbiters of group decisions.

Boehm calls the egalitarian system of male relationships found among nomadic hunter-gatherers a "reverse dominance hierarchy." The term recognizes that any would-be alpha would be dominated by a coalition of males. Others prefer the term "counter-dominant hierarchy," since an alpha male defeated by a coalition becomes part of it, rather than having his position reversed.[38]

The revolution that first brought down the alpha bullies during the Mid-Pleistocene had given the new leaders extraordinary power. By discovering that they could control even the most imposing fighter, the previously subordinate males found that they could further their goals in other ways, too. Did they use their new ability selfishly? The familiar dictum of the historian and politician Lord Acton surely applies: "Power tends to corrupt, and absolute power corrupts absolutely." Some three hundred thousand years ago, males discovered absolute power. They had surely been individually dominant to females before the onset of capital punishment, just as chimpanzees are. Afterward, however, the dominance of males over females took a new form. It became a patriarchy in the special sense of male dominance based on a system. The system was a network of mature males protecting their mutual interests.[39]

In societies where executioners held sway, death was a sword of Damocles hanging over deviants. In those circumstances, selection favored individuals who minimized the risk of being viewed as social outcasts. That meant that everyone needed to know which behaviors were "right" and which were "wrong." Getting it wrong could be fatal. Our ancestors, the ones who successfully negotiated this dangerous world, were the ones who got it right.

In light of that perspective, the three puzzles that I have cited seem soluble.

The first is why are we more prosocial, or more generous, than could be expected from the theories of kin selection and mutualism. Selfish individuals in the Pleistocene, grabbing at resources possessed by others, would have risked getting into fights. A coalition of militant egalitarians was in a position to cut them down. Selection would accordingly have favored those whose spontaneous generosity and noncombativeness protected them from such a risk by minimizing their selfish urges and increasing their tendency to help others.

Based on his knowledge of hunter-gatherers, Boehm sees this process benefiting the whole social group. "The punishing of deviants occurs," he writes, "because people feel individually threatened or dispossessed by social predators, but also, in a larger sense, because socially disruptive wrongdoers so obviously lessen a group's ability to flourish through cooperation."[40]

Boehm's argument here is that the coalition might make their collective decision based on their evaluation of group benefits. To judge from contemporary hunter-gatherers, those who were able to decide about capital punishment would mainly have been married men. Sometimes their decisions about who was "socially disruptive" would indeed have benefited the group as a whole, as Boehm suggests. The suppression of theft, fighting, and antisocial play would tend to be good for everyone.

Again, however, it is also plausible that more selfish motivations applied. Men could have enforced patriarchal norms that allowed them to trade women, use them as sexual and political pawns, and beat them. So, although the coalition promoted prosociality by punishing deviants, it did not necessarily advance the welfare of the whole group.

Either way, anyone in the group could be censured for antisocial acts. Prosocial behavior would therefore have been strongly rewarded.

The second puzzle is how we reach moral decisions. What causes us to classify some behaviors as right and others as wrong? And why are we guided by our curious internal biases, rather than following general rules for deciding whether a behavior is moral?

The Inaction, Side Effect, and Noncontact biases lead to different outcomes, but each of them puts distance between a moral actor and his or her act: "I did nothing." "That wasn't my goal." "I never touched them." The claims seem designed as defenses against an accusation of wrongdoing.

Self-protection makes sense in a world of uncertain moral criteria and heavy costs for wrong decisions. Imagine an individual faced with a decision but unsure how the dominant coalition will judge his or her act. Anyone who does something "wrong" risks being seen as a nonconformist, or deviant. Plausible deniability then becomes an important criterion for the best decision. The ideal moral act, in this self-interested view, is the one that protects an individual from possible censure by the tyrannical cousins.

Jonathan Haidt put it this way: "The first rule of life in a dense web of gossip is: Be careful what you do. The second rule is: What you do matters less than what people think you did, so you'd better be able to frame your actions in a positive light. You'd better be a good 'intuitive politician.' "[41]

The three biases can thus be seen as self-help mechanisms designed to avert criticism. Each of them helps the actor to deny having done something that might be considered unpopular. The biases would have been favored by selection because they protected individuals from being ostracized for inadvertently doing the wrong thing—meaning, the thing that the coalition dislikes. We retain the ancient unconscious intuitions. The penalties for minority choices when we are tested in a moral experiment in Psychology 101 are trivial, but atavistic instincts prompt us to act as though the consequences were still enormous. They encourage us to avoid making unpopular decisions.

When we have time to think about our moral decisions, we turn to our conscience. H. L. Mencken called it "the inner voice that warns us somebody may be looking." He seems to have got it right. According to the psychologists Peter DeScioli and Robert Kurzban, conscience is a mechanism of self-defense. "Through natural selection," they write, "humans became equipped with an increasingly sophisticated moral conscience for steering clear of moral mobs. These cognitive mechanisms would prospectively compare the individual's potential actions against the set of moral wrongs in order to avoid actions that could trigger coordinated condemnation by third parties." Conscience protected our ancestors from the kind of behavior that could lead them to be accused of being deviants. Once again, self-defense explains our moral motivation.[42]

The third puzzle is why humans are so sensitive to right and wrong that we not only try to do the right thing ourselves, but also monitor one another's behavior and punish those who we judge to do wrong. The answer seems clear. Individuals need to protect themselves from being regarded as nonconformists.

We have already seen why troublemakers get punished: members of a conspiring coalition have absolute power, and they benefit by eliminating a problem. Power is absolute, because punishing is relatively easy. The simple predictive formula is that, by carrying out a coordinated plan, a large coalition can dispatch a lone social outcast at an extremely small risk to themselves of being physically harmed. There are often complexities in the process of forming the coalition, or in deciding that a kill is the appropriate action, but the killing itself is not risky. So executions can happen for a wide variety of offenses, some of which seem trivial to anyone not steeped in the culture. Once that system is put into practice, it means, of course, that offenders are expected to work hard to avoid the ultimate penalty. Accordingly, a few words from a senior member of the group should be enough to remind anyone of the importance of conformity. Our sensitivity to right and wrong is understandable as an evolved response to the extreme danger of being in the wrong.

Such sensitivity to moral values also became biologically embedded into novel emotional responses. Prominent human emotions not known to occur in animals include shame, embarrassment, guilt, and the pain of being ostracized, all of which are human universals. Involuntary and painful, they have been convincingly explained as mechanisms that show an individual's commitment to a social group after his or her social standing has been jeopardized.

In expressing shame, people admit to some kind of failing, such as cheating, physical weakness, incompetence, or even being diseased. By doing so, such a person recognizes that they are less worthy than previously supposed. Because shame signals acknowledgment of having violated a social norm, it offers the restorative power of an ingratiating apology. Shame thus seems designed to protect from the ostracism that can lead to social or physical death.[43]

The same kind of argument applies to embarrassment, which likewise admits a social blunder. Emotionally, embarrassment is as painful a feeling as shame. Behaviorally, it is signaled by several finely choreographed displays. Within less than a second of having been offensive (which is normally unintentional), the embarrassed individual starts a signal that lasts for two or three seconds. He looks downward, turns his head (mostly to the left), smiles, controls the smile such as by sucking his lips, gives furtive glances, and often touches his face. Meanwhile, a blush emerges over a longer period, peaking fifteen seconds after the signal began. As with shame, the intensity of embarrassment depends on the individual's perception of what others think. The elaborate programming of the response testifies to its having been important in the evolutionary past.[44]

A well-supported explanation proposed decades ago by the sociologist Erving Goffman finds that embarrassment serves to restore social relationships that have gone awry. Individuals who do not show embarrassment after a social faux pas are more likely to be viewed negatively, whereas those who readily blush redeem their standing. The execution hypothesis offers an explanation of why social standing matters so much. To be viewed positively is worthwhile, but to be viewed negatively is a potential disaster. Someone who is unapologetic after accidentally insulting his or her superior risks the dreaded fate of becoming an outcast.[45]

Guilt is another painful emotion that serves to mend social relationships. Defined as "a painful affect arising from the belief that one has hurt another," guilt involves admission of a wrong. The acceptance of blame is supposed to inhibit self-assertive aggression toward others, turning it inward on oneself. The associated expressions of remorse again pave the way to being forgiven.[46]

The pain of being ignored or excluded was investigated by the social psychologist Kipling Williams after an incident when he was relaxing in a public park. Two people who were strangers to Williams were nearby, throwing a Frisbee to each other. The Frisbee happened to land near him. He threw it back. A few more throwing exchanges ensued.

Then, however, the strangers kept the Frisbee to themselves. Williams felt ostracized. Thinking as a social psychologist, he realized that he was experiencing an intensity of social pain out of all proportion to the significance of the encounter. He inferred that the pain of being ostracized reflects an ancient adaptation to a more intense social world. This led him to make a series of studies using an online game that he invented, called cyberball.[47]

Experiments by Williams and others showed that a mere two or three minutes of play with strangers, followed by being excluded, led predictably to sadness, anger, and a series of negative effects including feelings of alienation, depression, helplessness, and even a reduced sense of meaning in life. The effects did not depend on the subject's personality or whether he or she felt similar to the ostracizers. Experimental subjects experienced elevated activation of a part of the brain, the dorsal anterior cingulate cortex, that is also activated by physical pain. To be ostracized, in short, engages a swift and strong series of neurally encoded responses that are very unpleasant. In the Pleistocene, the ostracism presumably involved familiars more than strangers. Once again, selection favored a strong emotional reaction because of the potential dangers of becoming socially isolated.[48]

Just as our emotions today are adapted to an earlier world in which people were perpetually walking a social tightrope, the way we currently think is adapted to saving us from ancient mortally dangerous faux pas. We have to use our thinking ability, because what a culture determines to be "right" and "wrong" can vary, as we saw with Kullabak's filicide: automatic responses to moral questions would have been inadequate. Our ancestors had to learn what kind of behaviors the culture considered appropriate. The evolved system for learning cultural norms is called a "norm psychology."

Norms are "learned behavioral standards shared and enforced by a community"; in other words, they are rules that everyone is expected to obey. A norm psychology is "a suite of cognitive mechanisms, motivations and dispositions for dealing with norms."[49] According to Joseph Henrich, our norm psychology evolved to protect us from social pitfalls.[50] Once humans became aware that a social rule militates against selfishness, selection would have favored mentally internalizing the norm, recognizing norm violators, and responding appropriately (such as by ostracizing the violators). That is why even a three-year-old, seeing another child, or a puppet, using a pencil "wrongly" (differently from what he has been told is "right") will point out the mistake.

Following Christopher Boehm, the core idea in this chapter is that our moral psychology was forged during a period when being a social outcast was even more dangerous than it is for most people today. The essentially social nature of moral behavior is clear to all scholars, and the belief that much moral behavior is concerned with avoiding and giving censure is also widespread. However, alternative scenarios for establishing the origins of morality, based on the benefits of cooperation, have not normally charged social ineptitude with costs as high as the execution hypothesis does. When the risk of being a nonconformist is that you will be executed, it is easy to imagine intense selection in favor of moral sensibilities that maintain you as part of the in-group.

Think of Kullabak's son. The risks he took were indeed large, and he lost. Deviants have been handled similarly everywhere in the past. Such effective social control makes Boehm's theory of moral origins unnervingly appealing.

If Boehm is right, the new mentality of which humanity is rightly proud had darker origins than we normally like to think. The force that bred conscience and condemnation into our ancestors began in the revolution of males competing for a new kind of power. It ended in a tyranny of the cousins with two major kinds of social effect.

On the one hand, in Haidt's words, it binds and builds. It constrains society to follow moral principles that promote cooperation, fairness, and protection from harm. It has brought a new kind of virtue to the world. On average, everybody benefited.

On the other hand, it also brought a new kind of dominance, because the limited power of a single alpha became the absolute power of a male coalition.

11

Overwhelming Power

IN 1886, IN the wake of Darwin's argument that humans are descended from apes, Robert Louis Stevenson published the first-ever story about a split personality. *The Strange Case of Dr. Jekyll and Mr. Hyde* presented a psychic conflict between the allure of being good and the temptations of being bad. The story suggested that our tendencies to behave well come from our being human, whereas our tendencies to behave badly come from our inner ape. A critical feature was missing, however. The aggression that was so prominent in Hyde, and suppressed in Jekyll, was almost all reactive. Proactive aggression was hardly to be seen.

The fictional Jekyll was a London doctor with everything going for him—a wealthy Fellow of the Royal Society, good-looking, hardworking, ambitious, respectful, and a morally guided thinker. He was the "very pink of the proprieties."[1]

His alter ego, Mr. Hyde, was "pale and dwarfish, . . . with a sort of murderous mixture of timidity and boldness." Hyde lost his temper easily. He hit children at will, and in a fit of rage he killed an old man. Hyde was "hardly human! Something troglodytic, shall we say?"[2] He had hairy hands, jumped "like a monkey,"[3] and attacked "with ape-like fury."[4]

"The terms of this debate," in Jekyll's words, "are as old and commonplace as man."[5] The story portrayed the kind of struggles that we all have with our baser motivations, and the lesson was encouraging. In the end, Jekyll defeated Hyde: goodness won. The moral seemed to be that if we try hard enough we can all live up to the highest standards. No wonder the novel resonated with the public. It sold forty thousand copies in the first six months. It was read by Queen Victoria and the prime minister. It inspired psychodramas by Oscar Wilde and Arthur Conan Doyle. It was seen as a "profound allegory" and a "marvelous exploration into the recesses of human nature."[6]

The debate between good and evil may have been "old and commonplace," but Stevenson's story broke new ground for the general public. Fourteen years earlier, in 1872, Darwin had published *The Expression of the Emotions in Man and Animals.* Now Stevenson was taking up the challenge of Darwin's evolutionary ideas. *The Strange Case of Dr. Jekyll and Mr. Hyde* implied that human tendencies for good behavior had evolved from a more wicked, nonhuman past.

Stevenson was right, of course. In contemporary terms, we can agree with his thesis by acknowledging several ways in which the special goodness of humans is rooted in biology. Our large brains allow us to exert cortical control over subcortical emotional stimuli. Self-domestication explains our being less easily aroused than apes are. Without our evolutionary history of self-domestication, temptation would be stronger, and accordingly harder to resist. In addition, the evolution of moral senses has added new feelings that directed our civility. We have become more fearful of giving offense, readier to conform, and more eager to help than if we were merely intelligent self-domesticated apes.

The results of these processes perfuse our lives, from soft conversation around a campfire to global offers of help in response to natural disasters. The view of the meritorious aspects of humanity triumphing over those of our wilder past is legitimate. We have evolved to be a remarkably kind and cooperative species whose selfish urges are less insistent than they were in the past. We are blessed with better abilities to resist temptation than a chimpanzee, or a Mid-Pleistocene *Homo.* We have achieved much virtue.

To the extent that Stevenson's mythic tale provides an allegory about the evolution of good and evil, however, it has a problem: the story is incomplete. Alongside the reduction in reactive aggression, proactive aggression ought to play a starring role. The omission of proactive aggression from *The Strange Case of Dr. Jekyll and Mr. Hyde* might be considered disappointing because it limits the novel's statement about humanity, but it is not as unfortunate as the same omission from scenarios of human social evolution. Yet in a parallel failing, proactive aggression has been given little attention by evolutionary anthropologists. We have already seen two examples of the importance of capital punishment, in domesticating humans and giving us our moral senses. Execution is only one of many human practices of proactive violence that permeate our societies and make our social lives crucially different from those of other animals.

I have explained how proactive aggression appears to be responsible for virtue. In this chapter and the next, I consider its contrasting consequences, those that show its responsibility for making humans an especially violent and despotic species. Proactive aggression gives us a key for solving the goodness paradox. Even though it made us into a calmer, gentler species, it also brought evil.

Coalitionary proactive aggression is a particularly important behavior, and will be a common phrase in this chapter. It sounds simple enough, but it has implications beyond its intuitive meaning, because it is shorthand for a longer and clumsier phrase. So I want to be clear about what I intend by it.

"Coalitionary" implies that several individuals join forces in a coordinated act of violence. "Proactive" is also used according to its standard definitions, referring to an act that is "planned or conscious, not spontaneous or related to an agitated state," and "used in pursuit of attaining a goal." So, at one level, the phrase is easily understood: a group of individuals combine in a deliberate attack.[7]

The further implication that I referred to adds another layer of meaning. A planned or premeditated aggressive act makes sense only if it has a reasonable chance of being successful: no one should plan to lose. Ordinarily, therefore, an act of coalitionary proactive aggression should carry with it the implication that the aggressors are being proactive because they know they can win. Accordingly, the additional hidden message included in the phrase "coalitionary proactive aggression" is that there is a large imbalance of power in favor of the aggressors. The aggressors would not plan their surprise attack unless they could know they have absolute power.

So a more accurate, longer phrase would be something like this: "coalitionary proactive aggression with such a large imbalance of power in the aggressors' favor that they are confident that they are going to win." Given how cumbersome the phrase is, I refer to it simply as "coalitionary proactive aggression."

War violence is normally dominated by coalitionary proactive aggression. In a classic style, one side uses surprise to attack another. Later, the other side retaliates, again with surprise. Reciprocal incidents of one-sided aggression continue. I discuss the evolution of warfare in the next chapter.

Within polities, the maintenance of civil society also depends on coalitionary proactive aggression. States use it to round up criminals, terrorists, gangs, or rivals for power. The power of the state is social oil. Without it, a state grinds quickly into chaos of competing militias, as Libya reminded us after the 2011 death of President Qaddafi, or Yugoslavia after the 1980 death of President Tito, or the eastern Congo after the 1997 death of President Mobutu.

Proactive aggression has a big impact in the lives of humans, chimpanzees, wolves, and a few other species, but in many species it occurs rarely or not at all. That makes it different from reactive aggression.

Imagine taking a walk in a tropical forest. In Uganda's Kibale National Park, there are few greater pleasures than to stop for a minute, close your eyes, and simply listen. At almost any time of day, you may hear the trills of warblers and insects, persistent cuckoos and tinkerbirds, and, every so often, the cawing of hornbills, the grumbling of colobus monkeys, even occasionally the hoots of chimpanzees. After dark, the calls of frogs, bats, and nightjars provide a backdrop for cicadas, bush babies, and owls. Peace seems to reign. "Soft stillness and the night / Become the touches of sweet harmony."[8]

Alas for innocence. The babble of the woods brings solace only if our ears are not properly attuned. The sounds that soothe are mostly male. Overwhelmingly, they tell of typical male actions: showing off, defending territories, threatening neighbors, calling allies, attracting females. They speak of color, weapons, and readiness for aggression. The human listener may be relaxed, but the callers are not. Jacked on testosterone, the males are loud, rough, and pushy. The sweet harmony is a testament to the pervasiveness of reactive aggression.

Yet hardly any of those melodious species exhibit proactive aggression.

Proactive aggression is so comparatively rare that at one time it seemed to be entirely absent in animals other than humans. In his celebrated book *On Aggression,* published in English in 1966, Konrad Lorenz claimed that evolution stops animals from deliberately killing one another. He noted that a wolf that has been defeated in a fight rolls over and exposes its vulnerable neck to the winner. The winner is then inhibited from further attack. Lorenz suggested that wolves offer a general lesson about the evolution of violence. Natural selection, he thought, leads to inhibitions against killing members of the same species. Lorenz suggested that the reason why humans kill relatively easily is because weapons allow us to kill at a distance; so armed humans are not exposed to the inhibiting signals of submission. Lorenz's account suggested that deliberate killing by humans is an unfortunate consequence of our technological advances.[9] The argument makes some sense. It is doubtless easier to send poisoned chocolates through the mail than to hand them to a victim, or to drop a bomb than to shoot someone pleading for his life.

But then people started watching mammals in the wild more intensely. They found that, contrary to the implication of Lorenz's idea, proactive killing of members of the same species is not confined to humans.

One context in which proactive aggression is observed among non-human mammals is infanticide, or the deliberate killing of an infant by adults. Infanticide by both male and female adults occurs but with different reproductive effects. Among primates, infanticide was first documented in a wild monkey, India's Hanuman langur, shortly after the publication of Lorenz's *On Aggression.* Hanuman langurs are easily observed in the wild because they occupy relatively open habitats, and they live in large groups that spend a lot of time on the ground. Groups contain many females, often more than ten, whereas they have only a single breeding male. The breeding male is an immigrant who achieves his position by defeating the previous male in combat. He remains as the breeding male by fighting new challengers, until finally he too is defeated and leaves the group. In July 1969, the primatologist S. M. Mohnot was studying a group in a near-desert environment in Jodhpur. An immigrant adult male who had become the new alpha was sitting on his own about ten meters (thirty-three feet) from a group of females. One female had approached and groomed him, but the male ignored her. His attention lay elsewhere.

> About 9:50 a.m. the male, with a sudden bound, was among the females. He grabbed the infant from the lap of Ti [the infant's mother], clasped it in [his] right arm, held its left flank in his mouth and ran fast. [Ti] and two other females . . . rushed at the running male. [Ti] twice blocked his passage but could not recapture [her] infant; the other two females also failed. All the while [her] infant was screeching. . . . After running for 70–80 metres, the male stopped for a moment, took a quick bite at the infant's left flank with his canines [producing a 6-centimeter gash through which part of the infant's intestines emerged], dropped him on the ground and sat near the bleeding infant. When the mother approached him, he barked loudly, tossed his head, bared his teeth and stared at her. It was all over in less than three minutes.[10]

This was no tragic accident. It was a hunt-and-kill.

Observations like Mohnot's cropped up with increasing frequency and launched a discussion about why infanticide occurred. There were critical issues at stake. Some commentators were convinced that human violence was not adaptive, and were therefore unwilling to see any exceptional violence as natural, even in a monkey. This line of thinking suggested that infanticides might be nonadaptive pathologies in a few disturbed individuals. Others thought that, if infanticide was indeed adaptive, the benefits must accrue to the group as a whole, not to the killer. For instance, the male who killed an infant could be benefiting the group by keeping the number of individuals in the group low enough so that everyone would have enough to eat. The idea had some political appeal. To social-justice warriors of the early 1970s who wanted a biological rationale, it implied that human behavior might have evolved for the good of a larger group.

The major alternative explanation was that infanticide is a selfish action that increases the male killer's chance of fathering an additional infant. The death of an infant fathered by a different male would lead the mother to come into estrus sooner than she would if her infant lived; the infant-killer might then be the father of her next offspring. The concept, laid out for Hanuman langurs by Sarah Hrdy in 1977, is called the sexual selection theory of infanticide, and has been richly supported since then. Neverthless, many people find it difficult to accept that a behavior we humans judge as morally abhorrent might have some evolutionary adaptive value.[11]

Even in the abstract world of scholarly journals, the tone of the discussion sometimes deteriorated.[12] In the 1990s, when the primatologists Susan Perry and Joseph Manson attempted to interpret three apparent infanticides by males among white-faced capuchin monkeys in Costa Rica, they were met with disdain. "All five referees," they reported, "using abusive and contemptuous language, urged the journal editors to reject our manuscript."[13] One reviewer compared the widespread appreciation of the sexual selection theory of infanticide to enthusiasm for eugenics before World War II. But nothing was wrong with Perry and Manson's data. The young scientists published their results, similar data continued to accumulate, and their study became a classic.

The debates were fierce, and for a time during the 1970s and 1980s, particularly in the United States, infanticide denial was a thorn in the flesh of the ethologists. Those supporting the hypothesis of adaptive infanticide were sometimes accused of having a right-wing political agenda. The accusations of bias were largely untrue, but the mudslinging rapidly became irrelevant as the observational evidence grew overwhelming. The percentage of infant deaths due to infanticide varied widely, and rose as high as 37 percent in a population of mountain gorillas, 44 percent in chacma baboons, 47 percent in blue monkeys, and a remarkable 71 percent in red howler monkeys.[14] In 2014, the behavioral ecologists Dieter Lukas and Elise Huchard surveyed 260 species of mammals studied in the wild and reported that infanticide had been found in almost half of them. Infanticide, Lukas and Huchard observed, occurs most often in species where males have something to gain from the killing. It is usually a selfish reproductive strategy used by males to bring females into breeding condition as soon as possible. With regard to primates, among eighty-nine wild species, Lukas and Huchard found infanticide in sixty (67 percent), including chimpanzees and gorillas.[15]

In a few primate species (such as chimpanzees), infanticide occurs for reasons other than sexual selection. Male chimpanzees who encounter mothers from neighboring communities tend to attack them and can severely wound or kill their small infants. In this case, the protagonists are unlikely to meet again, so there is little chance of the killer's fathering the female's next infant. The traditional sexual selection theory, therefore, does not apply. Possibly, the killers benefit by intimidating the female into avoiding the area, leaving more food for the killers' community. Alternatively, the attackers might gain by killing male infants that would otherwise grow up in the neighboring community to become future opponents.[16] Further observations will eventually test such ideas.

Female primates can also kill infants. Among marmosets and tamarins, groups contain up to four females. Normally only the alpha female breeds. If a lower-ranking female gives birth, her offspring are likely to be killed by the alpha female. The killing is adaptive for the alpha female because additional infants compete for care by adults, and thereby jeopardize the survival of her own young.[17]

With regard to the sexual selection hypothesis, the infanticide data were rich enough to show that the behavior is thoroughly strategic. In the best-known species, such as Hanuman langurs and lions, observers have clearly established that males attack only infants that they could not have fathered; that the victims are young enough for their death to hasten the time when the mother will become sexually and reproductively available; that killers attack only when the chance of success is high; and that they later mate with the females. Any male who might have been a father to the victimized infant tries to defend it, so variables such as the number of possible fathers in the group and how well the mothers can defend their young, influence whether infanticide attempts are likely to be made. Such factors vary not only between species, but also over time and among populations within a species. The rate of infanticide varies accordingly from population to population.[18]

In short, in numerous primates and for various reasons, adults use their power to kill infants of their own group deliberately. Most often the perpetrators are adult males. Sometimes, as in tamarins and marmosets, or as Passion and Pom illustrated in chimpanzees, they are adult females. In different ways, natural selection has favored deliberate attacks on members of the killer's own species. Infanticide forced scientists to confront the fact that even when behavior is highly flexible and context-dependent it can evolve for deliberately selfish purposes. Mother Nature, in the words of the evolutionary biologist George Williams, can be a Wicked Old Witch.[19]

The discovery of adaptive proactive infanticide in the last half-century showed for the first time that natural selection could lead mammals to kill members of their own species deliberately. However, this did not speak directly to the most troubling human practices. Killing a helpless infant is a much simpler behavior than the kind of killing that adult humans do to one another, and the sexually selected infanticide mostly performed by animals is not readily obvious in humans. So, although the infanticide story showed that premeditated aggression could be favored by natural selection, it did little to bridge the gap between animal behavior and human homicide.[20]

It took a second major revelation to bring home the lesson that a well-adapted evolutionary psychology in animals could include a tendency to kill adults. Chimpanzees provided an early mammal example: their cases of lethal aggression against adults of neighboring communities.

I happened to see the first hint that chimpanzees deliberately kill one another. It came on August 13, 1973. Yasini Selemani was a research assistant working for Jane Goodall in Tanzania's Gombe National Park. Late in the evening, he was following three adult males, named Godi, Sniff, and Charlie, through thick brush. They were walking toward a grove of leafy trees where they would make their night nests, but they made a strange detour en route. The object of their diversion proved to be the corpse of an old female chimpanzee. It was fresh, perhaps a day or two old. The males inspected her body briefly, then returned to their path.[21]

Next day, I joined Selemani to follow the three males. They did not revisit the dead female, but while the males were up a tree, eating fruits, shortly after dawn, Selemani took me to see her. We crawled under a network of dry vines to reach her contorted body. It lay on a steep slope. The brush was too thick for her corpse to have fallen to the point where she lay. Her left shoulder rested uncomfortably on a springy sapling that she must have pushed down as she collapsed. She had evidently died while being attacked. Her left hand was uppermost, her death grip wrapped around the stem of a shrub in a final effort to stop herself from being dragged downslope. Her arm and body stretched straight below. Puncture wounds on her back appeared to be due to chimpanzee canines. The only plausible conclusion was that she had been violently attacked by one or more chimpanzees. There were no other easy explanations: no leopards had been seen in the area for years, and she was uneaten. Most probably, her killers included Godi, Sniff, and Charlie, since those males knew where her body was. They were very unlikely to have found it by chance. Her case has gone down as an "inferred" killing.

Six months later, the hints of a murderous streak were confirmed. The victim was Godi. He was a member of the Kahama community. In January 1974, he had the misfortune to be caught on his own by six adult males from the neighboring Kasekela community. Watched by an adolescent male and a childless female from their own community, the Kasekela males stole up to Godi, grabbed him, and beat him for ten minutes. He escaped alive, but in appalling condition. He was never seen again. He probably died within a day or two.

Similar observations mounted over the next few years and have continued to accumulate. To students of chimpanzees, the cases were understandable, conforming to much behavior that seemed designed for intergroup attacks: Males would make regular patrols to the boundary areas of their territories, typically in a silent column, and only in parties with a substantial number of other males. Females and low-ranking males were less likely to visit the border areas than the powerful high-ranking males. In a boundary area, a patrolling party might climb trees that held no food, and spend up to twenty minutes looking outward, into the ranges of neighboring communities. They would show signs of fear on the patrols, reacting quickly to unexpected sounds. Nevertheless, they sometimes advanced as far as a kilometer (about half a mile) into the neighboring range. Their silent progress would be toward places where they appeared to have detected lone strangers, but if they heard calls suggesting that they were approaching a big party, they would sprint back to the safety of their home area.

Although these kinds of behavior prepared observers of wild chimpanzees for the idea that interactions would be hostile, the discovery of killing was still surprising because such extreme behavior was unknown in any other primate except humans. Still, it made adaptive sense. The average number of attackers per victim is eight, enough to explain why the attackers hardly get hurt at all. Several males can each hold a hand or a foot of the victim, making him or her vulnerable to any damage that others try to impose. The victim can be killed on the spot, or bruised, bitten, and torn so badly by a few minutes of punishment that he survives only briefly.

A stranger male killed by eight males in the community that I am studying, the Kanyawara chimpanzees of Uganda's Kibale National Park, revealed what a massive imbalance of power can do.[22] The Kanyawara males found the stranger in the northern part of their territory late in the evening. Not long later, he lay on his back, limbs spread, with numerous wounds all over the front of his body. His dorsal side was undamaged except around an elbow, where an attacker had seemingly gripped the skin with teeth and reared back, tearing it off. His thorax was torn out. One testis lay a few meters away, while the other was under his back. He had been killed fighting for his life, a hugely powerful animal in his prime. But none of the attackers showed so much as a scratch. Television footage of such interactions, taken in

Kibale by David Watts and in Tanzania's Gombe National Park by Bill Wallauer, has now allowed viewers to see for themselves how gruesome, chaotic, and effective these deliberate killings are.[23]

Anthropologists unfamiliar with apes were not prepared for these events. Once again, for similar political reasons as with infanticide, the field data were strongly resisted. Writers such as Margaret Power, Robert Sussman, and Brian Ferguson argued that the killings must be due to human interference, such as disturbing a pristine habitat by providing extra food to the chimpanzees, a practice that had happened in two studies. Logging, hunting, and the introduction of human diseases were also proposed as factors that could in theory throw a population off balance and cause a strange new maladaptive behavior to emerge. Skeptics feared that, if killing among chimpanzees proved to be a natural behavior, there might be a ripple effect on the way people thought about human killing: it would strengthen the supposedly worrying idea that violence and warfare were evolved behaviors. Such concerns appear to explain why hefty accusations of scientific naïveté and political bias were directed toward people such as me who reported the chimpanzee findings and argued that they represented adaptive behavior.[24] The skeptics' main objection was that the killings must be unnatural consequences of human disturbances of the coexistence of naturally peaceful groups.

As with infanticide, the debate is now settled. The coalitionary killing of adults is not common, but it is a characteristic feature of chimpanzees, independent of human activity. In 2014, the primatologist Michael Wilson compiled data from the eighteen chimpanzee communities that had been studied for longest in the wild. The results were unambiguous. The chimpanzee populations varied greatly in terms of how frequently killing occurred. More killings were committed by members of communities that included more adult males and lived in denser populations. The statistics indicated that the western chimpanzee subspecies, which lives from Senegal to Nigeria, proved less aggressive than others. But, as a whole, chimpanzees shared a proclivity for violent coalitionary attacks, and the variation in frequency bore no relation to variation in measures of human disturbance.[25]

Instead, the killing was explicable as a biological adaptation. The behavior often begins without provocation, males setting out for the boundary area simply, apparently, because they have time and energy to do it. The attacks cost little for the attackers, but by eliminating rivals they benefit their own community. In Kibale's Ngogo community, John Mitani and David Watts's team recorded instances when males killed or fatally wounded eighteen members of neighboring communities during a period of ten years. The Ngogo community then expanded their territory into the area where most of the kills had occurred.[26] In Gombe, Anne Pusey and her colleagues have shown that, when the territory occupied by a community increases in size, community members are better fed, breed faster, and survive better.[27] Kill some neighbors, expand the territory, get more food, have more babies—and be safer at the same time, since there are fewer neighbors who might be able to attack you.

Similar evidence of an evolved function for coalitionary proactive aggression has come, ironically, from wolves. Lorenz could hardly have been more wrong when he suggested that wolves evolved to be inhibited from killing one another, because the rate at which wolves kill other adults has proved exceptionally high. Lorenz erred by assuming that relationships within packs would be similar to those between packs. Within packs, as he observed, ritualized signals of submissiveness indeed normally inhibit a dominant individual from killing. Between packs, the story is different.[28]

In the United States, wolves reintroduced into Yellowstone National Park in Montana and Wyoming have been studied closely. In a twelve-year sample of 155 wolf corpses found inside the park, an estimated 37 percent were killed by other wolves.[29] Shortage of space, rather than shortage of food, predicted aggression between packs. Likewise, in Denali National Park in Alaska, where wild wolves were monitored mostly by helicopter, about 40 percent of 50 adult deaths were caused by adults from other packs.[30] How many of the kills were due to proactive aggression as opposed to reactive aggression getting out of control is unknown, but direct observations of fights between packs show that they were sometimes proactive. In April 2009, five members of Yellowstone's Cottonwood Creek pack illustrated the point.[31] The five Cottonwood Creek wolves approached a female from another pack, who went into her den. As the intruders approached, they were distracted by the resident female's mate, who was running nearby. Four times the intruders chased the female's mate, as far as three hundred meters (about a thousand feet) at a time, but kept returning to the den where the female was hiding. On the last occasion, they reached the den, attacked the female, and killed her. They killed at least two pups as well. The female died from bite wounds on her head, neck, stomach, and groin; the rock walls on the inside of her den were bloody. The attackers stayed in the area for five hours, and then moved off. Wolves tend to benefit from such attacks by later expanding their territories.

Chimpanzees live in large communities of multiple breeders, sleep scattered throughout their territory, and mate promiscuously. Wolves live in small groups, sleep in central dens, and form monogamous relationships. Despite such differences, the logic of killing is the same in the two species. Defenseless victims can be found; large parties can kill safely; and killers get extra territory.

It might seem surprising that this effective killing behavior is rare in the animal kingdom. The explanation is simple. In most species, the costs of attacking members of your own species are too high because you might get hurt. Only a few species happen to live in societies in which gangs of allied individuals can form, and the gangs can regularly find vulnerable loners of another group to beat up on with minimal risk of being hurt. Among mammals, these coalitions are so far known to occur only among social carnivores and primates.[32]

Coalitionary proactive aggression that kills adults in other groups of the same species is rare, but where it occurs it appears to be a natural and adaptive behavior that benefits the killers. It is known in wolves, lions, spotted hyenas, chimpanzees, white-faced capuchins, various ants, and a few other species. The ways in which animals use coalitionary proactive aggression bear sufficient similarity to those found in humans that an obvious question emerges. Are animal intergroup violence and human intergroup violence explicable by the same principles? Mobile hunter-gatherers provide a critical part of the evidence. They show that coalitionary proactive aggression has been practiced regularly by humans in raids and ambushes, in a style resembling intergroup violence in wolves and chimpanzees.

Some people think of hunter-gatherers as so peaceful that coalitionary fighting would hardly be part of their lives at all. That idea is generally right for a specific type of hunter-gatherer—namely, those who lived alongside farmers or pastoralists (herders of mobile animals such as cattle, sheep, and goats). Classic examples are the Hadza of Tanzania and the Ju/'hoansi of southern Africa (who have also been called !Kung Bushmen, San, or Basarwa). They both live in the same area as pastoralists, with whom they also intermarry, and have done for hundreds of years. The pastoralists are militarily superior to the hunter-gatherers. Although there are historical records of warfare among those peoples, in recent years peace has reigned in both groups of foragers. If fights arose between members of the two cultures, as they sometimes did, the hunter-gatherers would be soundly defeated, as they were. The reason why recent studies have found little evidence of recent intergroup combat in such places seems clear. Humans are smart enough to know that, if they are likely to lose, it is better not to start a fight.[33]

To assess how hunter-gatherers might have behaved during the Pleistocene, when their neighbors would have been more equally matched, we need to find modern cases where different societies of hunter-gatherers still lived alongside one another, with no farmers or pastoralists nearby. Only six such areas are known: Australia, Tasmania, the Andaman Islands, Tierra del Fuego, western Alaska, and the Canada–Great Lakes region. In every case, the same story emerges. Relations between neighboring forager groups were often peaceful, particularly within ethnolinguistic societies. Yet on occasion hunter-gatherers did fight with one another. When they fought, violence was typically initiated in the form of ambushes, raids, and sneak attacks. Coalitionary proactive aggression was the primary technique, and the aim was to kill.

Australia's entire population were hunter-gatherers before European contact in the seventeenth century led to the destruction of traditional life. A 1940 estimate indicated there were almost six hundred different linguistic groups, or societies, in Australia.[34] Intergroup conflict was found to occur in every climatic zone throughout the continent, from the lush regions of the north and southeast to the harsh central deserts.[35] Occasionally it was directly observed.

In a 1910 book focusing on relationships between societies, the anthropologist G. Wheeler summarized the encounters: "A common procedure in such warfare is to steal up to the enemy's camp in the dead of night, and encircle it in the earliest dawn. With a shout, the carnage then begins."[36] Night attacks appear to have been universal. Herbert Basedow detailed the logic: "The aggressors know that the most radical method to extinguish the enemy is to take them unawares, and to slaughter them before they can retaliate. For this purpose, it is best to either steal on them in the earliest hours of morning . . . or to lie in ambush at a place . . . where the enemy is sure to call."[37]

Similar conclusions emerged in each of the regions in which hunter-gatherers lived without farmers or pastoralists nearby.[38] In Tasmania, "the combats usually took the form of ambushes or personal fights," and "the perfection of their war was ambush and surprise."[39] In India's Andaman Islands, isolated in the Bengal Sea, "the whole art of fighting was to come upon your enemies by surprise, kill one or two of them and then retreat. . . . They would not venture to attack the enemy's camp unless they were certain of taking it by surprise. . . . If they met with any serious resistance or lost one of their own number, they would immediately retire. . . . The aim of the attacking party was to kill the men. . . ."[40] In Tierra del Fuego, at the southern tip of South America, "the first intimation of attack was the swish of hostile arrows and the first move was to run for shelter until the strength and nature of the attack could be determined."[41] The Inuit of the Far North were equally given to raiding. "Surprise attacks were of two basic types: ambushes and night-time raids. Both were reported from all parts of western Alaska. . . . The objective of surprise attacks and battles was to kill as many of the enemy as possible."[42] People living in the Canada–Great Lakes region were no different. "Their attacks are all by stratagem, surprise, and ambush."[43]

Attacks were made by men. Often the attacking party would be small, around five to ten, though there are occasional reports of assemblages of up to several hundred warriors.

The frequency of hunter-gatherer intergroup conflict, as well as its context and its rate of mortality, have often been hotly contested issues. One school of thought has held that battles and "real war" were trivially rare before agriculture. Supporters of this view argue that nomadic hunter-gatherers had low rates of intergroup violence from time immemorial until they encountered farmers, at which time they had to start defending themselves. Until the agricultural revolution, these scholars tend to suggest, there was no need to fight: if a dispute arose between forager groups, one group could always move elsewhere. The motivation for taking this perspective was sometimes explicitly political. For instance, the anthropologist Douglas Fry wrote, "One important, general, contribution that anthropology holds for ending 'the scourge of war' lies in demonstrating that warfare is not a natural, inevitable part of human nature."[44]

On an opposite side of the debate, it can be noted that other primates invariably populate their habitats fully, which leaves groups with few options when conflict occurs. When competing groups cannot find vacant land, they fight. It would be surprising if human groups could regularly find empty, well-resourced space to occupy, or if humans could relate to their neighbors without aggression. Thus the claim that groups of hunter-gatherers before the agricultural revolution had generally peaceful relations and could move to unoccupied resource-rich land is implausible. Furthermore, many scholars point to archaeological evidence of frequent warfare prior to agriculture in the form of fortified settlements, armor, and high levels of violent trauma exhibited in skeletons and skulls. Finally, while known death rates from between-group fighting among hunter-gatherers vary widely, they are rarely zero. A standard sample of twelve hunter-gatherer societies found a median death rate from intergroup conflict of 164 deaths per 100,000 per year, compared to a median for twenty small-scale farming societies of 595 deaths per 100,000 per year. By comparison, the worldwide death rate from battle has been below 10 deaths per 100,000 per year since 1960, whereas it reached a maximum of just under 200 deaths per 100,000 per year during World War II. In 2015 the worldwide homicide rate within countries averaged 5.2 deaths per 100,000 per year. Hunter-gatherer societies therefore appear to have experienced significant risks of death from intergroup combat.[45]

The debate about the frequency of war, however, is a different question from whether interactions between groups led to killing. As the political scientist Azar Gat showed in a comprehensive review in 2015, both sides of the debate agree that killing occurred.[46] The discussion is confusing, partly because the word "war" is sometimes restricted to particular kinds of killing. Douglas Fry wrote that Australian hunter-gatherers were "unwarlike," even though he acknowledged significant killing rates. Killing occurred in "feuds" rather than what Fry called "war."[47] The anthropologist Raymond Kelly wrote a book about hunter-gatherers called *Warless Societies and the Origins of War.* Despite the implication of the title, his "warless" societies focused on the Andaman Islanders, among whom Kelly was explicit in recognizing much intergroup killing. In fact, he reported for the Andaman Islanders that "peace was unattainable in external war" (that is, armed conflict between different ethnolinguistic societies, of which there were eleven in the Andaman Archipelago). "External war is unremitting," Kelly continued, "and constitutes a condition of existence that defines the boundaries of the niches exploited by two populations." Kelly probed the primary sources more carefully than anyone before him. He reported how, when groups met, there was "a policy of attacking whenever a . . . hunting party has the advantage of surprise."[48] In other words, Kelly's careful work showed unambiguously that, whether or not you called the fights among the ethno-linguistic societies of Andaman Islanders "war," those hunter-gatherers used coalitionary proactive aggression to kill one another in a very chimpanzee-like style.

The archaeologist Brian Ferguson was one of the founders of the "tribal zone" theory, which proposed that hunter-gatherers were peaceful until they encountered colonial states. By 1997, however, the evidence of war prior to colonialization led him to write in his contribution to an edited book on violence and warfare in the past: "If there are people out there who believe that violence and war did not exist until after the advent of Western colonialism, or of the state, or of agriculture, this volume proves them wrong."[49] In short, whether or not intergroup killing was as frequent among mobile hunter-gatherers as it has been among farmers and pastoralists, the evidence that intergroup killing did occur is clear; and the killing happened often, and almost certainly mainly, by coalitionary proactive aggression.

Evidence from the rest of the world points to the same conclusion. Complex hunter-gatherers such as the Kwakiutl of the American Northwest Coast[50] or the Asmat of New Guinea[51] used boats to make predatory raids for taking trophy heads and scalps, destroying villages, or capturing slaves. Slash-and-burn farmers such as Brazil's Mundurucu[52] or Venezuela's Yanomamö[53] would plot in their home villages, then walk for hours or days to kill one or more unsuspecting victims before running away without, they hoped, having to defend themselves. Mounted pastoralists from the Mongols to the Moors would sweep into farming settlements and rape, kill, and pillage at will. The air forces of contemporary states plan bombing raids. Confrontational battles can certainly occur within every society, and are much more common in those with a specialized military. The ideal of raiding groups and military commanders throughout history, however, has always been the surprise attack with overwhelming force that leaves the opponents defeated, captured, or dead—and the attackers unharmed: the very model of coalitionary proactive aggression.

No one knows for certain how coalitionary proactive aggression came to be so prevalent a part of human behavior. Two reasons suggest an origin in intergroup aggression.

First, between-group aggression is where coalitionary proactive aggression is most common in animals. Within groups, by contrast, there is little evidence for it.[54]

There are cases of adult murders within chimpanzee communities. In Michael Wilson's 2014 compilation, five killings of adults had been observed within groups (and an additional thirteen incidents had been inferred or suspected). All were coalitionary. One that was documented from the start did show premeditation. Since chimpanzees cannot talk, however, it was not coalitionary proactive aggression conducted in the human style of making a plan in the absence of the victim.[55]

The primatologists Stefano Kaburu, Sana Inoue, and Nicholas Newton-Fisher described the incident in detail.

Pimu was the twenty-three-year-old alpha male of the M-group in Tanzania's Mahale Mountains National Park. His final hours began when he did something surprising and stupid: he suddenly bit the hand of a male who was grooming him, a twenty-year-old rival for power called Primus. The unprovoked bite launched a one-to-one fight that was caught on video. For at least thirty-five seconds, Pimu and Primus rolled on the forest floor, punching and biting each other. Primus eventually broke off and ran toward a party of other males some thirty meters (one hundred feet) away. He screamed as if seeking support, climbed a tree, and disappeared. He was not seen by the primatologists for almost a week.

Four of the solicited males now approached Pimu, joined almost immediately by two others. Pimu was already bleeding from a big cut on the side of his head and another on a hand. The ensuing mêlée was confusing. Mostly, the first four attacked Pimu, while the two others made limited attempts to defend him by threatening the attackers. Bouts of violence lasted for ten to fifteen seconds at a time, separated by pauses that could go on for more than a minute. Oddly enough, the most persistent aggressor was an old former alpha male, Kalunde, who until then had been Pimu's most frequent partner, helping him in coalitions against other males. After about two hours, all six males withdrew and climbed trees at least ten meters (thirty-three feet) away. Pimu was left incapacitated and bleeding. He died within half an hour.[56]

The four males had not been involved in the fight between Pimu and Primus, and by the time they attacked Pimu, Primus was already gone. Their aggression was therefore not in direct support of Primus. Nor was it reactive, since they had not been involved in the original fight and they took a long time to become involved. The attack was certainly coalitionary and appeared to be premeditated. Kaburu and colleagues thought that it was best explained as a challenge for social dominance provoked by the fact that Pimu had been wounded and was suddenly vulnerable.

So the coalitionary aggression that sometimes occurs within chimpanzee communities can in a limited sense be proactive, when individuals who have been given an opportunity to attack take their time deciding whether to join. However, there is as yet no evidence that it can be premeditated in the sense of a coalition deliberately searching for a particular victim to attack. The obvious reason for this limitation is that chimpanzees have no ability to discuss whom to target for killing.

Certainly, whatever prompts coalitionary proactive violence within communities, the behavior is much more common between communities.

The second reason for thinking that coalitionary proactive aggression developed primarily in the context of intergroup interactions is that it makes fewer cognitive demands than the same behavior expressed within groups.

Intense violence within groups calls for a difficult decision, because an actor must decide whether to join a coalition against a victim who could, at a different time, be a useful ally. That is one reason why attacks like the killing of Pimu are surprising: the advantages of retaining strength in numbers (in fights against neighboring communities) would normally seem to outweigh the benefits of eliminating a rival.

Furthermore, there is the problem of planning. How is a target decided, and how does any one individual know who is going to join a coalition? There seems to be no way for chimpanzees to share in advance their intention to attack a particular individual. Chance brought an opportunity for the killers of Pimu, and even then, four were against him but two were not. Humans solve this coordination problem through language. Plotters gossip with increasing confidence that someone should be killed, yet uncertainty remains up to the moment of a joint attack. The assassins check one another's commitment, lest someone turn traitor.

In short, complex calculation and refined communication are needed to make within-group killing a successful adaptive strategy. By contrast, in intergroup interactions the formula is simple: side with your friends against the enemy.

The obvious evolutionary route for coalitionary proactive aggression is that it was first expressed against members of other groups, as it is in chimpanzees, wolves, and some other mammals. That might have happened among our ancestors of 6 to 8 million years ago, the last ancestors we shared with chimpanzees. But we cannot be sure. The behavior might have been invented earlier, or more recently. Later, during the Pleistocene evolution of the genus *Homo,* increasing cognitive ability, and in particular the development of sufficient language skill, would have allowed coalitionary proactive aggression to be employed more selectively within the group. Further back in evolutionary time, before proactive aggression was coalitionary, it would presumably have been practiced by individuals acting alone, much as male Hanuman langurs do.

Coalitionary proactive aggression in humans, therefore, is most simply understood as an elaboration of ancient tendencies. Though it is uniquely elaborated in our species, it likely originated in the cognitively simple context of between-group interactions. A plausible scenario holds that killing of conspecifics by chimpanzees and humans has the same evolutionary origins in intergroup raids.

Advances in neurobiology might ultimately help test that idea. We saw earlier that, in rodents and cats, reactive and proactive aggression are controlled by different pathways within the brain's "attack circuit." In humans also, it is clear that proactive aggression is regulated differently from reactive aggression. The future holds unknown possibilities. In a 2015 review, the behavioral physiologist S. F. de Boer and colleagues presented evidence that offensive (proactive) aggression is supported by specialized types of neurons that respond to particular molecules. Within a few years, it may be possible to find out how similar or different the neural mechanisms are that underlie proactive aggression in humans and chimpanzees. Those kinds of comparisons would provide an exciting opportunity for further understanding the evolutionary biology of this rare and potent behavior.[57]

By the time in prehistory when coalitionary proactive aggression was directed not just at excessively aggressive rivals but at any supposed deviant, the degree of social power it bestowed on those who used it well was unprecedented. Much as a male langur monkey can pluck an infant from its mother and slice into its guts with little fear of the consequences, or as a gang of chimpanzees can tear out the thorax of a rival male without getting hurt, so a group of committed humans could aggress against a chosen victim while remaining personally immune from harm or retribution. The advanced skills of proactive aggression in humans licensed forms of despotism not found in other primates. Obedience and sovereignty are potent examples.

Obedience is a uniquely human relationship. Nonhumans such as dogs can learn to obey, but they cannot give orders. A system of obedience depends on punishment. Within families or small groups, the mechanism of punishment can be emotional manipulation or physical beatings, but in the politics of large-scale groups, a proactive coalition provides the power. An order given to a subordinate is essentially a threat to use aggression unless the order is obeyed. If the threat depended merely on the fighting ability of the leader, it would rarely be convincing. No leader could risk repeated fights. Even an alpha chimpanzee tries to avoid fighting when possible. However, a human leader does not have to fight personally; a coalition of supporters guarantees the value of the leader's threat, and the likely danger from an aggressive encounter is low for the supporting coalition because their overwhelming power can be brought to bear on the subordinate. Knowing this, the subordinate must obey or suffer the consequences. In the authoritarian courts of medieval European monarchs, Chinese emperors, twentieth-century fascist regimes, or Mafia families, a signal from the leader could be enough to cause the execution of a disrespectful member. The obedience of courtiers, slaves, prisoners, or unwilling soldiers shows us the consequence of hierarchical power in its rawest form. Those who attempt to challenge, escape, disobey, or desert are subject to being killed.

In democratic states, the system works so much more gently that you might think that its success depends entirely on willing cooperation. The political philosopher Michel Foucault has argued convincingly, however, that, although democracies are less violent than authoritarian states, even benign social institutions such as factories, hospitals, and schools depend ultimately on power enforcement for their success. Rule breakers can try to negotiate their way out of a conflict, but if worse comes to worst, they face imprisonment—courtesy of the coalitionary proactive aggression of police and other regulatory institutions. Even nonviolent states depend on decisive force for dealing with miscreants.[58]

Sovereignty over substantial areas of land is also uniquely human. One view of sovereignty has been that the rulers and the ruled were equally involved in controlling the area in which they lived. However, that sunny perspective ignores the realities of how power is distributed. In practice, real sovereignty, according to the anthropologists Thomas Hansen and Finn Stepputat, is "the ability to kill, punish, and discipline with impunity,"[59] which of course is concentrated in the powerful. "Sovereign power,"[60] they write, "is fundamentally premised on the capacity and the will to decide on life and death, the capacity to visit excessive violence on those declared enemies or on undesirables."[61] In 1954 Adamson Hoebel had come to much the same conclusion in his survey of the functions of law. "The really fundamental *sine qua non* of law in any society—primitive or civilized—is the legitimate use of physical coercion by a socially authorized agent."[62]

Sometimes the underlying violence is exhibited deliberately. When the East India Company decided that the punishments meted out by local princes were inadequate and ineffective, they erected public gallows and opened new prisons across their Indian colony. Monarchies have often displayed the heads or bodies of executed offenders. Violence is also patent in the multiple sovereignties that exist outside the rule of law. These might include "pirates, bandits, criminals, smugglers, youth gangs, drug lords, warlords, Mafiosi, traitors, terrorists." The underworlds that these liminal groups control occur worldwide, from fragmented postcolonial countries to the richest and most powerful states. Their domains are normally small, but, as elsewhere, their "sovereign power has always depended on the capacity for deployment of decisive force."[63]

Obedience and sovereignty are not found among mobile hunter-gatherers to nearly the same extent as among settled peoples, and arguably not at all. The archaeologist Brian Hayden suggested that hierarchical relationships among families began during the Upper Paleolithic, a few thousand years before agriculture was invented, when humans found a way to produce surplus food. Those who owned the surplus could use it to buy labor or goods from those who needed it, thus creating an advantage to producing as much food as possible. One of the consequences seems likely to have been fealty. One can imagine a wealthy man buying loyalty with food, and in so doing he could extend a kin-based coalition to create a long-term fidelity of allies.[64]

Robert Louis Stevenson implied that we inherited our violence from apes, but his moral tale did not mention proactive aggression. The impunity resulting from coalitionary proactive aggression, with its consequences that reach to obedience and sovereignty, has given successful individual humans power beyond the imagination of nonhuman primates. To the powerless it has given unprecedented misery. A more evolutionarily telling version of Dr. Jekyll and Mr. Hyde would have Jekyll as a wealthy, good-looking, hardworking, ambitious, kind, and cooperative thinker who was also a premeditated killer enforcing obedience and sovereignty over his social world.

Coalitionary proactive aggression is responsible for execution, war, massacre, slavery, hazing, ritual sacrifice, torture, lynchings, gang wars, political purges, and similar abuses of power. It permits sovereignty as a right over life, caste as a system of casual domination, and guards who make prisoners dig their own graves. It makes kings of wimps, underlies fidelity to groups, and gives us long-term tyrannies. It has battered our species since the Pleistocene. Along the evolutionary way, it brought us our great capacity for good, but it has also brought enormous harm.

It is therefore cheering to remember that in sane individuals proactive aggression is a highly selective behavior that is delicately attuned to context. Male langurs do not kill at random; they do not kill at all if their potential victims are adequately defended. Chimpanzees reserve their proactive attacks for when they have overwhelming force. Hunter-gatherers living with pastoralists do not get into wars. Proactive aggression is not produced by individuals in a fit of rage, or in an alcoholic haze, or out of a testosterone-induced failure of cortical control. It is a considered act by an individual or coalition that takes into account the likely costs. It has a strong tendency to disappear when it does not pay.

The evolutionary psychologist Steven Pinker testified to that effect. In his *Better Angels of Our Nature,* Pinker documented in detail the multiple ways in which violence has declined in the most recent decades, centuries, and millennia. Almost all of the iniquities that Pinker wrote about stemmed from coalitionary proactive violence. If we continue to improve the protections in our societies, the level of damage will continue to recede. But we should never forget the alarming potential made possible by the exertion of extreme power. The human species has yet to record a peace that lasts for millennia; and in a nuclear world, the frequency of violence might be less important than its intensity.[65]

12

War

Humans are often said to be intensely tribal. Indeed we are. But if "tribal" refers to a sense of solidarity with a large social group, the same is true of most primates. Tribalism does not distinguish us, nor does reactive aggression. It is coalitionary proactive aggression that makes our species and societies truly unusual. Among our ancestors, coalitionary proactive violence directed at members of their own social groups enabled self-domestication and the evolution of the moral senses. Now it enables the functioning of states. Unfortunately it also gives our species war, caste, the butchery of helpless adults, and many other forms of irresistible coercion.

The reason coalitionary proactive aggression enables these despotic behaviors is straightforward. A coalition of proactive human aggressors can choose when and how to be aggressive to their victims in such a well-planned way that they can achieve their goal with ovewhelming force and without risking their own safety. As long as the victim cannot assemble a defense, the ability to plan with clinical detachment gives a coalition extraordinary power. Success in removing opponents is predictable and cheap.

In theory, the path to resistance is obvious. "When bad men combine," the British parliamentarian Edmund Burke wrote in 1770, "the good must associate; else they will fall one by one."[1] But of course the "bad men" arrange matters to stop "the good" from "associating" in any meaningful manner. The S.S. were well organized. They chose to make their arrests when their victims were helpless. Prisoners being transported into concentration camps were given no opportunities to organize balances of power in their favor. They had no chance to fight back in a way that would give them the advantage. Collaborative planning enables murders to be conducted with cold efficiency.

It is no wonder that "Power tends to corrupt, and absolute power corrupts absolutely." As Acton also wrote: "Great men are almost always bad men." The cost-benefit dynamics of coalitionary proactive aggression make murderous violence an alluringly easy tool. "Here are the greatest names coupled with the greatest crimes," wrote Acton. Queen Elizabeth I could ask a jailer to kill Mary, Queen of Scots. William III could order his minister to extirpate a Scottish clan. Adolf Hitler could call for the physical extermination of the Jews. Regardless of their personal physical weakness, individual human leaders use coalitions to kill with unprecedented ease.[2]

Given the pervasive effects of coalitionary proactive aggression, its origin and effects are core issues for understanding our social evolution. There is no anthropological consensus about where it comes from. At one extreme, aggression is seen as unconnected to our evolutionary biology. The anthropologist Agustin Fuentes's view is widely shared by scholars mistrustful of evolutionary thinking: "Human aggression, especially in males, is not an evolutionary adaptation."[3] That kind of claim is often linked to the idea that if warfare and related forms of violence are found to be important evolutionary adaptations, politicians and the general public will treat them as inevitable: pessimism will reign and efforts for political improvement will be thwarted. According to Richard Lee: "By constantly asserting the dominance of the side of human nature that emphasizes war over peace and competition over cooperation, the dominant forces in the modern world order can more plausibly maintain a permanent war economy, justify the obscene profits of multinational corporations and their CEOs, and affirm the inevitability of winners and losers in life's sweepstakes."[4] As I will later explain, such fears seem to me both exaggerated and counterproductive. Still, they remind us that an evolutionary analysis is fraught with the potential for emotionally and politically sensitive responses. Care needs to be taken, when discussing the evolution of violence, to state the limits of its implications.

The alternative view sees proactive aggression in humans as an evolutionary adaptation, an elaboration of abilities found in other animals. We are just another mammal species, albeit with some unusual traits. That view is supported by similarities and differences in aggression between us and other species.

The publication of *On the Origin of Species* in 1859 inspired much discussion on the evolution of war. Darwinism raised the possibility that war, like any other behavior, was adaptive. For thinkers in the Hobbesian tradition, such as Thomas Henry Huxley, the idea was right. Humans had obviously evolved as a fighting animal. For Rousseauians, however, such as the Russian philosopher Peter Kropotkin, the idea that primitive man was inherently warlike was an illusion, not to mention politically dangerous.[5]

Nowadays most evolutionary anthropologists agree with Huxley's perspective that hunter-gatherers engaged in significant war, and that tendencies for warfare are strongly influenced by psychological adaptations that evolved in the Pleistocene. For these anthropologists, the important questions have shifted from whether humans have a violent past, to how our psyches are adapted to it. The distinction between proactive and reactive aggression has not yet been widely applied to the problem, however. In this chapter, I therefore introduce some thinking about it.[6]

Before doing so, I first address a classic difficulty. For some Rousseauians, the view that humans have an evolutionary history of war is unacceptable, whether or not it is theoretically compatible with our having an evolutionary history of docility. The skeptics insist that warfare was absent in the Pleistocene, or else that it was at most a rare activity that did not affect our ancestors' biological adaptations. As a result, they say, our contemporary genetic predispositions are irrelevant to the practice of war.

Many of the Rousseauians' concerns center on the notion that biology is destiny, an idea captured in the phrase "biological determinism." "Biological determinism" is a slippery concept, but the core idea is that it would lead to our behaving somehow robotically, in unthinking obedience to the programs of our DNA. The Rousseauian archaeologist Brian Ferguson characterized the vision as "the image of humanity, warped by bloodlust, inevitably marching off to kill."[7] For reasons that I will explain, I regard the problem of biological determinism as much less worrying than the Rousseauians' fears would indicate. Nevertheless, biological determinism is important, because ever since Darwin, and right up to today, it has cast a shadow on discussions of the past, present and future of war. A key question is whether, if our ancestors were adapted for war in the Pleistocene, we are biologically driven to conduct war today. As I will explain, my answer is that while war is not inevitable, conscious effort is needed to prevent it.

Almost everyone hopes for a future without war. As a result many people (such as me) feel cheered by the data showing a decline of war deaths over historical and prehistorical time. We seem to be going in the right direction. Yet not everyone agrees with that simple idea. Even though Rousseauians hope for a peaceful world, many dislike the notion of a trend toward nonviolence. They note that a long-term drop in the frequency of warfare means that in the past humans were killing each other at higher rates than today; and that idea of ancient violence is an obstacle to positive thinking. Douglas Fry explained the concern. If war is ancient, he worried, it must be natural. And "if war is seen as natural, then there is little point in trying to prevent, reduce, or abolish it." His claim that adaptations for war must mean that war is inevitable is a classic statement of biological determinism.[8]

Social psychologist Erich Fromm offered an equally determinist explanation for the supposed apathy that comes from seeing war as natural. He suggested that a sense of doom is comforting. In *The Anatomy of Human Destructiveness,* he asked: "What could be more welcome to people who are frightened and feel impotent to change the course leading to destruction than a theory that assures us that violence stems from our animal nature, from an ungovernable drive for aggression?"[9]

Such feelings are widespread among Rousseauians. They seem odd.

Why should the idea of war as natural mean that "there is little point in trying to prevent, reduce, or abolish it"? We do not apply that formula to other unpleasant natural things. We try to stop diseases even though they are clearly biological in nature. We try to intervene when men harass women, or when bullies throw their weight around, or when children fight one another. The fact that we think such behaviors have evolved does not inhibit us from trying to reduce their effects.

Furthermore, I know of no factual basis to support the claim that a belief in ancient war induces fatalism. In my experience, the claim is quite wrong. Chimpanzees were first observed to practice warlike violence in Gombe National Park, Tanzania, in the 1970s. Three senior scientists were involved in the research. Each of them was rocked by the implications, but they were inspired to action rather than being frozen into fear. All three became important ambassadors working in their own way to avert the dangers of war.

Jane Goodall, shocked by the revelations that chimpanzees committed infanticide, rape, and murder, published a book of supreme optimism called *Reason for Hope.* She wrote movingly about the disturbing discoveries, and became a tireless advocate for positive thinking and a sustainable world.[10]

The behavioral biologist Robert Hinde was Jane Goodall's graduate adviser, and later mine too. He spent much of his subsequent career working to reduce the risk of war, including serving as chair of the British Pugwash Group, which won a Nobel Peace Prize for the promotion of nuclear disarmament. Hinde wrote extensively about the deep prospects for international peace and moral improvement, including such books as *War No More: Eliminating Conflict in the Nuclear Age*[11] and *Ending War: A Recipe.*[12]

David Hamburg was originally an academic psychiatrist. After writing about the evolution of chimpanzees' lethal aggression and its significance for understanding human violence, he worked for decades to help reduce global threats.[13] As president of the Carnegie Corporation, he commissioned the 1997 report *Preventing Deadly Conflict,*[14] co-written with former U.S. Secretary of State Cyrus Vance. Hamburg was inspired by the observation that, since genocides require much planning, they can be anticipated and prevented. His books *No More*

Killing Fields[15] and *Preventing Genocide*[16] are rich in detailed practical solutions for war-preventing mechanisms and international peacemaking, using existing institutional systems and imagining even better ones.

Goodall, Hinde, Hamburg, and many others did not consider that, if aggression is adaptive, war must be inevitable. They understood that violence responds to circumstance, not to unstoppable genetic instructions. They assessed that chimpanzee and human psychology include some similarly dangerous motivations, and as a result they were moved to action.

Why, then, have many Rousseauians inferred that pessimism will be the primary response to the same ideas that can motivate such positive action? The Rousseauians' negativity is clearly associated with their claim that, if warring tendencies evolved, humans must have an "ungovernable drive for aggression." They imply that an evolutionary understanding of war as an adaptive behavior forces people toward biological determinism. Yet few if any of the scholars who believe that war has an adaptive basis regard war as inescapable.

The unjustified attribution of determinist thinking to scholars who interpret human violence as adaptive is a long Rousseauian tradition. The historian of science Paul Crook wrote a gripping history of how the relationship between evolution and war was discussed between 1859 (when Darwin's *Origin of Species* was published) and 1919. Claims of biological determinism were rife. Rousseauians employed them then, as some do now, to accuse opponents of simplistically implying that aggression and war were inevitable. As now, the accusations were misplaced. Twenty leading scholars reviewed by Crook credited human violence partly to evolved biological influences. Contrary to their opponents' claims, nineteen of them were explicit in arguing that culture could tame biology. They were big names of the day: Auguste Comte, George Crile, Charles Darwin, William James, Vernon Kellogg, Ray Lankester, Henry Marshall, William McDougall, Peter Chalmers Mitchell, Lloyd Morgan, G. T. W. Patrick, Ronald Ross, Charles Sherrington, Herbert Spencer, J. Arthur Thomson, Wilfred Trotter, Alfred Russell Wallace, Graham Wallas, and Lester Ward. The only hard-line "biological determinist" was the neurobiologist Karl

Pearson, who was also a eugenicist. Overall, the accusations had barely any truth.[17]

Abundant evidence shows that violence is socially influenced and socially preventable. History, after all, has long told us that societies can be at peace for generations. Evolution of a behavioral tendency does not mean that the behavior has to be inevitable, inflexible, or in some other way independent of human will. Genes affect the size and sensitivity of different brain regions, the nature and activity of the physiological stress systems, the production and fate of neurotransmitters, and on and on. Genes create a system, and the system responds to context. A primate that invariably produced aggression as predictably as it went to sleep or felt hungry or pulled away from a smelly cadaver would quickly fail in the evolutionary game. The secret to successful aggression is appropriate behavioral flexibility.

The different forms of aggression vary in how predictably they are expressed. Reactive aggression is harder than proactive aggression for an individual to control, but even reactive aggression is still subject to cortical inhibition. The reason men are more dangerous after drinking alcohol is that their ordinary controls have been loosened. In the absence of alcohol, tempers are more easily regulated. Ordinarily, in other words, inhibitions are active.

As I described in the previous chapter, proactive aggression is enacted only when the protagonists are able to assess a high likelihood of cost-free success. If that circumstance never arises, proactive aggression is not expected to occur. That is why chimpanzees kill each other only occasionally, and why people can live for long periods at peace.

In a revealing contrast to their reasoning about aggression, Rousseauians have been relatively comfortable with the idea that tendencies for attachment, romantic feelings, or cooperation have evolved. They readily acknowledge that cognitive and neuroendocrine systems differ between humans and the great apes in ways that foster our special human versions of empathy or altruism. When dealing with those positive aspects of human behavior the arguments about determinism are (quite rightly!) forgotten.[18]

A willingness to recognize the role of genetic adaptation in positive behavior suggests that the problem with recognizing its equivalent role in negative behavior does not come from a failure to appreciate the sophisticated nature of behavioral biology. Other motives appear to be at work. What else can explain why even accomplished evolutionary biologists like Stephen Jay Gould have reacted to the concept of adaptive killing with accusations of biological determinism? Gould satirized the idea of humans having evolved adaptations for intergroup aggression as the suggestion that "our accursed genes have made us creatures of the night."[19]

In short, some Rousseauians claim that belief in an evolutionary history of adaptive violence must commit one to the view that aggression and war are inevitable; but the claim is wrong. It is nonsensical to believe that if humans practiced war in the Pleistocene, our species must now be "warped by bloodlust, inevitably marching off to kill."[20] Humans do not have "an ungovernable drive for aggression." The theory of biological determinism that some Rousseauians appear to believe and that others have foisted onto scenarios of the evolution of aggression belongs in the dustbin. We can explore the evolutionary significance of warfare without being bound to the idea of an irresistible urge for violence.

How, then, if humans are decision-making beings rather than automatons, does an evolutionary theory based in the separate natures of proactive and reactive aggression help to explain warfare? Our species' potential for intergroup killing is not in question. In Richard Lee's words: "It is important to point out that the historically nomadic foragers are not nonviolent. They fight and sometimes kill."[21] Following the standard assumption that all human populations have the same basic psychology,[22] therefore, we must agree with Frederick the Great: "Every man has a wild beast within him."[23] The question is what releases the beast.

Two forms of warfare, simple and complex, are so different from each other that they require separate consideration. Complex war, involving military organization, group-on-group attacks, and battles in state-level societies, is addressed below.

The more evolutionarily relevant style is the simple war that occurs in small-scale prestate societies and is more similar to intergroup aggression in some animals. Its skirmishes are so brief, and it is so relatively unmilitary in its organization, that some anthropologists prefer not to apply the word "war" to this style of violence. It consists mainly of brief surprise attacks. Simple war is the only type of war in societies where men (ordinarily, the adult married men, or what Gellner calls "the cousins") have egalitarian relationships, and no men work for or hold authority over others. All except the infirm are warriors, without military hierarchy. When conflict arises with another group, the warriors can discuss a plan together, but no one coerces another's participation. Each man makes up his own mind about whether to join an attack or stay at home. These patterns are found both in mobile hunter-gatherers and some horticulturalists such as the Mundurucu of Brazil or the Yanomamö of Venezuela.

The patterns of aggression in simple war consist mainly of a party of males approaching the enemy by stealth. After wounding or killing one or more victims, ideally the attackers leave so rapidly that they escape without being drawn into an escalated encounter. Battles are accordingly rare. When opposing groups of warriors find themselves confronting each other, men on both sides tend to disperse.[24]

Although simple warfare is thus more complex than chimpanzee intergroup aggression, its adaptive logic is nevertheless much the same. Aggressors organize their raids so as to minimize their risk of getting hurt. Usually the main aim is to kill. Reducing the number and proximity of rivals is likely to lead to benefits. The benefits could include a reduced risk of being attacked in the future, or more access to neighboring resources. Whatever motivates an attack, if the neighboring group is weakened, the killers will likely end up doing better.

While the most likely outcome of a raid is total success, because raids are undertaken only when the attackers have overwhelming force, there is always a chance of failure.[25] The raiders' approach might be detected, the victim group might be unexpectedly well prepared, or there could be traps such as spikes embedded in the ground around the enemy village. The raiders therefore need courage, and in many cases a willingness to tolerate intense physical exertion. To help overcome their nerves, warriors often work themselves into a state of excitement before leaving home. Ritual practices may be used in prepartion for the attack. Raids might accrue benefits such as capturing women or taking heads; anticipated rewards may take the form of prestige or goods.[26] In some societies, cowards may be punished.[27]

The harder behavior to explain is when the act of killing is its own reward. In the New Guinea highlands, an Enga horticulturalist told the anthropologist Polly Wiessner how his people felt about killing in small-scale war.

> "Now I will talk about warfare. This is what our forefathers said: When a man was killed, the clan of the killers sang songs of bravery and victory. They would shout Auu! ('Hurray!' or 'Well done!') to announce the death of an enemy. Then their land would be like a high mountain and that is how it was down through the generations."[28]

Similar accounts, in which warriors perceive no benefits other than the thrill of making a kill, are rife. From an evolutionary perspective, we can explain their action as we can among animals.

Why do they kill? The unnerving answer that makes biological sense is that they enjoy it. Evolution has made the killing of strangers pleasurable, because those that liked to kill tended to receive adaptive benefits.

At first glance, this idea might seem absurd. None of us would ordinarily find pleasure in killing a stranger. But the strangers that any of us is likely to meet are very different from the strangers met by a warrior in a small-scale society living in an anarchic world. Using cues such as the stranger's weapons, dress, and dialect, the warrior can tell at once whether or not the stranger is part of his society. A true stranger, a member of a hostile neighboring society, will probably be regarded as nonhuman, and is liable to be as dangerous to the warrior as the warrior is to him. To enjoy a successful attack makes grisly sense. When every group relies on its own strength for protection, reducing the power of the neighbors brings rewards.

The rewards do not have to be anticipated consciously. All that is needed is enjoyment of the kill. Sexual reproduction works in a parallel way. A chimpanzee, or wolf, or any other animal, cannot be expected to know that an act of mating will lead to babies. Why do they mate? They enjoy it. Evolution has made sex pleasurable because those that liked to mate tended to have offspring.

The notion that humans evolved to enjoy killing unknown enemies is unpleasant and inimical to our ordinary view of humanity. We can hope that it will become increasingly irrelevant to the human future, given that our species is already largely unified by worldwide social connections: unconnected enemies are already rare. Where the social divide is sufficient, however, occasional outbursts of killing still seem to manifest a deep joy of killing. The historian Joanna Bourke wrote of atrocities on all sides in the Second World War. A Japanese solder remembered Nanking. "When we were bored, we had some fun killing Chinese. Buried them alive, or pushed them into a fire, or beat them to death with clubs, or killed them by other cruel means."[29] In Europe in the 1940s, members of the collaborationist Croatian Ustaše movement enthusiastically killed Jews, Serbs, and Gypsies, "often hacking them to death with primitive implements."[30] Slaughter was estimated at 40,000 Gypsies and 400,000 Serbs in Croatia and Bosnia. A priest, the Reverend Dionizije Juričev, explained: "In this country, nobody can live except Croatians. We know very well how to deal with those that oppose conversion [to Roman Catholicism]. I personally have put an end to whole provinces, killing everyone—chicks and men alike. It gives me no remorse to kill a small child when he stands in the path of the Ustase."[31] There are probably very few wars in which equivalent histories could not be found.

Revenge is a frequent motivator of war and violence.[32] In laboratory experiments, the feelings of pleasure derived from revenge are associated with enhanced neural activity in a particular part of the brain, the caudate nucleus.[33] In rats and monkeys, the caudate nucleus is involved in the processing of anticipated rewards, including those derived from cocaine and nicotine. In humans, individuals who experience higher activation of the caudate nucleus are more willing to punish others. Activation in the caudate nucleus is the kind of neural process that could have evolved to help motivate an interest in killing.

The satisfaction of killing a hated enemy is only one of many reasons to kill, and in recent wars it has probably played a relatively small part.

Moral pressures also explain why ordinary people become murderers. For many people in World War II, writes Joanna Bourke,

> It was not the fear of punishment that made people kill, but group pressure that ensured that those who resisted killing were cast as outsiders and suffered a dramatic lowering of their esteem within the group. The perpetrators of genocide were no different from those who did not participate in the bloodletting. What distinguished them was the situation in which they found themselves. This explanation is disquieting, implying as it does that the capacity for exceptional violence lies within each of us. We are all potentially "evil."[34]

A similar dynamic doubtless sometimes operates in simple warfare.

Revenge motivations and moral pressures are only two of many unique features of our species that influence the practice of simple warfare. Others include advanced weaponry, language, social norms, docile psychology, training of warriors, and the ability to devise a shared plan. But the mere occurrence of simple warfare depends on none of those explanations, given that the human pattern is strongly similar to intergroup aggression in some other species. In humans practicing simple warfare, as in chimpanzees and wolves, proactive aggression is the norm; the goal is to be safe; and killing tends to lead to long-term benefits for the killer. The traits that decorate human warfare beyond these elements are rococo additions, not necessary features. The essential facts of simple warfare of humans are barely more puzzling than the intergroup aggression of other animals that kill their neighbors when they get a chance to do so.[35]

We should therefore not find it surprising that small groups of men with disregard for national laws easily form gangs that take advantage of local imbalances of power to kill. Similar behaviors emerge among chimpanzees and hunter-gatherers as among freedom fighters, street gangs, or the underworld. Selection favors safe killing whenever it can be carried out at sufficiently low risk to the attacker, sometimes even when there is no obvious benefit is at stake. In an anarchic world, the satisfaction of killing an enemy can have its own rewards.

The cold planning of coalitionary proactive aggression can be responsible not only for execution of selected individuals but also for the deliberate killing of larger groups. Nataruk, an archaeological site in northern Kenya, reveals an incident of apparent warfare from ten thousand years ago. Among the bones of twenty-seven individuals were twelve relatively complete skeletons. Ten showed evidence of violent death. The hands of four victims appeared to have been bound before they were killed. We cannot be sure what happened so long in the past, but history and ethnography tell of innumerable cases of the dispatching of defeated warriors and noncombatant captives much as appears to have occurred at Nataruk.

In the ultimate example, coalitionary proactive aggression enabled concentration camp employees to shoot or gas millions of Jews, Romanies, Poles, homosexuals, and others during World War II with hardly a single killer being hurt in the act.[36] We are inclined to label callously planned violence such as the Holocaust as "inhuman." But phylogenetically, of course, it is not inhuman at all. It is deeply human. No other mammal has such a deliberate approach to mass killing of its own species.

Natural selection has doubtless honed our ability to use coalitionary proactive aggression to advantage during our long evolutionary history of simple warfare. Whether the much shorter history of complex warfare also affected our evolutionary psychology is unknown. The earliest strong evidence for complex warfare comes from Qermez Dere in northern Iraq, where defensive walls and skeletons associated with arrowheads and heavy clubs have been dated to about ten thousand years ago. From then onward, evidence of complex warfare has been relatively common.[37] Ten thousand years is certainly enough for biological adaptation to occur. Pastoralists have been consuming milk products from herd animals for no more than eight thousand years, and during that time they have evolved genetic adaptations for more efficiently digesting the milk-sugar lactose. Even though no psychological adaptations to complex warfare have yet been detected, they cannot be ruled out; but here I will treat them as absent.[38]

Complex warfare occurs in societies that have political leadership and involves two major classes of actors. Commanders decide independently what to do, while soldiers are under orders. The combination of commanders and soldiers makes complex warfare so much better organized for combat that it is described as being a truly military system—"above the military horizon," in the words of the anthropologist Harry Turney-High—unlike societies practicing simple warfare.[39]

Like simple warfare, complex warfare can involve proactive aggression, in the form of raids in which small groups aim to achieve a goal and return alive. A platoon that is ordered to take out a machine-gun nest, or a bombing crew aiming for a military installation, will do their best to avoid being detected until the action starts. Proactive raids have obvious military advantages. Sometimes they appear to engage the same enthusiastic psychology of attack as happens in simple warfare.[40]

Alongside these similarities, however, complex warfare involves features that contrast strongly with simple warfare, particularly in the context of organized battles. In battles of complex warfare, soldiers have no choice about whether to participate and may be thoroughly unenthusiastic about doing so. Their situation can be emotionally traumatic and often (when the soldier is coerced into action) highly maladaptive for the individual. In some battles, soldiers are required to approach a body of armed opponents in an action exposing themselves deliberately to a high risk of injury or death. The question is what makes them do it. Motivations forged over evolutionary time are not the answer.[41]

All men on a battlefield are afraid, according to U.S. General S. L. A. Marshall's study of battle behavior in the Second World War.[42] The military scholar Ardant du Picq noted examples of how soldiers responded to being ordered into battle.[43] Sometimes the whole army turns and runs. This occurs, in the words of the military historian John Keegan, "not because [the army] has been physically shaken but because its nerve has given."[44] At other times, "fainthearts" dribble away in small numbers until, by the time the two sides engage, hardly anyone is left. Or the attackers on both sides come to a halt before they are in range of their opponents' weapons—to the fury of their commanders, of course. Another outcome is for the opposing armies each to edge leftward as they approach each other, eventually sliding past each other without contact. The soldiers' fear of engagement is so strong that Keegan summarized the main task of officers as suppressing it, partly by being ready to kill deserters.[45] That was Frederick the Great's formula.[46] He was said to insist that the common soldier must fear his officer more than he fears the enemy. The commanding officer not only must bring the soldiers to the killing zone at the front, he must make them stay there.

Clearly, the battle behavior of soldiers in complex warfare bears little relation to the relative eagerness of a hunter-gatherer warrior attacking an enemy camp. Soldiers in complex war have to be trained to reduce their fear, in Keegan's words, so as "to perceive a face of battle which, if not familiar, and certainly not friendly, need not, in the event, prove wholly petrifying."[47]

According to Keegan, commanders tend to think that soldiers go into battle because they are obedient, whereas in reality two other reasons are more important. One is found when fighting can actually improve the chance of survival; in some circumstances, being left behind can be the worst outcome.

The other reason for bravery is to avoid incurring the contempt of close companions. Military organizations intentionally foster close bonds. Soldiers are typically organized into small groups, often around five to seven men. They have bonded with one another through training, prior action, often ritual exercises such as hazing, possibly hostility toward their officers, and endless hours of boredom. Their desire to maintain one another's respect is sometimes suggested to have come from developing a false sense of kinship. It seems to me more likely to have come from a moral sense that evolved, ultimately, in response to the threat of execution in the Pleistocene, as described in chapter 10. By this view, the solidarity that promotes unit effectiveness comes from not wanting to let one's peers down; and the respect for peers comes from the evolution of self-protective moral responses. Respect also comes from being hazed, a practice that emphatically shows a recruit that he is subject to the coalitionary power of his group members. A soldier exposed as a coward can be in danger from his own unit.[48]

A feeling of duty to close companions will promote military effectiveness. Other emotions forged in our evolutionary past will not.

The cognitive assessment that underlies proactive aggression is expected to discourage men from joining confrontational battles.

Proactive aggression is successful when it involves attacking at low risk of being hurt. Confrontational battles, by contrast, entail a high risk of wounding or death. Du Picq's and Keegan's lists of ways for soldiers to escape confronting the enemy should not surprise us. Human psychology is not well adapted to being a soldier. That is why the most successful armies are those that have most completely worn down the self-interested tendencies in their troops, whether through discipline or through inspiration. In war, claimed Napoleon, "three-quarters turns on personal character and relations; the balance of manpower and materials counts only for the remaining quarter."[49]

The reason that battles happen at all, of course, is that the top commanders insist on them. Commanders' motivations for aggression may be either proactive or reactive at any given point in time. A commander coolly planning a surprise attack with a specific goal in mind exhibits a purely proactive kind of aggression. Alexander the Great illustrates the possibilities. During the thirteen years from 336 to 323 BCE, Alexander led an army that conquered most of the Middle East, including the Persian Empire and kingdoms as far as western India. His substantial military actions include nine sieges, ten battles, and a major campaign. He never lost a fight. He inspired his troops by repeatedly fighting at the front and sometimes personally leading a charge. He was occasionally wounded, and when he died in bed in Babylon at the age of thirty-two, it is possible that an injury sustained in India contributed. Overall, however, his military ambitions were tuned with high precision to the power of his coalition.[50]

The political scientist Dominic Johnson and the mathematician Niall MacKay showed that the history of warfare is dominated by engagements that are similarly asymmetric, meaning that as in Alexander's battles, the attacking side is greatly superior to its opponent. Commanders who initiate aggression do so when their forces are overwhelmingly stronger than the enemy's, and as a result, until recently those who begin battles (or wars) have tended to win them.[51]

The success of commanders who take judicious initiative is easily understood in terms of an evolutionary history of proactive aggression, because selection should of course favor the ability to judge correctly the chance of achieving victory. Evolution is not the only theory that can explain routine victories, however. Commanders' triumphs could just as well be attributed to high intelligence. Evolutionary theory has more obvious value explaining military incompetence, a more surprising phenomenon.

"Military incompetence" was the term used by the psychologist Norman Dixon to refer to protagonists' losing even when they expect to win. Such losses are characteristic of battles in which the forces are relatively evenly balanced (as opposed to asymmetric attacks), but unexpected losses also appear in recent asymmetric warfare. The political scientist Ivan Arreguín-Toft tallied war victories from 1800 to 1998 according to whether one opponent was "stronger" than the other. To be judged as relatively "strong," a side had to have a material power at least ten times greater than the other. The chance of the stronger side's winning declined steadily from 88 percent before 1850 to 45 percent after 1950. Commanders are thus apparently no longer very good at anticipating victory. According to Johnson and MacKay, among other difficulties, the possibilities for counterinsurgency and guerrilla tactics have become too numerous.[52]

Military establishments of course want to eliminate failures in decision making. In an effort to understand factors affecting the chances of winning a battle, Dixon was given unparalleled access to a century's worth of British Army files from the 1853–56 Crimean War onward. He found four main symptoms of incompetence governing the outcome of battles: overconfidence, underestimation of the enemy, the ignoring of intelligence reports, and wastage of manpower.[53]

Groupthink, Dixon found, exacerbated the problem by contributing six additional symptoms: a shared illusion of invulnerability, collective attempts to maintain shaky but cherished assumptions, an unquestioned belief in the group's inherent morality, stereotyping the enemy as too evil for negotiation (or too weak to be a threat), a collective illusion of unanimity in a majority viewpoint (based on the false assumption that silence means consent), and self-appointed censors to protect the group from information that might weaken resolve (such as reports from spies).

The net result is that when forces are fairly equal in power, whether decisions are made by individuals or groups they are based on assessments by the attackers that commonly overestimate their own military strength and underestimate the strength of the opponent. Roughly half the time, the result is disastrous.

Take the Bay of Pigs. On April 17, 1961, a CIA-led brigade of 1,400 Cuban exiles was ordered by President John F. Kennedy to invade at Cuba's Bay of Pigs. They were defeated in three days by vastly superior forces. In retrospect, the decision to invade seemed extraordinary. Evidence of the strength of the Cuban armed forces had been abundant, and none supported the touted claims of an organized resistance of 30,000 who would "make their way through the Castro army and wade the swamps to rally to the liberators."[54] "How could I have been so stupid as to let them proceed?" President Kennedy repeatedly asked later.[55] Most of his team were similarly puzzled about their assessment failures.

According to Peter Wyden, whose 1979 book on the topic remains a classic, the answer was clear. It was arrogance, "egos so tall that the eyes and ears can shut out whatever one prefers not to see or hear."[56] Kennedy, the final decider, desperately wanted to avoid being called "chicken," had unbounded confidence in his own luck, and was surrounded by people who echoed his feelings. "Everyone around him thought he had the Midas touch and could not lose,"[57] wrote Arthur Schlesinger. Richard Bissell, the CIA deputy director of plans who pushed for the invasion, was such an ambitious and confident risk-taker, says Wyden, that he was unable to give up on his superman task in the face of increasingly clear evidence of risk. Even after the fiasco, Bissell clung to his view that they had done the right thing. The CIA's secret internal report on the affair, published in 1998, "painted a picture of an agency shot through with deadly self-deception."[58]

Delusions of this type are so routine in war that they are a mainstay of theories about military failure.

The first two delusions in Dixon's list, overconfidence and underestimation of the enemy, come from positive illusions.[59] The third, the ignoring of intelligence reports, allows the illusions to be maintained.[60] The fourth, the coldly stated "wastage of manpower," is the awful result. Armies pride themselves on their efficiency, and therefore might have been expected to have systems for ensuring accurate assessments of enemy strengths, but the opposite occurs: counter to commonsense intuitions that we might have, systems arise for ensuring inaccuracy. A core problem is the emergence of positive illusions. People overestimate the positive.

Tendencies for positive illusion occur not only in military interactions but also in intergovernmental relations. The historian Barbara Tuchman concluded that, regardless of place or period, governments routinely pursue policies contrary to their own material interests even when these policies are decided by a group, and even though feasible alternatives are available and openly discussed. Such tendencies, she found, were universal across three thousand years, unrelated to history or the type of political regime, nation, or class. They reflect "a rejection of reason" in the face of "ambition, anxiety, status-seeking, face-saving, illusions, self-delusions, fixed prejudices."[61]

Wherever groups compete, even without fighting, the same positively biased judgments recur. Mark Twain got it right: "Nations do not *think*," he wrote,

> they only *feel* . . . each nation *knowing* that it has the only true religion and the only sane system of government, each despising all the others, each an ass and not suspecting it, each proud of its presumed supremacy, each perfectly sure it is the pet of God, each with undoubting confidence summoning Him to take command in time of war, each surprised when He goes over to the enemy, but by habit able to excuse it and resume compliments—in a word, the whole human race content, always content, persistently content, indestructibly content, happy, thankful, proud, *no matter what its religion is, nor whether its master be tiger or house-cat.*[62]

The sense of confidence even extends to the general public. "Heroes, not horsetraders, are the idols of public opinion," wrote the international-relations theorist Hans Morgenthau,[63] so much so that the public sometimes drives policy as conflicts approach. The declaration of the First World War "was greeted with enormous popular enthusiasm in the capitals of all combatant countries,"[64] despite the statesmen's foreboding. "Sons of mine, I hear you thrilling / To the trumpet call of war,"[65] wrote British poet William Noel Hodgson in 1914, expressing a widely held sentiment.

Reality changes the perception. By the end of the war, after his son was killed, Rudyard Kipling wrote, "If any question why we died, / Tell them, because our fathers lied."[66] Kennedy recognized that he had lied to himself, and puzzled why. A victory over Castro would have been important in the Cold War era, and would have cemented his glory. But the attempt to damage a Soviet ally was too wild a gamble, relying on a misreading of Cuban spirit and ability. Its failure ended the Kennedy euphoria, lost Bissell his career, reduced U.S. credibility as a world leader, and, according to Che Guevara, handed a great political victory to Cuba, transforming it from an aggrieved little country into more of an equal.

The positive illusions that carried the United States into a war of their choosing were clearly a disaster. But they are typical of the thinking that tends to shape the approach to escalated battles, and they raise the evolutionary question: why do we have them?

The psychology of reactive aggression suggests an answer. Unlike asymmetrical encounters, the decision to go into battle against a roughly equal opponent is made by opposing commanders, each of whom expects serious resistance. In this context, confidence matters. The more confidence, the more likely it is that a combatant will win. The principle applies to animals as it does to humans. Two reasons stand out: focus and bluff.

First, the focus afforded by self-belief allows total commitment to winning. Courage, "the fixed resolve not to quit,"[67] beats caution every time. If, as Hamlet says, "the native hue of resolution / Is sicklied o'er with the pale cast of thought, / . . . enterprises of great pith and moment / With this regard their currents turn away, / And lose the name of action."[68] In an even match, a rational opponent who thinks like Hamlet would accurately perceive himself as having a 50 percent probability of defeat. Accordingly, he might think about protecting himself in the event of failure, whether by devising escape plans, avoiding damage, or attempting to reassess the opponents' strength.

Such attention to the possibility of loss would lead to anxiety (a sure predictor of defeat) and, more generally, to distraction. So, because in a contest between equals 100 percent effort beats 90 percent effort, arrogant blind confidence will predict the winner. "Championship thinking" is irrational and wasteful and half the time deluded, but in an even match, it brings more psychological resources to the fight and increases the chance of winning.

Self-confidence can involve beliefs that are patently irrational to the nonbeliever. In Borneo in 1997, long-simmering tension between native Dayaks and immigrant Madurese erupted into conflict. The Dayaks believed that their magic made them invulnerable to bullets. This made them fearless, and therefore particularly effective. Such self-deception is common in battles. In the eastern Congo in the 1990s, the Mai-Mai fighters believed that bullets hitting them would turn to water. In Uganda in the 1980s, Alice Lakwena's rebel groups were ferocious fighters because they thought themselves safe from bullets. A belief that one is magically protected from harm is a wonderful illusion for inspiring unrestrained aggression against the enemy. It works particularly well for commanders who are not at the battlefront.

The second benefit of self-confidence is that it can create fear in the enemy; often a good bluff is sufficient. The Dayaks inspired fear in various ways. Tales of their cannibalism and parading of severed heads meant that "they were possessed, they weren't acting normally," and their enemies were duly scared.[69]

By giving focus and promoting a successful bluff, positive illusions help people win. Their false confidence can also have collateral benefits, such as inspiring confidence in potentially wavering allies, but their key adaptive feature is the promotion of victory.

In the grand scheme of life, it is an irony of nature that selection for the ability to win brings with it a failure of assessment, and resulting "wastage of manpower." The great poet and dictionary-writer Samuel Johnson understood how absurd it was if both opponents in a confrontation believed that they would win. "Of victory, indeed, every nation is confident before the sword is drawn; and this mutual confidence produces that wantonness of bloodshed, that has so often desolated the world. But it is evident, that of contradictory opinions, one must be wrong." Johnson expressed his view in order to persuade England not to go to war with Spain. It is a reminder that could be useful when conflicts occur today. Unfortunately, the reminder that opponents are vulnerable to self-deception about how easily they will win comes up against the fact that they are selected to ignore such advice.[70]

The role of self-confidence in fighting is probably much the same in animals as it is in humans. In a relatively even contest, individuals on both sides have to be committed to winning. Losers therefore pay higher costs than a rational analysis might predict. The problem with the incorporation of overconfidence into human warfare is that, above the military horizon, a leader's overconfidence can have disastrous effects for the soldiers who have been coerced into fighting.

Complex warfare, in sum, involves soldiers whose evolutionary psychology very often tells them not to participate, to fight battles organized by commanders whose evolutionary psychology leads them to an extravagant commitment of resources, using coercive tactics honed by coalitionary proactive aggression. The result is bloodshed beyond what would be predicted were the system designed for mutual benefit. Unfortunately, natural selection favors mechanisms that lead to winning; and they include the positive illusions that exacerbate the wastage of war.

Complex war does not depend on "an ungovernable drive for aggression," in Erich Fromm's phrase, or on a joy of killing, or an outside source of evil. It is a complicated result of the interaction of propensities to use proactive and reactive aggression. Very often soldiers have to fight despite their desire not to do so. Most important, complex war is made possible by our refined ability for coalitionary proactive aggression, which contributes obedience and hierarchy. It is also driven partly by ego; and as crises loom, players become increasingly poor at rational assessment, for reasons forged in the evolution of reactive aggression.

This understanding is a mere preliminary sketch of some initial relationships between evolutionary psychology and military action. The aim has been to imagine how a view of war psychology can be illuminated by simpler models from behavioral biology; and to emphasize how, above the military horizon, some of our evolutionary adaptations for aggression can perversely reduce, rather than increase, the chances of winning at war, when precise assessment really matters.

The aim of this book has been to understand better how evolution has shaped humanity into the best and worst of species. The aim was not to say how the story ends.

We can at least limit the pessimism, however. As I emphasized earlier, an account of how aggression was adaptive in the Pleistocene does not lead to the conclusion that wars will continue in the Anthropocene.

The evidence is overwhelming that there has been a long-term decline in the proportion of deaths that are due to violence. Among other reasons, societies have grown larger over time; and in larger societies, a lower proportion of the population is directly involved in war. The decline is understandable. People work hard to make themselves safer.[71]

How long, and how completely, the decline will continue is an open question. At the end of the Pleistocene, just before the beginning of agriculture, *Homo sapiens* occupied most of the world, either as mobile or settled hunter-gatherers. At the time, there were probably some tens of thousands of different societies, perhaps around thirty-six thousand,[72] each with sovereignty over its home area. Because all men were hunters and potential combatants the opportunities for violent death in interactions between societies were legion. Today there are 195 nations, within each of which the state takes responsibility for controlling violence. As the number of independent societies has declined, so has the frequency of wars. Unfortunately recent data indicate that the longer the time spent in peace, the more casualties tend to result when war eventually breaks out.[73] Still, other things being equal, the probability of dying from violence should continue to decrease if the average size of nations continues to increase. In the distant future humanity could become a single nation: extrapolations from past trends suggest a date between 2300 CE and 3500 CE for a World State to be established.[74] A World State could then be expected to minimize the rate of death from anarchic violence, though the possibilities of tyranny could enable other kinds of homicide to flourish.

On the other hand, as long as the number of nations remains stable or increases, continuing intense efforts will be needed to regulate international relations. The challenge is hard. In 1928, leaders of sixty-two nations pledged not to resort to war as an instrument of policy. The Kellogg-Briand Pact that they signed was not perfect. It failed to prevent Japan from military expansion into China in 1931. It did not stop the aggressive nationalism of Germany or Italy that led to the Second World War. Within a decade, every signatory except Ireland was at war. Other wars involving member nations included the Korean War, the Arab-Israeli conflict, the Indo-Pakistani wars, the Vietnam War, the Yugoslav Civil War, and the wars in Syria and Yemen. Yet, despite all these failures, and in the face of much skepticism, the pact was in fact a success according to the law scholars Oona Hathaway and Scott Shapiro, because it changed the rules of war. Between 1816 and 1928, most wars had been fought to acquire territory. Such wars of conquest became illegal under the Kellogg-Briand Pact. As a result, annexation of territory became rarer, and nations turned increasingly to trade.[75]

There will be bumps along the road, but if international law is pursued with sufficient vigor and cunning it at least has the potential to avert catastrophe. The more challenging difficulty for our species is that, as resource distributions change, new coalitions are expected to form repeatedly to challenge existing sovereignties. All human societies are composed of competing subgroups. Some subgroups will ignore existing laws by attempting to carve their own territories out of prior nations, as ISIS did in Iraq in 2014. Handling such efforts nonviolently will surely always be a challenge. The global response to ISIS illustrates the intensity of violence that can readily emerge when a new ideology defies existing mores.

Blind optimism about the prospects for a decline in the frequency of war would therefore be as foolish as apathetic pessimism. Our species swings between the desire for peace and the temptations of power, and faces the contradiction that although the risk of dying from violence has fallen, the risk of nuclear holocaust has risen. The great merit of proactive aggression, from the perspective of a nonviolent philosophy, is that a well-adapted animal does not attack if it expects to get hurt.[76] A good defense should be a good deterrent—as long, of course, as it is not so effective that it tempts its possessors into safely attacking rivals.[77]

The idea that warfare has evolved, and that even today it is facilitated by adaptive features of our psychology, does not make it inevitable. It does mean, however, that we are a dangerous species. In the face of our tendency for positive illusions about the merits of war, we will always need strong institutions and alert engagement to temper the rise of militaristic philosophies, the spread of excessively optimistic pacificism, and the abuse of power.

13

Paradox Lost

ROUSSEAUIANS SAY WE are a naturally peaceful species corrupted by society. Hobbesians see us as a naturally violent species civilized by society. Both perspectives make sense. To say that we are both "naturally peaceful" and "naturally violent," however, can seem contradictory. The mismatch created by that combination represents the paradox at the heart of this book.

The paradox is resolved if we recognize that human nature is a chimera. The Chimera, in classical mythology, was a creature with the body of a goat and the head of a lion. It was neither one thing nor the other: it was both. The thesis of this book is that, with respect to our tendency for aggression, a human being is both a goat and a lion. We have a low propensity for reactive aggression, and a high propensity for proactive aggression. The solution makes Rousseauians and Hobbesians both partially right, and raises the two questions I have discussed: why did this unusual combination evolve, and how does the answer contribute to understanding ourselves?[1]

First, what evolutionary stimulus pushed human aggression into two contrasting directions, decreasing reactive aggression and increasing proactive aggression?

To judge from the few relevant species, a high propensity for coalitionary proactive aggression is normally associated with a similarly high propensity for reactive aggression. Chimpanzees are the primate species that most often uses proactive aggression to kill adults, and they also have a high rate of reactive aggression within communities. Wolves are the species of carnivore whose proactive aggression against members of their own species is best known to be frequently lethal. Even though, as with chimpanzees, relationships within wolf groups are generally benign and cooperative, they are not so calm as they are in dogs. Lions and spotted hyenas are also wolflike in these respects. In those species, proactive and reactive aggression seem to occur at roughly parallel high levels.[2]

Something different happened in the human lineage. Reactive aggression became suppressed, while proactive aggression stayed high. According to the evidence in this book, our propensity for reactive aggression fell due to a process of self-domestication that started certainly by 200,000 years ago, and possibly with the first glimmerings of *Homo sapiens* a little more than 300,000 years ago. Language-based conspiracy was the key, because it gave whispering beta males the power to join forces to kill alpha-male bullies. As happens in small-scale societies today, language allowed underdogs to agree on a plan, and thereby to make predictably safe murders out of confrontations that could otherwise have been dangerous. Genetic selection against the propensity for reactive aggression was an unforeseen result of eliminating the would-be despots. The selection against alpha personalities led to males becoming, for the first time, egalitarian. Across some twelve thousand generations, the tenor of life became increasingly calm. Our species, though not ideally peaceful, is now more Rousseauian than it has ever been.

Blumenbach called us "the most completely domesticated animal," but there is no reason to regard our domestication as complete. How much more domesticated we could become if we were tamed for another twelve thousand generations, say, is an open question. Given sufficient sanctions against reactive aggressors, in another 300,000 years humans could in theory become as hard to rile as lop-eared rabbits at a petting farm, which remain gentle even when stroked repeatedly by dozens of eager children. Equally, however, if would-be despots were to escape sanctions, the process could go into reverse. The relationship between the propensity for reactive aggression and reproductive success will continue to depend on power inequities, but how power will be distributed, and what effects its distribution will have on reproduction, depend on too many unknowns to allow us to forecast how our emotions will evolve.[3]

The same ability to perform capital punishment that led to self-domestication also created the moral senses. In the past, to be a nonconformist, to offend community standards, or to gain a reputation for being mean became dangerous adventures; this is still true today, to some extent. Rule breakers threatened the interests of the elders, so they risked being ostracized as outsiders, sorcerers, or witches. Execution could follow. Selection accordingly favored the evolution of emotional responses that led individuals to feel and display unity with the group. Conforming was vitally important for everyone.

The moral senses of individuals thus evolved to be self-protective to a degree not shown by other primates. The strongly conformist behaviors produced by the new tendencies provided a safe passage through life, and they had a second effect as well. By reducing competition and promoting respect for the interests of others, individual conformity brought benefits to the group of moral enforcers and their supporters. This process seems to explain why humans show unexpectedly high concern for the welfare of their own groups. Group selection is commonly invoked to explain our species's interest in nonrelatives and our occasional willingness to sacrifice our own interests on behalf of a larger good.[4] Group selection theory, however, has never quite been able to explain how benefits at a group level override those of individuals.[5] The theory that the moral senses evolved to protect individuals from the socially powerful suggests that group selection may be entirely unnecessary for explaining why we are such a group-oriented species. Our deference to the coalitionary powers within our own groups leads to a reduced intensity of competition, enabling groups to thrive.

As for proactive aggression, according to the reconstruction sketched in the preceding chapters, a predisposition for premeditated violence was in place in our *Homo* ancestors by at least 300,000 years ago. How much earlier it was present is not marked by anything so concrete as the domestication syndrome. Based on inferring the behavior of our ancestors, however, a high propensity for coalitionary proactive aggression probably operated at least through the 2.5 million years of the Pleistocene, and possibly earlier.

The reason for this claim is the antiquity of hunting. *Homo erectus*, the first ancestor of *Homo sapiens* that was committed, like us, to living on the ground, evolved around two million years ago. Cut marks that *Homo erectus* left on meat-bearing bones show that they butchered animals the size of large antelope. By one million years ago, ambush hunting is suggested (humans repeatedly reused a site of that age at Olorgesailie, Kenya, a place where animal prey were limited to narrow travel routes and could therefore be killed easily); this, too, implies cooperation. Even stronger evidence of hunting of large deer and bovids by individuals based in a residential camp comes from Gesher Benot Ya'aqov about 800,000 years ago. However, only with *Homo sapiens* and Neanderthals, in the last few hundred thousand years, have we found sufficient evidence that *Homo*'s hunting had clearly become premeditated: using projectile points, catching small animals apparently by setting snares, and hunting from elevated positions. A conservative interpretation therefore might limit proactive hunting to the Mid-Pleistocene, but ambush hunting is still a plausible explanation for how *Homo* obtained much of their animal food as early as two million years ago.[6]

After our ancestors became good hunters, they could have killed strangers; hunting is a transferable skill. Hunting and simple war both require searching and safe dispatching, and both benefit from long-distance travel and well-honed coordination. Wolves, lions, and spotted hyenas use coalitionary proactive aggression not only to get food but also to kill rivals in other groups. Chimpanzees are social hunters that are likewise killers of their own species. Bonobos, by contrast, are not known to be social hunters (despite their liking for meat), and to date they have not shown clear evidence of planned aggression. Among humans living in small-scale societies, the anthropologist Keith Otterbein found a similar association to that among social carnivores: societies relying more on hunting tended to have more frequent war. The same correlation between hunting for food and killing the competition is found in the neural pathways of aggression in rats and mice. For all these reasons, human hunting of prey seems likely to have been associated with the ability to kill rivals in neighboring groups for two million years. Much as in chimpanzees and wolves seeking opportunities to attack strangers, once our ancestors had achieved the ability to kill safely a motivation to kill would probably have been present too. There seems no reason to excuse our ancestors from the links between hunting and violence found in other mammals.[7]

Dale Peterson and I have argued that the killing of strangers probably went back to our common lineage with chimpanzees and bonobos, when our Central African ape ancestor was most probably a chimpanzee-like hunter and killer.[8] The evidence is admittedly inferential, given that no fossils have confirmed the nature of the last common ancestor. The uncertainty about when the propensity to kill strangers evolved gets more acute during our ancestors' long era as Australopithecines, from about 7 million years ago to around 2.5 million years ago.[9] We have little basis for reconstructing our australopithecine ancestors' social behavior or organization during that period.

Regardless of when coalitionary proactive aggression began against strangers, the impact of such killing within groups was limited until humans' development of language. Much changed after individuals became able to share ideas with one another. People could then form alliances based on shared interests that they could articulate. With the arrival of planned and communally approved executions, the bullying of an alpha male was exchanged for the subtler tyranny of the previous underdogs. The newly powerful coalitions of males became the set of elders who would rule society—a system that largely continues today, albeit with more laws, threats, and imprisonment than execution.

Both our "angelic" and "demonic" tendencies, therefore, depended for their evolution on the sophisticated forms of shared intentionality made possible by language—an ability that undoubtedly also contributed to much prosocial behavior. A chimpanzee-style form of shared intentionality launched the process at least seven million years ago. It took the mysterious dawning of a language facility, sometime between 500,000 and 300,000 years ago, to shake us into a new world.

Language created our chimeric personality in which high killing power lies alongside reduced emotional reactivity. A unique communicative ability gave us a uniquely contradictory psychology of aggression.

The second question provoked by seeing both our good and our bad sides as being rooted in biology concerns our sense of self: what does the resolution of the goodness paradox do for our understanding of our nature?

The argument that human nature is a chimera is challenging because it is hard to hold two superficially contradictory concepts in mind at the same time. It is easy to think, as the Hobbesians and Rousseauians have wrongly argued, that only one side of our species's split personality is embedded in our biology; and if so, many people would find it emotionally easier to imagine that only our "good" side, our low reactive aggression, is the product of evolution. Nevertheless our "bad" side, in the form of the high proactive aggression that has often been responsible for deeds of evil, needs also to be charged to our evolutionary past. To understand what this means for contemplating the human future, I believe it helps to remember two things about evolution.

First, as I have emphasized, an evolutionary history is an account of the past; it is not predictive, does not say what kind of future awaits. Nor is it a political platform, or a justification for an ethical stance, or a recommendation that we return to some imagined delightful past. It does not change what we already know about the power of humans to adapt. It is just a story.

By "just a story" I do not mean to diminish its power as a cosmological narrative. Evolutionary stories could hardly be more captivating. It is astounding to learn that we ultimately derive from simple chemicals that became arranged into complex molecular patterns some four billion years ago, leading first to cells, then to animals, mammals, primates, apes, humans, and eventually *Homo sapiens.* The science of evolutionary biology still has gaps and uncertainties, but it gets more robust and exciting with every decade. The essentials will not change. Out of nonlife, life! Out of instincts, consciousness. Out of materialist brains, spirituality and laughter and joy and an understanding of the meaning of life. Out of darkness, a species that sees itself for what it is, a glint of mentality in a vast, mostly sterile universe.

So, when I say "just a story," I do not mean for a second to downplay the grandeur of an evolutionary perspective. I mean merely that it is a story without prescriptions and with few limits for the future. The social systems that we see today are mostly very different from those that existed a few hundred years ago. The capacity for social change is obvious. The system of nations that has been in place since the Treaty of Westphalia in 1648 might feel permanent, but it has already started to change, and anything is possible in the future. History is far more important than evolutionary theorizing as a reminder about human potential, because the historical evidence of change is so much more vivid. We know that over time society sometimes improves in quality, and sometimes decays. What we cannot know is which direction our descendants will take.

My second point, despite the open nature of the future, is that evolution has left us with biases that affect our behavior in predictable and sometimes disturbing ways, and we would do well to acknowledge those biases.

The great problem with the purest Rousseauian visions is that they are easily interpreted as implying that a state of anarchy would be peaceful. Take away capitalism, patriarchy, colonialism, racism, sexism, and other evils of the modern world, they seem to suggest, and an ideal society of love and harmony will emerge. The idea that humans have evolved merely with a Rousseauian tolerance, and not also with a Hobbesian selfishness, is problematic to the extent that it encourages people to let their guard down.

Consider the relationship between men and women. In small-scale societies, as I discussed, egalitarianism is primarily a description of relationships among men, particularly married men. Just as happens in every society throughout the world, in the public sphere men dominate women. This observation says nothing about the private sphere. Within marriages, wives often dominate their husbands. Personality is the strongest influence, but in a significant number of marriages women also use physical force to bully their men. In the public domain, however, where coercive alliances regulate societal rules, conflicts between the interests of men and women consistently end in men's favor. Patriarchy in this sense is currently a human universal.[10]

No evolutionary rule says that society has to stay this way, however. Recent political changes in Rwanda and Scandinavia show that the conventional tradition of legislative bodies numerically dominated by males can be overturned. Similar changes are possible at every level.

But they will not happen easily. It takes positive action and elaborate organization to make sure that such changes actually occur. They will not happen if we simply create anarchy—in other words, create a society without rules. Destroy the old institutions without replacing them, and violence will predictably emerge. Men will rapidly use alliances to compete for dominance: militias will bloom and fight. Male groups can confidently be predicted to use their physical power of coalitionary proactive aggression to dominate in the public sphere. History and evolutionary anthropology tell the same sad story.

The more general evolutionary lesson from an understanding of the human trajectory is that groups and individuals will always be interested in vying for power. They will not necessarily go to war. There will not necessarily always be patriarchy, bullying in schools, sexual harassment, street crime, or the flexing of power by those at the top for financial gain. Equitable and violence-free social arrangements are entirely possible, perhaps much more in the future even than in Iceland or other relatively egalitarian and peaceful countries of the present time.

The one guarantee that an evolutionary analysis can offer, however, is that it will not be easy for fairer and more peaceful societies to emerge. They will take work and planning and cooperation. Mobile hunter-gatherers had a system for protecting themselves against deviants and bullies. Every society has to find its own protection. To avert episodes of violence we should constantly remind ourselves of how easily a complex social organization can decay, and how hard it is to construct.

—

On a sunny day in July 2017, surrounded by well-fed people in casual summer dress, I walked around Auschwitz. I could feel the chimera at its best and worst.

Cooperation and prosocial feeling filled the air. I had come with a small group of tourists whom I had met in Kraków that morning. The camp was so full that at times we had to wait for a few minutes to be ushered into the next location. Everyone was patient, quietly chatting.

We saw where a camp orchestra had played. The music helped to keep prisoners in step, for easy counting. We saw the block where hundreds of women had been held for sterilization experiments from 1943 to 1944. We saw the courtyard between Blocks 10 and 11, where thousands were shot for clandestine activities, and others were flogged or hanged, after they had obediently removed their clothes. We crowded into the cramped chamber where up to two thousand naked victims at a time were gassed with Zyklon B. We saw the nicely treed garden surrounding the house where the first commandant of Auschwitz, Rudolf Höss, lived with his wife and children a few yards from the prisoners' blocks. We saw the gallows where Höss was hanged. In the parking lot, original handmade models of the Auschwitz and Birkenau gates were available from smiling traders.

So much cooperation. We sometimes think that cooperation is always a worthwhile goal. But, just like morality, it can be for good or bad.

The important human quest should not be to promote cooperation. That goal is relatively simple and firmly founded on our self-domestication and moral senses. The harder challenge is reducing our capacity for organized violence.

We have started on the process, but there is a long way to go.

Afterword

THE COMPLEXITIES THAT characterize human behavior are paralleled by the question of the moral attitudes that we adopt toward capital punishment. Capital punishment is a hot-button issue in the United States (where I live). Some are passionately in favor of it; a larger number are against it. The theory presented in the preceding pages suggests that our ancestors inadvertently created a more peaceful version of themselves partly by killing the most aggressive males. This means that capital punishment is a natural behavior that produced morally attractive results. Does that imply any societal recommendations? Does it mean we should embrace executions as a way to improve society?

My answer is a resolute no. Whatever the contributions of capital punishment once were, they are irrelevant to the question of justifying its use today. The administration of the death penalty by the powers of the state is very different from its use in small communities. Consensus is no longer required, nor is the killing performed, as it often was, by close relatives. Conditions have changed: prisons offer alternative forms of social control that our ancestors did not have. I believe that judicial execution is an outmoded punishment that should no longer have a place in the world. Capital punishment is generally found to be ineffective, in that it does not lead to a reduction in crime. It is more expensive for society than imprisonment. In some countries, such as the United States, it is strikingly unjust, because it targets the poor and disadvantaged. And it makes mistakes: innocent people have often been put to death. We can understand our past, but in this respect we should not admire it. The evidence that capital punishment has had a long and creative prehistory is irrelevant to contemporary societal questions.

The world is increasingly agreed that capital punishment belongs to the past. In December 2007, 104 member states of the United Nations voted to adopt the principle that "the use of the death penalty undermines human dignity" and to call "upon all States that still maintain the death penalty to establish a moratorium on executions with a view to abolish the death penalty." This international resolution was voted on again in 2008, 2010, 2012, and 2014. Each time, the number of votes in support rose. In December 2014, there were 117 votes in favor, 38 against, 34 abstentions and 4 absent. But in December 2016, the numbers were similar, and disappointingly unimproved. There were again 117 votes in favor, with 40 against, 31 abstentions, and 5 absent.[1]

I hope that, very soon, every country will abolish capital punishment, just as most countries have outlawed other ancestral behaviors such as cannibalism, slavery, and marital rape. Whether something is natural says nothing about whether we should give it a place in our lives today. In the 1951 movie *The African Queen,* Katharine Hepburn's character had the right idea when she reprimanded the unsophisticated Charlie Allnut, played by Humphrey Bogart, for his crude behavior: "Nature, Mr. Allnut, is what we are put in this world to rise above."

Nevertheless, we can appreciate capital punishment for what it did. Until recently, it has been celebrated for the wrong reasons. The consequences of socially approved killing have been too stealthy to be obvious, so its appeal was limited to the base instincts of the mobs and the tyrants. But if we step back and cherish our rough past, we can thank our cruel ancestors for making us *sapiens.* Ironically, executioners seem to have brought us to the beginning of wisdom.

Acknowledgments

First and foremost, I have been especially fortunate to be mentored throughout my career by David Hamburg and the late Robert Hinde, equal fonts of wise counsel about the evolution of violence. Their ability to combine scholarly insights with humane, practical perspectives continues to make them lasting role models.

With regard to self-domestication, many ideas described here began in the late 1990s in conversations about apes with Brian Hare, whose beautiful experiments and iconoclastic ideas continue as a source of inspiration to the present. David Pilbeam taught me how to see the wood for the trees in paleoanthropology. Victoria Wobber was a brilliant student whose painstaking research pushed forward our understanding of the effects of domestication. Natalie Ignacio responded superbly to the challenge of working in Novosibirsk. Adam Wilkins and Tecumseh Fitch have been wonderful colleagues with whom to explore the role of neural-crest cell migration. Christopher Boehm continues to lead the way in thinking about the evolution and control of male-male relationships, and has been endlessly generous in sharing thoughts and discoveries.

With regard to aggression, I have been lucky to have Luke Glowacki, Martin Muller, and Michael Wilson as collaborators in comparisons of chimpanzees and humans; and by sharing her psychological experiments on human aggression, Joyce Benenson introduced me to a rich new world. The acute scholarship of all these friends has sharpened my understanding at every turn. It has been a privilege to work with them.

Jane Goodall first gave me the opportunity to study chimpanzee behavior and continues to be inspiring. Daniel Lieberman has been a mine of information, ideas, and cautions about human evolution. I thank them as well as Terry Capellini, Rachel Carmody, Peter Ellison, Joe Henrich, Maryellen Ruvolo, and Noreen Tuross for their multiple kinds of biological advice.

More than anyone, Anne McGuire helped me reach the finishing line. In addition to reviewing every chapter in detail Anne was a constant source of deep thinking about the book as a whole. I could not be more grateful for her extraordinary contributions and support.

For comments on the first draft, I greatly appreciate Joyce Benenson, Tommy Flint, Chet Kamin, Daniel Lieberman, Martin Muller, David Pilbeam, Manvir Singh, and Adam Wilkins. Their efforts have substantially improved the book. I only wish that I had been able to respond to all their observations. For reviewing individual chapters or sections, I similarly thank Ofer Bar-Yosef, Christopher Boehm, Fiery Cushman, Madeleine Geiger, Marc Hauser, Karl Heider, Rose McDermott, Dale Peterson, Matt Ridley, Kate Ross, John Shea, Barbara Smuts, Ian Wrangham, and Christoph Zollikofer.

For advice on specific points, I am grateful to Johan van der Dennen, Paul Crook, Sylvia Kaiser, Steven Pinker, and Adrian Raine. For sharing unpublished data, I thank Cat Hobaiter, Nicole Simmons, Martin Surbeck, and Michael Wilson.

In addition to those mentioned above, conversations and correspondence with numerous other friends, family, and colleagues over the years have been critically helpful. My benefactors include Bridget Alex, Adam Arcadi, Robert Bailey, Isobel Behncke, Alex Byrne, Rachel Carmody, Napoleon Chagnon, Richard Connor, Meg Crofoot, Lee Dugatkin, Melissa Emery Thompson, Lee Gans, Sergey Gavrilets, Alexander Georgiev, Ian Gilby, Tony Goldberg, Joshua Goldstein, Stephen Greenblatt, Stewart Halperin, Henry Harrison, Kim Hill, Carole Hooven, the late Gabriel Horn, Nick Humphrey, Kevin Hunt, Carrie Hunter, Dominic Johnson, James Holland Jones, Jerome Kagan, Ewa Lajer-Burcharth, Kevin Langergraber, Steven LeBlanc, Richard Lee,

Zarin Machanda, Curtis Marean, Katherine McAuliffe, John Mitani, Mark Moffett, Michael Moran, Randolph Nesse, Graham Noblit, Kate Nowak, Nadine Peacock, Anne Pusey, Vernon Reynolds, Neil Roach, Lars Rodseth, Diane Rosenfeld, Elizabeth Ross, Graham Ross, Peter DeScioli, Lyudmila Trut, Carel van Schaik, Michael Tomasello, Robert Trivers, Vivek Venkataraman, Ian Wallace, Felix Warneken, David Watts, Polly Wiessner, Kipling Williams, David Sloan Wilson, Carol Worthman, David Wrangham, Ross Wrangham, Brazey de Zalduondo, and Bill Zimmerman.

Observations of chimpanzees in Kibale National Park, Uganda, and Gombe National Park, Tanzania, have enriched my understanding of human evolution. For codirecting the Kibale Chimpanzee Project I thank Martin Muller, Melissa Emery Thompson, and Zarin Machanda. For the financial support that made the Kibale studies possible, I am grateful to the National Science Foundation, National Institutes of Health, Leakey Foundation, National Geographic Society, MacArthur Foundation, and Getty Foundation.

It is a pleasure to thank Jeremy Bloxham for support at Harvard, Katinka Matson and John Brockman for their superb agency, Andrew Franklin for his help and backing at Profile Books, and Erroll MacDonald, Nicholas Thompson, and Terry Zaroff-Evans at Pantheon Books for shepherding this book to completion.

Most of all, I thank Elizabeth for her sharing of the journey, including her good humor at discovering that this book took three times longer to write than it should have done.

Notes

Introduction: Virtue and Violence in Human Evolution

1. Diary entry, May 1, 1958, in Payn and Morley, eds., 1982.
2. Although Rousseau has become the icon for human nature's being inherently nonviolent, in actuality he did not hold this view. See chap. 1, n. 10.
3. Dobzhansky 1973, p. 125.
4. Huxley 1863, p. 151.
5. Barash 2003, p. 513.
6. Kelly 1995, p. 337.
7. Darwin 1872, p. 1266.
8. Fitzgerald 1936, pp. 69, 70.

1. The Paradox

1. Bailey 1991; Grinker 1994.
2. Heider was accompanied by a small team (Heider 1972). Robert Gardner was a veteran filmmaker who stayed for five months. His resulting movie, *Dead Birds,* is perhaps the fullest visual account ever made of primitive war, a classic still very popular in anthropology classes. Michael Rockefeller was the sound recorder. He died in November 1961 on the Asmat coast, apparently killed by tribesmen (Hoffman 2014). See chapter 8, p. 145.
3. The figure of one hundred million deaths in the twentieth century includes those caused by famine and disease, and is one of the higher estimates. Keeley 1996, p. 93, extrapolates to 2 billion.
4. Heider 1997.
5. Barth 1975, p. 175.
6. Glasse 1968, p. 23.
7. Chagnon 1997.

8. Shermer 2004, p. 89.
9. Hess et al. 2010.
10. Lescarbot 1609, p. 264, cited by Ellingson 2001, p. 29. According to Ellingson, Rousseau did not attribute any natural moral goodness to living native peoples. Rousseau thought that even in an earlier time, when people supposedly lived without social groups, people would have been "savage . . . possessing few of the qualities that that would come with the advance toward civilization" (Ellingson 2001, p. 82). Rousseau was so far from accepting the idea of natural goodness that he criticized his contemporaries for deluding themselves that humans had once lived in a golden age of peace—ironically, much as contemporary Rousseauians are themselves criticized. Ellingson 2001, p. 22, reported that the phrase "Noble Savage" comes from Lescarbot 1609, who wrote, "The Savages are truly Noble." Lescarbot meant that, since every male hunter-gatherer hunted, and since in Europe hunting was a practice of the nobility, "savages" could be called noble. The association of natural goodness with Rousseau seems to have been made in 1861 by John Crawfurd, president of the Ethnological Society of London. The anthropologists E. B. Tylor and Franz Boas accepted Crawfurd's erroneous attribution. For example, Boas 1904, p. 514, referred to Rousseau's "naïve assumption of an ideal natural state which we ought to try to regain." Ever since, Rousseau has become a false icon for the belief in an ancient natural goodness that was subsequently corrupted. In this book, I use the term "Rousseauian" to refer to the popular view of Rousseau's ideas (i.e., that hunter-gatherers lived in a golden age of peace), rather than how Rousseau actually conceived of human nature.
11. Chinard 1931, p. 71, cited by Ellingson 2001, p. 65.
12. Davie 1929, p. 18.
13. Orwell 1938, chap. 14.
14. Pinker 2011; Goldstein 2012; Oka et al. 2017. The strong evidence that rates of violence have declined says little about the future. The power of modern weapons, the risk of the accidental use of nuclear bombs, and the trend for longer periods of peace to be followed by more violent wars are reminders that nothing can be taken for granted (Falk and Hildebolt 2017).
15. Wrangham et al. 2006.
16. Surbeck et al. 2012.
17. Shostak 1981. Shostak, a professional anthropologist, gave voice to a !Kung woman, Nisa, and set Nisa's life experience in the context of !Kung society generally. Shostak presented no quantitative data, but Nisa's account makes clear that !Kung women experienced levels of physical abuse that would be intolerable to women in a Western democratic society.
18. García-Moreno et al. 2005. Remarkably, most women who were surveyed thought that male violence against female partners was often justified—for

instance, if a woman went out without telling him, neglected the children, or failed to prepare his food. The proportion of women who thought that their men's action in beating them was sometimes justified ranged from 74 percent (Thailand) to 94 percent (Ethiopia). Even in North America, where only 1.5 to 3.0 percent of women reported one or more acts of violence in the previous twelve months, the frequency of intimate-partner violence is high enough to warrant intense efforts to reduce it. The same index tends to be much higher in poorer, more marginalized, and otherwise disadvantaged groups: in rural areas of Thailand, Tanzania, Peru, and Ethiopia it was 22 to 54 percent. A subsequent WHO study surveyed Bangladesh, Brazil, Ethiopia, Japan, Namibia, Peru, Samoa, Serbia-and-Montenegro, Tanzania, and Thailand (Pallitto and García-Moreno 2013).

19. U.S. data are from Black et al. 2011, based on 9,086 interviews conducted nationwide in 2010.
20. Pallitto and García-Moreno 2013, p. 2.
21. García-Moreno et al. 2013.
22. Goodall 1986.
23. Surbeck et al. 2012, 2013, 2015, and Surbeck, personal communication.
24. Herdt 1987.
25. Chagnon 1997.
26. Malone 2014.
27. Stearns 2011.
28. Keeley 1996.
29. Lee (2014) challenges the significance of Keeley's data.

2. Two Types of Aggression

1. Mashour et al. 2005, p. 412.
2. Wrangham 2018; Babcock et al. 2014; Teten Tharp et al. 2011.
3. Carré, et al. *Psychoneuroendocrinology* 36: 935–44.
4. See https://www.theguardian.com/uk-news/2016/mar/07/bailey-gwynne-trial-boy-16-guilty-culpable-homicide.
5. Wolfgang 1958. See Polk 1995 for an Australian example.
6. Craig and Halton 2009; Siegel and Victoroff 2009.
7. Byers 1997.
8. Clutton-Brock et al. 1982. Xu et al. (2016) report 2013 statistics on U.S. death rates. The annual total of U.S. male deaths was 1,306,034, of which 12,726 were homicides (table 12, p. 52)—i.e., a rate of 0.97 percent, (or 8.2 per 100,000 living males; see table 14, p. 64). Many of these homicides would have resulted from interactions other than character contests, including premeditated killings, spousal violence, and killing of infants. But even if we guess that character contests accounted for half of the male homicides

(likely an overestimate), such that 0.5 percent of deaths per year came from this form of reactive aggression, it is at least twenty times less than the 10 percent or more of male deaths from male-male contests found among red deer and pronghorn.

9. See ews.bbc.co.uk/1/hi/england/nottinghamshire/8034687.stm.
10. Hrdy (2009), pp. 3–4. Peterson 2011, p. 113, commented: "Before every commercial airlines flight, passengers are officially computerized, identified, interrogated, numbered, checked, x-rayed, metal-detected, searched, rechecked, organized, checked a third time, then buckled into an assigned seat and instructed not to move during the most critical times of the flight."
11. Craig and Halton 2009; Siegel and Victoroff 2009; Weinshenker and Siegel 2002.
12. Brookman 2015.
13. Wolfgang and Ferracuti (1967), p. 189.
14. Unclassified homicides: Wolfgang (1958).
15. Revenge killings: Brookman (2003).
16. Proportion of unsolved murders in U.S. more than 35 percent: Brookman (2015), footnote 2, data from the Federal Bureau of Investigation.
17. "Altercations appeared . . ." quote: van der Dennen (2006), p. 332, citing Mulvihill et al. (1969), p. 230. Van der Dennen (2006) reviewed five studies concluding that most murders are reactive (strictly, impulsive).
18. "Tempers flare . . ." quote: van der Dennen (2006), p. 332, citing Mulvihill et al. (1969), p. 230. Juvenile non-lethal aggression seems to follow the same pattern as adult homicide. Among children a relatively high frequency of non-lethal physical aggression is found to be reactive. Proactive aggression is more characteristic of indirect, non-physical aggression (e.g. gossip) (Frey et al. 2014, pp. 287–88).
19. Wilson and Daly (1985).
20. Fights over honor more frequent among lower class men: Polk (1995), Brookman (2003). Income inequality and reactive aggression: Daly and Wilson (2010).
21. Nisbett and Cohen (1996) showed experimentally that young men from the American south reacted with more aggression to a standardized experimental insult than those from other regions of the United States. They proposed that the southerners' high emotional reactivity resulted from their being derived from populations that had immigrated from parts of Europe where honor was especially highly valued. Daly and Wilson (2010) accepted the principle of a southern culture of honor, but showed that it could equally well be explained as resulting from high income inequality.
22. Keedy 1949, p. 760.
23. Ibid., p. 762.
24. Shimamura 2002.

25. LaFave and Scott 1986, p. 654, cited by Bushman and Anderson 2001, p. 274.
26. Bushman and Anderson 2001, p. 274.
27. Berkowitz 1993.
28. Dodge and Coie 1987. There remain some variants in the system of categorization. The National Institute of Mental Health launched a strategic plan in 2008 to understand the biological mechanisms underlying behavior. Their aim was to integrate clinical research with neuropsychology by creating categories of behavior that reflect the way brains and bodies work. They put aggression into three categories. Proactive (or offensive) and reactive (or defensive) aggression were two, but they also included "frustrative nonreward," which is aggression expressed by someone who cannot get what he or she wants. This third type is probably appropriately regarded as a subclass within "reactive aggression." See Veroude et al. 2015; Sanislow et al. 2010; https://www.nimh.nih.gov/research-priorities/rdoc/units/index.shtml.
29. Crick and Dodge 1996; Weinshenker and Siegel 2002; Raine 2013; Schlesinger 2007; Declercq and Audenaert 2011; Meloy 2006.
30. Raine 2013.
31. Raine et al. 1998a.
32. Cornell et al. 1996.
33. Neumann et al. 2015.
34. Coid et al. 2009.
35. Kruska 2014. Reductions in size of the limbic system have been described for all eight domesticated species that have been studied: pigs (44 percent loss), dogs (poodle) (42 percent), sheep (41 percent), guinea pigs (about 25 percent), mink (17 percent), rats (12 percent), llamas (3 percent), and gerbils (1 percent).
36. Umbach et al. 2015.
37. Dambacher et al. 2015.
38. Role of serotonin: Davidson et al. 2000; Siever 2008; Almeida et al. 2015. Although the relationship between low serotonin activity and impulsive, risk-taking and aggressive behavior is well established, a complicating feature is that in some circumstances excess aggression is associated with high, not low, levels of serotonin (de Almeida et al. 2015).
39. Weinshenker and Siegel 2002, for proactive aggression.
40. Almeida et al. 2015.
41. Flynn 1967; Meloy 2006.
42. Tulogdi et al. 2010, Tulogdi et al. 2015.
43. Shimamura 2002; Manjila et al. 2015. Muybridge's brain injury, consisting of damage to the prefrontal cortex, is similar to one received by a railway worker, Phineas Gage, in 1848, when a metal rod shot through his forebrain. Gage lived for almost twelve years after his accident and suffered a series of

personality changes similar to those experienced by Muybridge. His case led to important advances in understanding the functions of the prefrontal cortex (Damasio 1995).

44. Segal 2012.
45. Veroude et al. 2015.
46. Two main questionnaires are used: the Reactive and Proactive Questionnaire (RPQ) devised by Raine, and the Buss-Perry Aggression Questionnaire (BPAQ) (Tuvblad et al. 2009; Tuvblad and Baker 2011). Examples of questions are given by Tuvblad and Baker 2011.
47. Latest twin study of reactive vs. proactive aggression: Paquin et al. 2014. Prior studies finding higher heritability for proactive aggression: proactive aggression 32–50 percent, reactive aggression 20–38 percent, 9- to 10-year-old boys in California, Baker et al. 2008; proactive aggression 48 percent, reactive aggression 43 percent, 11- to 14-year-old boys in California, Tuvblad et al. 2011. Psychopathy heritability: Ficks and Waldman 2014.
48. Plomin 2014.
49. Ficks and Waldman 2014; McDermott et al. 2009.
50. Raine quote was cited by Adams 2013.
51. Nikulina 1991.

3. Human Domestication

1. Coppinger and Coppinger 2000, p. 44.
2. Hearne 1986.
3. Kagan 1994, p. 96.
4. Zammito 2006; Bhopal 2007. Painter 2010 and Gould 1996 (quote on p. 405) describe the irony that Blumenbach was a rare egalitarian with regard to his attitudes about human diversity, but at the same time his view that Caucasians were the original human race became the source for subsequent racist thinking.
5. Blumenbach 1795, p. 205; 1806, p. 294; 1811, p. 340.
6. As far as I know, Blumenbach never listed the criteria that persuaded him that humans were domesticated. However, he characterized domestication in nonhuman animals as a process of becoming tame; and he viewed humans as the most completely domesticated of all species. His views are shown in the following full quote from 1811 (Blumenbach 1811, p. 340): "Man is a domestic animal. But in order that other animals might be made domestic about him, individuals of their species were first of all torn from their wild condition, and made to live under cover, and become tame; whereas he on the contrary was born and appointed by nature the most completely domesticated animal. Other domestic animals were first brought to that state of perfection *through him*. He is the only one who brought *himself* to perfection."
7. Singh and Zingg 1942.

8. Blumenbach 1865; Candland 1993. Monboddo is cited by Singh and Zingg 1942, p. 191.
9. Nowadays Peter is thought to have suffered from Pitt-Hopkins syndrome, a condition caused by an abnormality in chromosome 18.
10. Blumenbach 1811, p. 340.
11. Ibid. 1806, p. 294.
12. Ibid., p. 903.
13. Brüne 2007.
14. Hutchinson 1898; Nelson 1970.
15. Hutchinson 1898, p. 115. Hutchinson (p. 116) relates an amusing story of trickery. The king met a tall girl while away from his palace. He gave her a note for his commandant: "The bearer is to be given without delay to Macdoll, the big Irishman. Don't listen to objections." The girl suspected what was happening and gave the note to an old woman, who was wedded at once to the disgusted Hibernian. The king later declared the marriage null and void.
16. Darwin 1871, p. 842.
17. Darwin 1845, p. 242.
18. Bagehot 1872, p. 38.
19. Crook 1994.
20. Brüne 2007.
21. The original, Lorenz 1940, is in German. Kalikow 1983 examined the sources of Lorenz's concerns about the decay of human civilization. She found that they came more from a long-standing German tradition than from Nazi politics. Lorenz appears to have been strongly influenced by the biologist Ernst Haeckel (1834–1919).
22. Lorenz 1943, p. 302.
23. Nisbett 1976, p. 83.
24. Haldane 1956 and Nisbett 1976 were two of several works that critiqued Lorenz's views on human domestication and racial purity. Haldane did so partly by challenging the idea that humans are domesticated. He pointed out that, unlike humans, domesticated animals tend to communicate poorly and to be physically specialized. He also argued that domesticated animals are produced by artificial selection, which humans could not have been.
25. Mead 1954, p. 477.
26. Boas 1938, p. 76.
27. Leach 2003; Boehm 2012; Frost and Harpending 2015; Cieri et al. 2014; Gehlen 1944 cited by Brüne 2007; Nesse 2007; Phillips et al. 2014; Lorenz 1940; Dobzhansky 1962; Clark 2007; Gintis et al. 2015.
28. Dobzhansky 1962, p. 196.
29. Leach 2003.
30. Ruff et al. 1993.
31. Brace et al. 1987; Leach 2003; Lieberman et al. 2002.

32. Frayer 1980.
33. Cieri et al. 2014.
34. Henneberg 1998; Bednarik 2014. The meaning of reduced brain size in humans is disputed because the reduction coincided with a fall in body weight, suggesting to some that smaller brains are a meaningless correlate of smaller bodies (Ruff et al. 1997).
35. Kruska 2014; Kaiser et al. 2015; Lewejohann et al. 2010.

4. Breeding Peace

1. Darwin 1868; Hemmer 1990; Price 1999.
2. Increasing cultural sophistication from 300,000 years ago: McBrearty and Brooks 2000, Brooks et al. 2018.
3. Oftedal 2012;. Gould and Lewontin 1979; Gould 1997.
4. Gould 1987; Herrera et al. 2015. Alcock 1987 notes that, even if the clitoris arose as an evolutionary by-product, it can subsequently take on an adaptive function.
5. Dugatkin and Trut 2017.
6. Wrinch 1951.
7. Adam Wilkins, personal communication.
8. Shumny 1987; Trut 1999; Bidau 2009. From 1959 to 1985, Belyaev headed the institute, which became the largest genetics-research center in the USSR.
9. Statham et al. 2011.
10. Trut et al. 2009; Dugatkin and Trut 2017.
11. Trut et al. 2009.
12. Ibid.; Dugatkin and Trut 2017.
13. Trut 1999.
14. Belyaev et al. (1981) p. 267.
15. Ibid., p. 268.
16. Trut 1999, p. 164.
17. Ibid., p. 167.
18. Darwin 1868.
19. MacHugh et al. 2017.
20. Kruska and Sidorovich 2003.
21. Sidorovich and Macdonald 2001.
22. Kruska and Sidorovich 2003; Kruska 1988; Groves 1989.
23. Lord et al. 2013. See also Hughes and Macdonald 2013.
24. Dugatkin and Trut 2017.
25. Kruska 2014; Kruska and Steffen 2013.
26. Künzl et al. 2003.
27. Plyusnina et al. 2011; Price 1999; Malmkvist and Hansen 2002; Bonanni et al. 2017. Range et al. 2015 argued that dogs are more hierarchical than

wolves, but Bonanni et al. 2017 note problems with their conclusion. See also Mech et al. 1998.

28. Simões-Costa and Bronner 2015.
29. Trut et al. 2009.
30. Wilkins et al. 2014.
31. Trut et al. 2009, Simões-Costa and Bronner 2015.
32. Wilkins et al. 2014, p. 801.
33. Crockford 2002.
34. At least four subspecies of the house mouse (*Mus musculus*) from different geographical regions (from Spain to India) independently came to associate with humans. These "commensals" have acquired some features of the domestication syndrome, including changes in hair color, shortening of the face and molar row, and possible reduction of body size (Leach 2003). House mice started living with *Homo sapiens* during the Late Pleistocene at least 15,000 years ago, several thousand years before the development of agriculture. They were apparently drawn by the houses and stored grain of hunter-gatherers living for the first time in long-term settlements (Weissbrod et al. 2017).
35. Kruska 1988; Trut et al. 1991 (article in Russian). Note that, because silver foxes have been kept in captivity in Siberia since they were imported from Canada in the 1920s, there was possibly some inadvertent selection for domestication before Belyaev's experiment began in 1958 (Statham et al. 2011). No one has yet compared Belyaev's selected lines of foxes with wild Canadian foxes. Experimental selection for low fear in red jungle fowl (the wild ancestor of chickens) led to smaller brains within five generations (Agnvall et al. 2017).
36. Creuzet 2009; Aguiar et al. 2014. "FGF" stands for "fibroblast growth factor."
37. Van der Plas et al. 2010; Feinstein et al. 2011; Chudasama et al. 2009; Stimpson et al. 2016; Brusini et al. 2018; Suzuki et al. 2014
38. Librado et al. 2017; Singh, N. et al. 2017; Pilot et al. 2016; Pendleton et al. 2017; Montague et al. 2014; Theofanopoulou et al. 2017; Wang et al. 2017; Alex Cagan, quoted by Saey 2017; Sánchez-Villagra et al. 2016. Carneiro et al. 2014 found no neural-crest effects in a genetic comparison of wild and domestic rabbits, but their data did not rule out such changes.
39. Theofanopoulou et al. 2017.
40. Ibid., p. 5.

5. Wild Domesticates

1. This was the domesticator imagined by Blumenbach's "very profound psychologist" (Blumenbach 1806, p. 294). See chap. 3, p. 54.

2. Clutton-Brock 1992, p. 41.
3. Schultz and Brady 2008.
4. The anthropologist Franz Boas may have been the first to cite a version of a domestication syndrome as part of a justification for seeing humans as being a domesticated species (1938, pp. 83–85). His list of traits that humans shared with domesticated animals focused on variation. Thus, he drew attention to pigmentation being either reduced or intensified, to hair being tightly curled or excessively long, and to body stature varying greatly. He also cited "changes in milk-secreting structures," and "anomalies of sexual behavior."
5. Asian elephants are strikingly easy to tame. Individuals that work with humans also often have a pale mottling suggestive of a failure of melanoblast migration, raising the possibility that they have been subject to either domestication or self-domestication.
6. Hare et al. 2012. See also Clay et al. 2016.
7. Furuichi 2011.
8. Goodall 1986, Muller 2002.
9. Muller et al. 2011.
10. Feldblum et al. 2014.
11. Wilson et al. 2014.
12. Kano 1992, Furuichi 2011.
13. Behncke 2015, p. R26.
14. Ibid.
15. Kelley 1995.
16. Raine et al. 1998b; Ishikawa et al. 2001.
17. Smith and Jungers 1997.
18. Hare et al. 2007.
19. Hare and Kwetuenda 2010.
20. Stimpson et al. 2016.
21. Divergence date of bonobos and chimpanzees: 875,000 years (Won and Hey 2005); 1.5–2.1 million years (de Manuel et al. 2016). Estimated dates vary widely because they depend on factors that are not precisely known, including mutation rates and generation times.
22. Van den Audenaerde 1984.
23. Coolidge 1984, p. xi.
24. Myers Thompson 2001 describes a more complex history of the naming of bonobos than I have given here. Although Coolidge was the first to name bonobos as a species, they had been classified as a subspecies of chimpanzee since the 1880s, based on collections and photographs. Coolidge was the first, however, to recognize the paedomorphic form of the skull, and to propose them as a separate species.
25. Gorilla as overgrown chimpanzee. Quote is from Mitteroecker et al. 2004,

p. 692. The similarities between gorillas and chimpanzees that lead to this concept were articulated clearly by Pilbeam and Lieberman 2017.

26. Pilbeam and Lieberman 2017. See also Duda and Zrzavý 2013.
27. Hare et al. 2012.
28. Shea 1989, p. 84.
29. Hare et al. 2012.
30. On age changes in sexual mounting in rhesus monkeys, see Wallen 2001.
31. Treves and Naughton-Treves 1997.
32. Palagi 2006.
33. Behncke 2015, p. R26.
34. Furuichi 1989 first noted that wild males were submissive to female coalitions, who sometimes attacked males. Parish 1994 supported the importance of female coalitions in zoo populations, comparing bonobos and chimpanzees.
35. Surbeck and Hohmann 2013 describe their detailed study in the wild (in Lui Kotale, Democratic Republic of the Congo). In nineteen months of observation of thirty-three to thirty-five bonobos, they recorded twenty-six female coalitions against males. In fourteen of these cases, the coalition was preceded by male aggression toward either a female (typically, one who was lower-ranking than he was) or, more often, toward an offspring of a female. Tokuyama and Furuichi 2016 give further details from Wamba, Democratic Republic of the Congo.
36. Data on five chimpanzee and two bonobo communities show clear species differences in patterns of association and cooperation, Surbeck et al. 2017. Routine cooperation among female bonobos is described by Tokuyama and Furuichi 2016.
37. Baker and Smuts 1994.
38. Smith and Jungers 1997.
39. Hare et al. 2012 describe this scenario by which the lack of competition with gorillas provoked a new series of selective pressures on bonobos.
40. Takemoto et al. 2015 show that the age of the Congo River is well dated.
41. Recall that the divergence time of bonobos and chimpanzees is estimated around 1 million years ago, Won and Hey 2005, Prüfer et al. 2012.
42. de Manuel et al. 2016.
43. Yamakoshi 2004; Wittig and Boesch 2003; Pruetz et al. 2017.
44. Limolino 2005, Losos & Ricklefs 2009.
45. Stamps and Buechner 1985.
46. Raia et al. 2010 review the connection to high population density, which they tested by studying lizards on an island that has an unusually low population density. In agreement with the population-density theory, the studied lizards failed to show the typical reduction of aggression found on most islands.

47. Rowson et al. 2010.
48. Nowak et al. 2008; Cardini and Elton 2009.

6. Belyaev's Rule in Human Evolution

1. Dirks et al. 2017; Berger et al. 2017; Argue et al. 2017.
2. Stringer 2012 and Lieberman 2013 provide excellent introductions to human evolution.
3. The number of distinct range extensions by African *Homo sapiens* into Eurasia, and their genetic contribution to living populations, continue to be studied. There is good evidence for at least two (Nielsen et al. 2017; Rabett 2018).
4. McDougall et al. 2005.
5. Jebel Irhoud: Hublin et al. (2017). The claim that the Jebel Irhoud population represents *Homo sapiens* is disputable partly because little is known about other populations of *Homo* living around the same time in Africa. The small-brained *Homo naledi,* which was discovered only in 2013, occupied parts of southern Africa. Other *Homo* populations from around 300,000 years ago likely remain to be discovered. Some might one day be found to have characteristics of *Homo sapiens* not found in Jebel Irhoud, raising the possibility that Jebel Irhoud *Homo* did not give rise to the full suite of characteristics found in *Homo sapiens* until after they had interbred with other, still unknown, populations (Stringer and Galway-Witham 2017).
6. Lieberman et al. 2002, Lieberman 2011, Brown et al. 2012.
7. Genetic evidence: Nielsen et al. 2017, Schlebusch et al. 2017. Changes in stone tool technology: McBrearty and Brooks 2009, Lombard et al. 2012. Olorgesailie: Brooks et al. 2018.
8. Stringer 2016.
9. The description is based on skulls and skeletons of African *Homo,* with a bit of help from some European fossils from around the same time, since there is more European material. See Cieri et al. 2014; Stringer 2016.
10. Behavior of Mid-Pleistocene *Homo* at Gesher Benot Ya'aqov. Plant foods: Melamed et al. 2016; flake production and possible hafting: Alperson-Afil and Goren-Inbar 2016; stone "boards": Goren-Inbar et al. 2015; fire use: Goren-Inbar et al. 2004, Alperson-Afil 2008; cooking and complex preparation of prickly water-lily: Goren-Inbar et al. 2014; butchery: Rabinovich et al. 2008. The "species that is still eaten today" is the prickly water-lily (or Fox nut) *Euryale ferox,* which produces highly nutritious edible seeds. Evidence of complex preparation comes from Bihar, India, where the seeds are collected underwater, then dried, roasted, and popped (Goren-Inbar et al. 2014).

11. Harvati 2007; Stringer 2016.
12. Cieri et al. 2014; Ruff et al. 1993; Frayer 1980.
13. Brain sizes: Schoenemann 2006, Hublin et al. 2015.
14. Skulls of *Homo sapiens* globular: Lieberman et al. 2002. The developmental patterns responsible for globularization are debated. Contrasting analyses suggest that critical changes occur before birth (Ponce de Leon et al. 2008, 2016), after birth (Gunz et al. 2010, 2012) or perhaps in both phases (Lieberman 2011).
15. Zollikofer 2012.
16. Neanderthals as models of the common ancestor with *Homo sapiens:* Williams 2013.
17. Reduction in brain size: Henneberg and Steyn (1993), Henneberg 1998, Allman 1999, Groves 1999, Leach 2003, Bednarik 2014, Hood 2014. Groves (1999, p. 10) claimed that reduced brain size was not due to smaller bodies, partly because the reduction does not coincide with a reduction in stature. The argument by Ruff et al. 1997 that brain size differences between Neanderthals and *Homo sapiens* are due entirely to differences in body size refer only to body mass, rather than to body stature. Hublin et al. 2015 agreed with Ruff et al. 1997.
18. Higham et al. 2014; Bridget Alex, personal communication.
19. Williams 2013.
20. Slon et al. 2017.
21. Prüfer et al. 2014.
22. Dates are subject to change due to continuing efforts to understand mutation rates and generation times better; see Moorjani et al. 2016.
23. Lieberman 2008, p. 55. Lieberman wrote those words before the Jebel Irhoud fossils had been properly dated to 315,000 years ago.
24. Marean 2015. The focus on cultural skills makes every sense, as Henrich 2016 showed.
25. Pearce et al. 2013. The skulls dated to less than seventy-five thousand years ago included thirty-two *Homo sapiens* and thirteen Neanderthals.
26. Melis et al. 2006; Tomasello 2016.
27. Asakawa-Haas et al. 2016; Schwing et al. 2016; Drea and Carter 2009.
28. Hare et al. 2007.
29. Joly et al. 2017.
30. Stringer 2016.
31. Weaver et al. 2008.
32. Lieberman 2013; effects of climate: Pearson 2000. Ideas explaining changes in skull and teeth are from Lieberman 2008. Lieberman 2011 discusses these issues in detail.
33. Leach 2003, p. 360.
34. Genetic roots of *Homo sapiens:* Schlebusch et al. 2017.

7. The Tyrant Problem

1. Animals exhibiting unselfish behavior: de Waal 1996, 2006, Peterson 2011. Greene 2013 discusses Darwin's concern to explain morality without invoking a deity.
2. Men more violent than women: Daly and Wilson 1988, Wrangham and Peterson 1996, Pinker 2011.
3. Darwin 1871, p. 875.
4. Ibid., p. 876. Although Darwin was impressed by the potential selective effect of execution and punishment, he made no effort to assess its importance compared with societal approval of morally positive behavior. Darwin's next sentence (after writing "the fundamental social instincts were originally thus gained") shows his ambivalence: "But I have already said enough, whilst treating of the lower races, on the causes which lead to the advance of morality, namely, the approbation of our fellow-men—the strengthening of our sympathies by habit—example and imitation—reason—experience and even self-interest—instruction during youth, and religious feelings." Given that this summary statement makes no reference to the effects of societal punishment that he had discussed in the preceding paragraph, it is not surprising that his argument about the evolution of a reduced propensity for violence has been largely disregarded.
5. Darwin 1871, p. 842.
6. Remarkably, Darwin even compared his scenario for the evolution of morality to the process of domestication. "In the breeding of domestic animals," he wrote, "the elimination of those individuals, though few in number, which are in any marked manner inferior, is by no means an unimportant element towards success." The "inferiority" that Darwin referred to is excessive aggression. He was saying that the human laws that constrain the survival of violent men have the same effects as human actions that constrain the survival of violent animals. The result in both cases is a reduction in aggressiveness—an effect that Darwin assigned to the evolution of morality, whereas nowadays, thanks to Belyaev, we can recognize it as a key component of domestication or self-domestication. Darwin elaborated the analogy even further. As with domesticated animals, he speculated that a result of humans being rendered biologically moral (thanks to executions of the aggressors) would be occasional genetic throwbacks. This might explain the mysterious appearance of exceptionally violent people in otherwise well-behaved families. "[W]ith mankind," he wrote "some of the worst dispositions, which occasionally without any assignable cause make their appearance in families, may perhaps be reversions to a savage state, from which we are not removed by very many generations" (Darwin 1871, p. 876). Here again, he envisaged that aggressiveness in humans had been reduced by a process that we can recognize as being very similar to domestication.

7. Ibid., p. 872. In this quote I have elided a phrase that I regard as a distraction: "and this would be natural selection." Although the added phrase could be interpreted to mean that Darwin thought natural selection responsible for the increase of patriotism, fidelity, etc., it is clear from other passages in this chapter (chapter V) of *The Descent of Man* that he did not think these traits spread through genetic change.
8. Bagehot 1883, p. 32. The "compact" tribes, for Bagehot, were those within which the society was strongly united. He added: "Man, being the strongest of all animals, differs from the rest; he was obliged to be his own domesticator; he had to tame himself. And the way in which it happened was, that the most obedient, the tamest tribes are, at the first stage in the real struggle of life, the strongest and the conquerors. . . ."
9. Hanson 2001.
10. Choi and Bowles 2007; Bowles 2009. Choi and Bowles 2007 included an explicit definition of parochial altruism referring to individuals incurring *"mortal risks"* or forgoing *"beneficial opportunities for coalitions, co-insurance, and exchange"* when *"the members of the actor's group benefit as a result of one's hostile actions toward other groups"* (p. 636).
11. Langergraber et al. 2011; van Schaik 2016.
12. Bellinger Centenary Committee 1963, p. 14. The reference to *"pitched battles"* among aboriginal Australians cited by Choi and Bowles 2007 was Lourandos 1997. Lourandos 1997 gave no description of a battle, citing Coleman 1982. Coleman (1982, p. 2) referred to a pitched battle involving 700 as follows: *"A fight witnessed by early settlers at North Beach near the Bellinger River involved men from the Macleay and Bellinger Rivers against men from the Clarence, a total of about 700 men."* (Bellinger Centenary Committee 1963). Women, children, and old people frequently accompanied the warriors, the fights then being followed by up to a month of feastings, weddings, and corroborees. The report from the Bellinger Centenary Committee (1963) is a pamphlet of 30 pages, almost half of which are full-page ads for farm materials such as agricultural machinery. It makes no pretense of being an academic publication: it has no references. Its theme is the performance of farms in the area. No author is named for the report of the battle, which was said to have been watched by a Mr. John Greer. *"[H]e and a number of other early settlers were privileged to watch from an area which the blacks specially set apart for them as a 'safety zone.' This was a practice invariably carried out at a tribal battle"* (p. 13). The battle is implied to have taken place sometime after 1862, when the first official settlers arrived in the Bellinger Valley area, in the northeast region of New South Wales.
13. Cited by Gat 2015, from Wheeler 1910, p. 148–149.
14. Cultural influences on self-sacrificial behavior: Kruglanski et al. 2018.
15. Darwin 1871, p. 870.
16. Alexander 1979.

17. Engelmann et al. 2012. See also Nettle et al. 2013; Engelmann et al. 2016.
18. Kurzban and Leary 2001 cite cases from fish (sticklebacks) to primates (baboons, chimpanzees) in which "bad cooperators" are shunned. For the stickleback, being infested with a parasite made an individual a "bad cooperator."
19. Bateson et al. 2006.
20. Gurven et al. 2000. See also Boehm 2012.
21. Nesse 2007, p. 146. See also Nesse 2010.
22. Cieri et al. 2014.
23. Feminization was reviewed by Cieri et al. 2014; Leach 2003; Lefevre et al. 2013.
24. Anderl et al. 2016; Stirrat et al. 2012; Carré et al. 2008, 2013; Haselhuhn 2015. Standardized facial breadth is measured as the bizygomatic width divided by the height of the upper face.
25. Hrdy 2009; van Schaik 2016; Tomasello 2016.

8. Capital Punishment

1. Anonymous 1821, pp. 15–16.
2. Banner 2003, pp. 1–2, described Clark's case, based on Anonymous 1821. Even now there are parts of the world that allow execution for nonviolent crimes. In Saudi Arabia, the death penalty can be sentenced for drug offenses, apostasy, heresy, and witchcraft. According to Amnesty International 2015, between January 1985 and June 2015, at least 2,208 people were executed in Saudi Arabia.
3. Bedau 1982; Fischer 1992; Banner 2003.
4. Fischer 1992, pp. 91–92.
5. Ibid., p. 193; Bedau 1982, p. 3.
6. Morris and Rothman, eds., 1995.
7. Hoffman 2014, p. 281.
8. Otterbein 1986, p. 107. There were fifty-three societies in his sample.
9. Boehm 1999, 2012.
10. Lee 1984, p. 96.
11. Workman 1964.
12. Boas 1988, p. 668.
13. Warner 1958, pp. 160–61.
14. White, ed., 1985, p. 132.
15. Gellner 1994, p. 7.
16. Bridges 1948, p. 410.
17. Knauft 1985. The division of sorcery vs. witchcraft that I describe here as volitional vs. innate is a sharply abbreviated summary of a rich discussion of the differences by Knauft 1985, pp. 340–45.

18. Ibid., pp. 98–99. Knauft paraphrased and condensed the conversations from complete transcripts of several actual accusations.
19. Vrba 1964, p. 115.
20. Des Pres 1976, p. 140. A similar ethos to the bread law was described in Andersonville Prison, also known as Camp Sumter, during the American Civil War, Futch 1999.
21. Weinstock 1947, pp. 120–21.
22. Kim 2015.
23. Marlowe 2005 found that the total population size for the 341 best-known ethnolinguistic hunter-gatherer societies averages 895 individuals.
24. Bridges 1948, p. 216.
25. Fried 1967; Woodburn 1982; Flanagan 1989; Boehm 1999. The least egalitarian mobile hunter-gatherers are Australian Aboriginals, whose society has been called "gerontocratic," referring to the monopolization of young wives by the elders, men about fifty years old (Berndt 1965; Meggitt 1965; Hiatt 1996). The elders derive inordinate power from having absolute control over religious beliefs, but "they are in no sense a council of chiefs. . . . They are more often the emblems of Aboriginal authority than formal leaders" (Liberman 1985, p. 65). As in other egalitarian societies, Aboriginals are said to "have no aspiration to issue orders to other Aboriginals." The reason seems clear: "It is not so much the absence of a desire for social power as it is a fear of the social consequences of their appearing to presume that they are better than their fellows" (Liberman 1985, p. 259).
26. Liberman 1985, pp. 27–28.
27. Shostak 1981.
28. Marlowe 2004, p. 77.
29. There are interesting exceptions to this generalization. Bonobos form separate and overlapping hierarchies among males and females, such that each sex has an alpha and either the male or female can dominate. A few primate species superficially echo mobile hunter-gatherers by having a system of male-male relationships that lacks any alpha position and has no obvious hierarchy. However, egalitarianism among such primates differs from that of humans by coming with little ambition: male primates that are egalitarian show minimal interest in competing with one another (such as tamarins; see n. 31 below).

    A primate species whose male hierarchy is more similar to that of hunter-gatherers is the hamadryas baboon, where breeding males rarely fight one another. Each forms a tight bond with one or more females and stays permanently close to them in a single family, or one-male unit. Several one-male units combine into a clan, within which the males respect one another's families: they do not try to steal one another's females. Two or three clans similarly combine into a band, within which, again, breeding males are mutually respectful. Only when different bands meet are males likely to show aggression, especially when competing for access to one another's females (Swedell and Plummer 2012). Male hamadryas baboons thus show no interest in trying to compete against males within their own clans or bands. By contrast, in humans, male hunter-gatherers compete for prestige in physical skills, hunting ability, sacred knowledge, and shamanism. They owe their egalitarianism to the "tyranny of the cousins," not to a lack of interest in competing.

30. Chapais 2015.
31. Groups of the tiny tamarin monkeys of South America often have only a single breeding female and two adult males. Both males mate, yet they are so rarely aggressive to each other that dominance has not been described between the males. Males groom each other often and share food. Twins are routine, and the two infants can have different fathers. (Goldizen 1989; Huck et al. 2005; Garber et al. 2016.)
32. Woodburn 1982, p. 346.
33. Boehm 1999, p. 68.
34. Inuit bully: Boehm 1999, p. 80.
35. Lee 1969.
36. Woodburn 1982, p. 440.
37. Cashdan 1980, p. 116.
38. Burch 1988, p. 25.
39. Briggs 1970.
40. Durkheim 1902.
41. Boehm 1999, p. 68.
42. Ibid., p. 83. Boehm found one exceptional case in which the bully was not killed. This bully was an Eskimo shaman in Greenland who killed rivals within his group, leaving the survivors fearful. The report of his behavior came from a group of Scandinavians who had visited the Eskimo band for only a few weeks. Boehm's survey convinced him that, had the visitors stayed longer, they would have recorded the shaman's execution.
43. Phillips 1965, p. 187.
44. Norm enforcement in Ju/'hoansi: Wiessner 2005. Data were from 308 conversations.
45. Muller and Wrangham 2004; Carré et al. 2011; Terburg and van Honk 2013.
46. Langergraber et al. 2012; Mech et al. 2016.
47. Knauft 1985.
48. These figures are calculated from tables 4 and 5 in ibid. Table 4 shows that 81 of 230 adult male deaths (i.e., 35.2 percent) were homicides, and 48 of 164 adult female deaths (29.3 percent) were homicides. Table 5 shows that, of 101 adult homicide victims, 70 (i.e., 69.3 percent) were killed as sorcerers, and of the 55 female homicide victims, 29 (i.e., 52.7 percent) were killed as sorcerers. Overall rates of being killed as sorcerers were thus 35.2 percent of

69.3 percent of adult males (i.e., 24.4 percent), and 29.3 percent of 52.7 percent of adult females (i.e., 15.4 percent). Other homicides were listed as "directly sorcery-related," "sorcery raid," "battle," "insanity related," and "Other / cause unknown."

49. Knauft 2009.
50. Kelly 1993, p. 548.
51. Nash 2005.
52. Benedict 1934.
53. Otterbein 1986, p. 38.
54. Boehm 1999, p. 79.
55. Beard 2015; Workman 1964.
56. Dediu and Levinson 2013; Hublin et al 2015; Prüfer et al. 2014; Tattersall 2016.
57. Otterbein 1986; Kelly 2005; Okada and Bingham 2008; Phillips et al. 2014; Gintis et al. 2015.
58. Steven LeBlanc, personal communication. LeBlanc heard this story from a Yanomamö man on a fund-raising tour for a missionary group in California in the 1990s.
59. Tomasello and Carpenter 2007, p. 121.
60. Tomasello and Carpenter 2007.
61. Morrison and Reiss 2018.
62. Tattersall 2016, p. 164. This source provides a helpfully brief introduction to the immense literature on the evolution of language. See also Klein 2017, suggesting a slow development with a critical increase in ability around fifty thousand years ago; Corballis 2017, relating language to other mental abilities that developed during the Pleistocene; Buckner and Krienen 2013, suggesting how larger brains can lead to emergent properties such as linguistic ability; Hauser and Watumull 2017, suggesting the specific cognitive abilities that precede and enable the evolution of language.

9. What Domestication Did

1. Evolution of humans from unicellular life: Dawkins 2005.
2. The oldest primate fossils date to almost 60 million years ago, in the Paleocene, but genetic data suggest that primates originated in the late Cretaceous (Francis 2015).
3. Lovejoy 2009. See also Hylander 2013; Muller and Pilbeam 2017.
4. Hare 2017 also noted that simply becoming larger led to larger brains. Larger brains have a disproportionately larger pre-frontal cortex, so more self-control could have followed as an incidental consequence of growing bigger, rather than being directly selected.
5. MacLean et al. 2014.
6. Herculano-Houzel 2016.
7. Francis 2015.

8. Geiger et al. 2017.
9. Trut et al. 2009; Evin et al. 2017; Sánchez-Villagra et al. 2017.
10. Neoteny means that the paedomorphic feature develops at a slow rate. In postdisplacement, the rate of development is the same between ancestral and descendant species, but the paedomorphic feature starts developing relatively late. Progenesis occurs when the animal reaches sexual maturity relatively early (Gould 1977).
11. Trut et al. 2009, p. 353.
12. Gariépy et al. 2001.
13. An experimental domestication program has also been conducted with red jungle fowl, the ancestors of chickens. Among other results supporting the view that selection for low fear toward humans produced the domestication syndrome, selection for tamer fowl led to increased circulating serotonin and smaller brains (Agnvall et al. 2015, 2017).
14. Buttner 2016.
15. Trut et al. 2009.
16. Paedomorphic effects vary among domesticated species. Whereas the levels of basal cortisol are much lower in silver foxes as a result of selection for low emotional reactivity, domesticated guinea pigs show no reduction in basal cortisol compared with their wild ancestors, cavies, nor do domesticated foxes, mallards, or mice. However, although guinea pigs do not have low basal levels of cortisol, in response to stress they echo selected foxes in having a greatly reduced surge of cortisol (Künzl and Sachsler 1999). Tamer red jungle fowl, on the other hand, produce as much cortisol in response to being handled as their unselected ancestors (Agnvall et al. 2015).
17. Trut 1999.
18. Lord 2013; Buttner 2016. Dogs and wolves are almost identical genetically, but two genes involved in stress regulation differ in the rate at which they are expressed in the hypothalamus (CALCB [calcitonin-related polypeptide beta] and NPY [neuropeptide Y]) (Buttner 2016).
19. Künzl and Sachsler 1999; Trut et al. 2009.
20. Hemmer 1990; Leach 2003; Raine et al. 1998b; Isen et al. 2015. The direction of causation may be more complicated than physical strength permitting success in aggression. Research on 2,495 children in the Minnesota Twin Study found that more physically aggressive boys at age eleven were not necessarily the stronger ones when they were younger (Isen et al. 2015).
21. Lieberman et al. 2007; Durrleman et al. 2012.
22. Wobber et al. 2010; Durrleman et al. 2012; Behringer et al. 2014.
23. Huxley 1939.
24. Naef 1926. (Article in German.)
25. Gould 1977; Bromhall 2003. In many ways, human development is indeed slow. The science journalist Richard Francis, who began his career as a

neurobiologist, compared humans and chimpanzees for both somatic and neural features. In humans, bones and muscles develop more slowly, growth overall is slowed, and the adolescent growth spurt starts later and lasts longer. In the brain, axons become myelinated later, which allows learning to continue for longer. Patterns of gene expression vary in different regions of the brain. In several areas, such as the prefrontal cortex, genes that are expressed in chimpanzees at a certain age are expressed later in life in humans. However, many of the changes from ape to human were not paedomorphic, such as our developing long legs and large feet. The concept of a "global paedomorphism," covering human biology as a whole with respect to apes, is certainly wrong (Francis 2015).

26. Liu et al. 2012.
27. Miller et al. 2012.
28. Hrdy 2014 gave examples of how the neural and behavioral development of humans is a mosaic of delay and acceleration, compared with apes. For instance, humans develop slowly in terms of their physical abilities, but quickly with respect to interacting with their caretakers.
29. Zollikofer 2012.
30. Williams 2013.
31. Hublin et al. 2015.
32. Marean 2015.
33. Liu et al. 2012. Recent evidence indicates that the delay in synaptic development that is characteristic of humans in general is disrupted in autistic individuals (Liu et al. 2016).
34. Kaiser et al. 2015.
35. Higham et al. 2014; Bridget Alex, personal communication; Prüfer et al. 2014; Wynn et al. 2016. Neanderthals have previously been dated in Gorham's Cave, Gibraltar, to as late as twenty-five thousand years ago, but there is now uncertainty whether such a late date is accurate (Higham et al. 2014).
36. Hayden 2012; Villa and Roebroeks 2014; Roebroeks and Soressi 2016.
37. Hardy et al. 2012. Despite indisputable evidence of Neanderthals using fire extensively in many locations, two (late) Neanderthal European sites indicate occupation for significant periods without fire. This observation has led to the idea that Neanderthals could not make fire (Dibble et al. 2018), a view challenged by Sorensen (2017). Anatomical evidence that all *Homo* species need cooked diets makes the idea that Neanderthals could survive on raw food surprising (Wrangham 2009).
38. Roebroeks and Soressi 2016, p. 6374.
39. Marean 2015; Wynn et al. 2016.
40. Hayden 1993; Marean 2015; Shea and Sisk 2010.
41. Marean 2015; Pearce et al. 2013.

42. Large eyes in Neanderthals: Pearce et al. 2013.
43. Tattersall 2015, p. 206.
44. Hayden 2012.
45. Prüfer et al. 2014.
46. Marean 2015.
47. Hayden 2012.
48. Hare et al. 2002.
49. Hare et al. 2005. I was Brian Hare's graduate adviser at the time of these investigations, and I was enthralled with the chance to see a test that would distinguish between the two ideas. Natalie Ignacio bravely launched the research by committing herself to carrying out her senior honors-thesis research in Siberia. Lyudmila Trut and her colleagues in Novosibirsk were wonderful hosts for our small research team.
50. Since bonobos appear to be self-domesticated from a chimpanzee-like ancestor, they should have better cooperative-communicative skills than chimpanzees. There is some evidence that bonobos indeed show greater understanding of human behavior than chimpanzees do. When humans look in a new direction, bonobos are more likely than chimpanzees to follow the humans' gaze. But, unfortunately, like chimpanzees, bonobos fail the object-choice test (Hare 2017). This may be due to the fact that selection against reactive aggression in bonobos has not involved humans.
51. Hare et al. 2005.
52. Buttner 2016.
53. Vasey 1995. A critical factor is whether mating ability is constrained by hormones. In lemurs and lorises, mating is supported by hormones, and homosexual behavior is absent. In monkeys and apes, hormones are not needed for the expression of homosexual behavior, which is widespread. See also Wallen 2001.
54. Bagemihl 1999.
55. Young and VanderWerf 2014.
56. Vasey 1995.
57. Furuichi 2011; Tokuyama and Furuichi 2016. Although female bonobos interact sexually very broadly within their communities (and even with members of neighboring communities), Clay and Zuberbühler 2012 found sexual interactions among the higher-ranked females to be rare in a captive community.
58. Hines 2011; Skorska and Bogaert 2015; Whitam 1983; Barthes et al. 2015.
59. U.S. data from 2000 census, Peplau and Fingerhut (2007). UK data from Knipe 2017.
60. VanderLaan et al. 2016. Another hypothesis proposes that same-sex attraction is favored because it is associated with high fecundity in the trait bearers' mothers, but a similar problem applies as for kin-directed altruism—namely,

that the benefits are insufficient to warrant abandoning personal reproductive efforts (Skorska and Bogaert 2015).

61. Muscarella 2000.
62. Roselli et al. 2011.
63. Ibid.
64. Roselli et al. 2004.
65. Valentova et al. 2014; Li et al. 2016..
66. Bagemihl 1999.
67. McIntyre et al. 2009. Samples of apes were small compared with the thousands studied in human research: seventy-nine chimpanzees and thirty-nine bonobos.
68. Wallen 2001.
69. Note that the timing of when, and how long, self-domestication happened includes various possibilities in addition to the simplest version, the one that I have given. If linguistic skills were not sufficient for forming murderous conspiracies until, say, a hundred thousand years ago, the first elimination of alpha males would have happened then. In that case, the loss of brow ridges, the smaller face, and the reduction of sexual dimorphism in the early phase of *Homo sapiens* would be, as traditionally argued, adaptations rather than correlated consequences. Further back in time, there might have been a form of self-domestication when *Homo* first evolved from australopithecines, between two and three million years ago; or when australopithecines evolved from forest apes, 6 to 9 million years ago; or even earlier still. Such conjectures await examination.

10. The Evolution of Right and Wrong

1. Freuchen 1935, pp. 123–24.
2. Haidt 2012, p. 190.
3. Darwin 1871; de Waal 2006; Tomasello 2016; Henrich 2016; Baumard 2016.
4. Boehm 2012.
5. DeScioli and Kurzban 2009 and 2013 make this point clearly. They note that much evolutionary analysis of morality has been confined to the way that Darwin framed the problem. For Darwin, "to do good unto others" was "the foundation-stone of morality" (Darwin 1871, p. 871). DeScioli and Kurzban (2009, p. 283) listed Alexander 1987, Darwin 1871, de Waal 1996, Ridley 1996, and Wright 1994 as following Darwin's lead. See Peterson 2011. Fiske and Rai 2015 exemplify the alternative view that moral behavior includes rationalizing much competitive and violent behavior.
6. Graves 1929 (1960), p. 56.
7. Des Pres 1976.
8. Davie 1929, pp. 19–20.

9. Hinton 2005, p. 2. See also Gourevitch 1998; Goldhagen 1996.
10. Fiske and Rai 2015, p. xxi.
11. DeScioli and Kurzban 2009 and Saucier 2018 justify this definition.
12. De Waal 2006.
13. Warneken 2018.
14. Hamlin and Wynn 2011.
15. Hauser 2006; Bloom 2012. However, religious believers tend to be more prosocial (to their ingroup) than nonbelievers (Cowgill et al. 2017).
16. Decety et al. 2015. Children were aged five to twelve years, living in Canada, China, Jordan, Turkey, the United States, and South Africa. Religions included Muslim (43 percent), Christian (24 percent), Jewish (3 percent), Buddhist (2 percent), and Hindu (0.5 percent). Twenty-eight percent were not religious. Both Muslim and Christian children were less altruistic than nonreligious children.
17. Bloom 2012.
18. DeScioli and Kurzban 2009.
19. Haidt 2007, pp. 998, 1000.
20. Gintis et al. 2015; Henrich 2016.
21. Jensen et al. 2007. Jensen et al. 2013 respond to a critique of their conclusion that chimpanzees do not show any other-regarding behavior when playing the Ultimatum Game.
22. Elkin 1938; Berndt and Berndt 1988.
23. Cultural norms raise closely similar problems to moral behavior about whether, if they are good for a group, they necessarily arose because of that fact. That a norm such as a selectively applied food taboo might have group benefits (for example, in preventing pregnant women from eating foods that might be risky for her infant) suggests to many people that it arose through a form of group selection. Other processes are plausible also however, enforcement of rules by more powerful individuals or subgroups. See Singh, M. et al. 2017.
24. Hauser 2006.
25. Cushman and Young 2011 called these three the "Action/Omission," "Means/Side-effect" and "Contact/Non-contact" biases. For the sake of clarity, I have renamed them to stress the effect of the bias on the moral response (respectively the "Inaction," "Side Effect" and "Noncontact" biases).
26. Cushman and Young 2011 found evidence that the Inaction Bias and Side Effect Bias influence people's solutions not only to moral problems, but also to problems without moral consequences. For example, instead of a moral dilemma that affected how many people died, a dilemma might affect whether a rock fell down a cliff. Given that the biases were not limited to the moral domain, they argued (p. 1068) that the biases were derived from "the architecture of cognition in nonmoral domains such as folk-psychological and causal cognition." It seems equally plausible that the direction of causa-

tion is the other way round—i.e., that the nonmoral biases are derived from the moral biases.

27. Hoffman et al. 2016.
28. Rudolf von Rohr et al. 2015.
29. Jane Goodall, personal communication.
30. Goodall 1986. Pusey et al. 2008 described further infanticides by another mother-daughter pair.
31. Darwin 1871, p. 827.
32. De Waal 2006, p. 54.
33. Boehm 2012, p. 161.
34. Hoebel 1954, p. 142.
35. Ibid., p. 70.
36. Hiatt 1996.
37. Hoebel 1954, p. 90.
38. Boehm 1993. Erdal and Whiten 1994 preferred the term "counter-dominant hierarchy."
39. Dalberg-Acton 1949, p. 364. The quotation comes from an April 1887 letter to Mandell Creighton.
40. Boehm 2012, p. 15.
41. Haidt 2007, p. 999.
42. Mencken 1949, p. 617; DeScioli and Kurzban 2013, p. 492.
43. Sznycer et al. 2016.
44. Keltner 2009.
45. Goffman 1956. See also Feinberg et al. 2012.
46. Giammarco and Vernon 2015, p. 97; Scollon et al. 2004; Breggin 2014.
47. Williams and Jarvis 2006.
48. Williams and Nida 2011.
49. Chudek and Henrich 2011, p. 218.
50. Henrich 2016.

11. Overwhelming Power

1. Stevenson 1991, p. 10.
2. Ibid.
3. Ibid., p. 32.
4. Ibid., p. 15.
5. Ibid., p. 141.
6. Mighall 2002.
7. Babcock et al. 2014, p. 253; Teten Tharp et al. 2011.
8. From Shakespeare's *Merchant of Venice,* act V, scene I, line 54.
9. Lorenz 1966. (The original version of *On Aggression* was published in German in 1963.) The idea that seeing the faces of those we are supposed to kill will inhibit our aggression has some support (Sapolsky 2017, p. 644).

Numerous examples of intense face-to-face cruelty, however, show that any such inhibitions can be easily overcome. Examples from hunter-gatherers are given by Burch 2005 (for Arctic Inupiaq) and Zegwaard 1959 (for New Guinea Asmat).

10. Mohnot 1971, p. 188.
11. Hrdy 1974, 1977.
12. The infanticide debate was discussed by Sommer 2000. Bartlett et al. 1993 was one of the last major attacks on the sexual selection hypothesis in primates. Pusey and Packer's 1994 data on other species were challenged by Dagg 2000.
13. Perry and Manson 2008, p. 198. This book about the study of white-faced capuchins led by Susan Perry since 1990 is a primatological classic. It is based on high-class observation and theory, and has helped revolutionize our understanding of the species, the largest-brained of all New World primates. The species has fascinating cultural quirks, such as group-specific ways of greeting and showing friendship, and complex political systems, but also much violence toward both infants and adults. Their earliest data on infanticide were published in Manson et al. 2004.
14. Palombit 2012 reviewed the primate data. Based on observations of infanticide in fifty-six populations from thirty-five species, he again showed that the sexual selection theory is strongly supported in most species, but that in some cases other explanations are needed. Infanticide frequency data come from Watts 1989; Henzi et al. 2003; Butynski 1982; Crockett and Sekulic 1984.
15. Lukas and Huchard 2014. Lukas and Huchard found that infanticide occurs mainly in social mammals in which reproduction is monopolized by a minority of males. "The evolution of infanticide is largely determined by variation in the intensity of male-male contest competition . . . and closely reflects variation in the intensity of intra- and intersexual conflict" (p. 843). They found infanticide in 119 (45.7 percent) out of the 260 mammal species.
16. Wilson et al. 2014.
17. Infanticide in marmosets: Beehner and Lu 2013, Saltzman et al. 2009. Infanticide in tamarins: Garber et al. 2016.
18. Palombit 2012; Borries et al. 1999. Infanticide is not the only context for proactive aggression. As I noted in chapter 2, in laboratory strains of mice and rats, males may attack conspecifics proactively. In one strain (CD-1), 19 percent of dominant male mice showed such strong interest in being able to attack subordinates that their aggressive behavior was considered "addiction-like," which suggested to the authors that this style of aggression had an evolutionary origin (Golden et al. 2017). Proactive aggression does not appear to have been studied in wild mice, however.
19. George Williams quoted by Boomsma 2016.

20. Surprisingly, the pattern of infanticide by men in the United States and Canada was shown by Daly and Wilson 1988 to conform to several predictions of the sexual selection hypothesis. Most often, the victims were fathered by a different male from the killer, the mother was in a sexual relationship with the killer, and the infant was young enough (normally less than two years old) that its death would be expected to allow the mother to become pregnant sooner. However, the behavior was hardly ever adaptive, since the killer was normally arrested and imprisoned.
21. The dead female was never identified but was probably a member of the Kahama community. She is listed in Wilson et al. 2014.
22. Muller 2002.
23. For example, *The Dark Side of Chimps,* National Geographic Television production, 2004, directed by Steven Gooder.
24. Power 1991; Sussman, ed., 1998; Ferguson 2011; Marks 2002.
25. Wilson et al. 2014.
26. Mitani et al. 2010.
27. Williams et al. 2004.
28. Cafazzo et al. 2016.
29. Cubaynes et al. 2014.
30. Mech et al. 1998.
31. Cassidy and McIntyre 2016.
32. Wrangham 1999.
33. Lee 1979 considered that the Ju/'hoansi and other Bushman groups in Southern Africa represent a pristine society relatively unaffected by farming cultures, whereas Schrire 1980 and Wilmsen 1989 argued that interaction with pastoralists had had diverse impacts on their lives. Marlowe 2002 reviewed the debate and described the Hadza situation. Lee 2014 argued that patterns of hunter-gatherer violence indicate that humans have evolved from a peaceful background. He claimed that hunter-gatherers had few things to fight over, and could easily disperse to avoid conflict, and archaeological prehistorical forensic data show little evidence of violence. He acknowledged that southern African hunter-gatherer groups were "famous in colonial history for their fighting prowess" (p. 219) but argued that this "was largely an artifact of their historical positioning" and was therefore irrelevant to understanding whether they would have had warfare as hunter-gatherers. His evidence does not match the numerous, often grisly accounts of hunter-gatherer warfare around the world documented in works such as Chacon and Mendoza, eds., 2007, or Gat 2015.
34. Tindale 1940 estimated that there were 574 "tribes" (ethnolinguistic groups) in Australia prior to European contact; Birdsell 1953 suggested that there were 251,000 to 300,000 inhabitants, averaging 437–523 per "tribe."
35. Gat 2015.

36. Wheeler 1910, p. 151.
37. Basedow 1929, p. 184.
38. See table 1 in Wrangham and Glowacki 2012.
39. Tindale 1974, p. 327; Roth 1890, p. 93.
40. Radcliffe-Brown 1948, p. 85.
41. Lothrop 1928, p. 88.
42. Burch 2007, pp. 19–20.
43. Bishop and Lytwyn 2007, p. 40.
44. Fry 2006, p. 262.
45. Archaeological evidence of frequent warfare prior to agriculture: LeBlanc 2003. Death rates of hunter-gatherers vs. farmers: Wrangham et al. 2006. Knauft 1991 proposed that hunter-gatherers have a lower death rate than farmers from intergroup conflict, an idea that has generally been supported. Worldwide death rates from battle: https://ourworldindata.org/war-and-peace. Worldwide homicide rate: https://data.worldbank.org/indicator/VC.IHR.PSRC.P5.
46. Gat 2015.
47. Fry 2006. Fry has been a particularly strong advocate of the view that mobile hunter-gatherers had little war (Fry and Söderberg 2013). His evidence was critically reviewed by Gat 2015.
48. Kelly 2000, p. 118 and p. 139.
49. Ferguson 1997, p. 321. Gat 2015 noted that Ferguson followed the quoted sentence with one that appeared designed to redress the balance: *"Equally, if there are people who believe that all human societies have been plagued by violence and war, that they were always present in human evolutionary history, this volume proves them wrong."* Gat's comment (p. 113) on Ferguson's follow-up was: *"However, the various claims in the second proposition were anything but 'proved.' At best, they remained unproved and open to further investigation."* I agree.
50. Lovisek 2007.
51. Zegwaard 1959.
52. Murphy 1957.
53. Chagnon 1997. Chagnon's studies of the Yanomamö have often been accused of being biased, illegitimate, or otherwise wrong. However, the accuracy and fidelity of his work have been strongly supported by in-depth investigations (Dreger 2011, Shermer 2014).
54. I have occasionally seen interactions among chimpanzees in which a pair of males spend time maneuvering themselves into a position to corner and safely charge a rival. Such efforts have always occurred when the rival was already present. I know of no cases in which observers have reported coalitions of chimpanzees setting a trap for, or going hunting for, a rival.
55. Boehm 2017 argued that coalitionary proactive aggression did occur within chimpanzee communities, on the basis that occasional coalitionary killing

of males could have been premeditated. But premeditation has not been evidenced, and the victims have had such varied status that no systematic pattern has emerged: victims have been young or old, and rising or falling in rank. The only alpha male seen to be killed was Pimu, but another Mahale alpha male, Ntologi, was suspected of being killed by a coalitionary attack (Nishida 1996, 2012).

56. Kaburu et al. 2013. See also Wilson et al. 2014; Boehm 2017. In the other within-group attacks reported among wild adult male chimpanzees that led to deaths, the initiation of the attack was not seen (Pruetz et al. 2017; Michael Wilson, personal communication; Nishida 1996; Watts 2004; Fawcett and Muhumuza 2000; Nicole Simmons, personal communication; Nishida et al. 1995).
57. De Boer et al. 2015.
58. Foucault 1994.
59. Hansen and Stepputat 2006, p. 296.
60. Ibid., p. 301.
61. Ibid., p. 309.
62. Hoebel 1954, p.26.
63. Ibid., p. 305.
64. Hayden 2014.
65. Pinker 2011.

12. War

1. Burke 1770, p. 106.
2. Dalberg-Acton 1949, p. 364. The fuller quote is: "I cannot accept your canon that we are to judge Pope and King unlike other men with a favourable presumption that they did no wrong. If there is any presumption, it is the other way, against the holders of power, increasing as the power increases. Historic responsibility has to make up for the want of legal responsibility. Power tends to corrupt, and absolute power corrupts absolutely. Great men are almost always bad men, even when they exercise influence and not authority: still more when you superadd the tendency or certainty of corruption by authority. There is no worse heresy than that the office sanctifies the holder of it."
3. Fuentes 2012, p. 153. Fuentes argued that the idea that humans, and especially males, have natural aggressive tendencies "enables a kind of inevitability in our communal sense of aggression and society." See also Fry 2006.
4. Lee 2014, p. 224.
5. Kropotkin quoted by Crook 1994, p. 194.
6. Gat 2006; Lopez 2016. See also Tooby and Cosmides 1988; Johnson and Thayer 2016; Lopez et al. 2011; McDonald et al. 2012; Johnson and Toft 2014; Johnson and MacKay 2014. These references all focus on selection

acting at the individual level. There is also an important tradition of arguing that group selection can influence the evolution of warring behavior. See, e.g., Bowles 2009; Zefferman and Mathew 2015.

7. Ferguson 1984, p. 12.
8. Fry 2006, p. 2.
9. Fromm 1973, p. 22. The anthropologists Robert Sussman and Joshua Marshack feared long-lasting effects. "If warfare and killing are seen as universal, primordial, adaptive and natural . . . these views may become virtually immutable in the collective unconscious, diminishing our impetus for positive change" (Sussman and Marshack 2010, p. 24). Gat 2015 (p. 123) discussed the problem, which he summarized as: "People habitually assume that if widespread deadly violence has always been with us, it must be a primary, irresistible drive that is nearly impossible to suppress."
10. Goodall 1999.
11. Hinde and Rotblat 2003.
12. Hinde 2008. Robert Hinde was the graduate adviser for many other primatologists, including David Bygott (the first to study aggression in chimpanzees), Dian Fossey (who discovered infanticide in gorillas), Dorothy Cheney and Robert Seyfarth (who documented baboon social behavior), and Alexander Harcourt and Kelly Stewart (who described mountain gorilla behavior in detail).
13. Hamburg and Trudeau, eds., 1981.
14. Carnegie Commission on Preventing Deadly Conflict 1997.
15. Hamburg 2002.
16. Hamburg 2010.
17. Crook 1994. Crook showed that Darwinism contributed as much to thinking about the evolution of peace as it did to the evolution of war. For an equivalent book about war and evolutionary theory during the period between the First and Second World Wars, see Overy 2009. The First World War reduced enthusiasm for the idea that war was beneficial, but the debate between social and biological perspectives continued.

    Pearson was a socialist, freethinker, and sexual radical who embraced the idea that genetics should be used to ensure success of the patriotic group. Crook cites Pearson's 1888 *The Ethic of Free Thought:* "Socialists have to inculcate that spirit which would give offenders against the State short shrift and the nearest lamp-post."
18. Rousseauians allowing positive behaviors to be genetically influenced: Sapolsky 2017 presents a nuanced account. I am grateful to Carole Hooven for this point.
19. Gould quotes: Gould 1998, p. 262. I suspect that I was personally involved with Gould's writing his critical essay. Gould was a colleague of mine at Harvard University. I knew him only a little, and I was aware of his stern criticisms of human sociobiology. In early 1996 I sent him a pre-publication

copy of *Demonic Males: Apes and the Origins of Human Violence,* by Dale Peterson and myself. I hoped that he would appreciate the value of a behavioral-ecology approach for understanding some problems of human violence. He never responded to me about it. A few months later, however, Gould published a critique of the approach taken in *Demonic Males.* His essay appeared first as "The Diet of Worms and the Defenestration of Prague" in *Natural History* magazine (September 1996). It was republished in Gould 1998.

20. Ferguson 1984, p. 12.
21. Lee 2014, p. 222, cf. Fry 2006, p. 5: "humans are capable of a great deal of violence." Instead of disputing the potential for violence, the main complaint of Rousseauian scholars such as these is that Hobbesians bias their compilation of the rates of violent death among hunter-gatherers in favor of high rates (e.g. Lee 2014).
22. Tooby and Cosmides 1990 and McCrae and Costa 1997 discuss the universality of human nature.
23. Frederick the Great's quote is from a 1759 letter to Voltaire. Frederick the Great might have been persuaded about human beastliness by the behavior of his father, Frederick William I, who had frequently beaten him as a child (when not engaged in trying to collect or breed giants for his army).
24. Gat 2015; Wrangham and Glowacki 2012. Note that, in comparing humans and chimpanzees, I am referring to the nature of the social interaction. I am not saying that mobile hunter-gatherers are any more like chimpanzees than are you or I.
25. Wrangham and Glowacki 2012.
26. Glowacki and Wrangham 2013, 2015.
27. Mathew and Boyd 2011.
28. Wiessner 2006, p. 177.
29. Bourke 2001, p. 67.
30. Ibid., p. 105.
31. Ibid.
32. Boehm 2011.
33. De Quervain et al. 2004.
34. Bourke 2001, p. 155.
35. I say "barely more puzzling" rather than "no more puzzling" to acknowledge that simple warfare certainly includes more complexities than animal intergroup violence. For example, simple warfare ranges from feuding to occasional battles; it sometimes involves personal relationships of hostility; victims might be sought because of perceived wrongs; it can occur in internal or external war; and it can involve alliances among groups. Cultural influences can come from shared perceptions of the need for revenge, traditions of military training, patriarchal ideology, variable degrees of linguistic difference between groups, etc. I do not mean to belittle the importance of

such factors in affecting rates and styles of violence. Nevertheless, in spite of the variation that they bring, these influences do little to affect the core logic of coalitionary proactive aggression in a system of anarchic warfare.

Note also that explaining why killing happens is not the only challenge in theories of small-scale war. For evolutionary anthropologists, an important problem concerns the fact that success in intergroup contests is a "public good." A series of raids might lead to the attacking group's expanding the size of its territory, in which case all group members will have access to extra resources. This means that there could be an incentive for some group members to be free-riders. Free-riders would be those who avoid joining raids, so that for them the benefit of the extra resources would be free. If everyone became a free-rider, there would be no war. Lopez 2017 describes this problem and notes that it applies more to offensive than defensive war.

To explain how this free-rider problem does not stop the practice of simple war, scholars have sometimes invoked uniquely human characteristics such as the capacities for gossip or culture that could reward or punish men appropriately for fighting. Rewards can indeed be important. Warriors practicing simple warfare can be publicly celebrated or given increased access to women or goods. In societies with a higher death rate in simple war, warriors were found to have access to a wider set of potential rewards (Glowacki and Wrangham 2013). Physical whipping of cowards has also been recorded among East African pastoralists, suggesting that punishment, too, can encourage participation in simple warfare (Mathew and Boyd 2011).

Animals show, however, that reward and punishment are not needed to solve the free-rider problem. Lethal intergroup aggression occurs among chimpanzees and wolves without such encouragements. Among chimpanzees, for example, no rewards or punishments have been found in relation to participation in intergroup aggression (Wrangham and Glowacki 2012). Certain males simply take the initiative more than others, leading the way toward enemy territories. Theory suggests that the more eager participants should be the higher-ranking individuals, since they have a greater stake in the group's success (Gavrilets and Fortunato 2014). The limited available evidence supports that idea, though other sources of variation in individual initiative are possible too (Gilby et al. 2013).

36. Lahr et al. 2016.
37. Otterbein 2004; Gat 2006; Flannery and Marcus 2012. Note that war among settled hunter-gatherers, such as the Northwest Coast Native Americans and the Asmat of New Guinea, had features of complex war, including hierarchical relationships among warriors. Since the European Upper Paleolithic, from about forty thousand years ago, is when hunter-gatherer societies were probably settled, the roots of complex war may go back to that time also.
38. Other examples of human biology evolving in response to recent new factors include adaptations to increased exposure to malaria and to increased alco-

hol intake. A variety of effects on intelligence and social behavior have also been suggested, but have yet to be firmly proved (Cochran and Harpending 2009).

39. Turney-High 1949.
40. Johnson and MacKay 2015 explore imbalances of power as a general principle in war, following the essential logic of safe proactive aggression.
41. Keegan 1976; Collins 2008, 2009.
42. Cited by Keegan 1976, p. 71.
43. Du Picq 1921.
44. Keegan 1976, p. 69.
45. Collins 2008, 2009; Grossman 1995.
46. Fraser 2000.
47. Keegan 1976, p. 20.
48. The psychological impact of male-male bonding in war appears to last a long time. Junger 2016 argues persuasively that the modern failure to maintain group unity after returning from war is a major source of PTSD among U.S. veterans.
49. Bonaparte 1808. Hanson 2001 shows the importance of esprit de corps over sheer numbers of soldiers.
50. Grainger 2007 notes that, although Alexander was a successful commander, he failed as a politician.
51. Johnson and MacKay 2015.
52. Arreguín-Toft 2005.
53. Dixon 1976, p. 400.
54. Kornbluh, ed., 1998, p. 1.
55. Ibid.
56. Wyden 1979, p. 326.
57. Schlesinger cited in ibid., p. 316.
58. Weiner 1998.
59. Johnson 2004 noted that confidence has many sources in war, of which positive illusions are merely one important contributor.
60. This is a form of biased assimilation, a more general psychological phenomenon (Lord et al. 1979).
61. Tuchman 1985, pp. 5, 380.
62. Twain 1917, p. 108.
63. Morgenthau 1973.
64. Keegan 1999, p. 71.
65. William Noel Hodgson, "England to Her Sons," quoted in Gardner 1964, p. 10.
66. Eliot, ed., 1941, p. 164.
67. Moran 2007.
68. From Shakespeare's *Hamlet,* act III, scene i, line 56.
69. Parry 1998, p. 117.

70. Johnson 1771 (1913), p. 62.
71. Pinker 2011; Goldstein 2012; Falk and Hildebolt 2017; Oka et al. 2017.
72. This guesstimate is based on the total habitable land area's being 63.8 million square kilometers (24.6 million square miles), hunter-gatherers' living at a global population density of 0.5 per square kilometer (1.3 per square mile), and an average society size of one thousand.
73. Longer peace, more casualties in the next war: Falk and Hildebolt 2017.
74. Forecasts of the likelihood and timing of a world state were reviewed by Carneiro 2004. 2300 CE was predicted by the trend in the declining number of autonomous political units since 1500 BCE. The figure of 3500 CE was extrapolated from increases in the size of the world's twenty-eight largest empires since the Akkadian Empire in 2100 BCE.
75. Overy 2009; Hathaway and Shapiro 2017.
76. Suicide attackers are a striking exception, since their behavior seems unlikely to be adaptive. They normally represent the consequences of intense cultural persuasion (Atran 2003).
77. In 1983, President Ronald Reagan proposed a Strategic Defense Initiative, popularly known as Star Wars. Reagan persuaded Congress to fund the SDI on the basis that it would provide an effective shield against nuclear attack by the USSR. The effort proved technically impossible, politically challenging, and hugely expensive, and was eventually shelved. It was dangerous partly because it contributed to the arms race, and also because, if it had worked, it could have tempted U.S. leaders into increased aggression on the basis of their being invulnerable to nuclear attack.

13. Paradox Lost

1. This resolution of the paradox is much as was sketched by Knauft 1991 and elaborated by Boehm 1999, 2012, with the additional recognition here of the distinction between reactive and proactive aggression. Earlier writers who assumed the general proposition that humans' good and bad qualities come from our biology have done little to specify how the combination of aggressive and non-aggressive tendencies can be reconciled.
2. Some social insects show a similar combination. Many species of ant predictably kill members of other colonies but cooperate harmoniously with members of their own colonies. The ants' contradictory behavior is explicable by a relatively simple mechanism. Individuals are highly aggressive unless they interact with an individual that they perceive to be a nest mate, in which case their aggression is suppressed and they exhibit indiscriminate altruism. The Argentine ant is the most extreme example. There are groups of billions, including millions of colonies cooperating unselfishly, yet on encountering other groups they commit themselves to murderous battles (Giraud et al. 2002; Starks 2003).

3. Blumenbach 1811, p. 340. Note that earlier episodes of self-domestication could have happened, too. Future research might find evidence of relevant anatomical changes in the evolution of an australopithecine into a *Homo*, or from a chimpanzee-like ancestor into an australopithecine.
4. Prominent advocates for applying group selection theory (or its close cousin, multilevel selection theory) to human behavior include Sober and Wilson 1998; Nowak et al. 2010; Bowles and Gintis 2011; and Wilson 2012.
5. See, e.g., Dawkins 1982; West et al. 2007; Coyne 2009; and Pinker 2012. Pinker 2012 is an excellent survey of thinking about group selection, including twenty essays by different authors pro and con in response to Pinker's target article.
6. Domínguez-Rodrigo and Cobo-Sánchez 2017 briefly reviewed hunting by *Homo erectus*. There is excellent evidence of butchery by two million years ago at Kanjera, Kenya, including bones of many small and some medium-sized species (wildebeest-sized) (Ferraro et al. 2013). Pickering 2013 argued from such evidence that ambush hunting started around two million years ago. More specific indications of ambush hunting come from Olorgesaile, Kenya (Kübler et al. 2015). For evidence that prey animals were snared sixty to sixty-five thousand years ago in Sibudu, South Africa, see Wadley 2010. Thanks to the paleoanthropologist Neil Roach for advice on these topics.
7. Otterbein 2004, p. 85. See also Wrangham 2018. Pickering 2013 offered a contrary view to the likely association between hunting prey and killing conspecifics in *Homo*. He argued that predation is decoupled from aggression in *Homo*, and suggested that premeditated aggression was made possible only by the development of weapons. This argument is surprising given that chimpanzees, wolves, and other species use coalitionary premeditated aggression to kill without weapons.
8. Wrangham and Peterson 1996.
9. Estimates for the date of the last common ancestor between chimpanzees and humans (and, therefore, the origin of the australopithecines) continue to be refined. They depend partly on assessment of mutation rates, which are still uncertain. The dates given here are from Pilbeam and Lieberman 2017, p. 53. They determined that the best estimate for the time when the chimpanzee-bonobo and australopithecine-human lineages separated was 7.9 million years ago, with a range of possibilities from 6.5 to 9.3 million years ago.
10. Works in feminist anthropology such as Rosaldo and Lamphere 1974, and Collier and Rosaldo 1981, provided useful introductions to this generalization. For example: "Everywhere, from those societies we might want to call most egalitarian to those in which sexual stratification is most marked, men are the locus of cultural value. Some area of activity is always seen as exclusively or predominantly male, and therefore overwhelmingly and morally important. . . . This observation has its corollary in the fact that everywhere

men have authority over women, [and] have a culturally legitimated right to her subordination and compliance." (Rosaldo 1974, pp. 20–21.) As I say in the main text, this fact of human society does not mean that patriarchy is inevitable. It does suggest, however, that reductions in patriarchal power will require strong institutions.

Afterword

1. World Coalition Against the Death Penalty, http://www.worldcoalition.org/The-UN-General-Assembly-voted-overwhelmingly-for-a-6th-resolution-calling-for-a-universal-moratorium-on-executions.html.

Bibliography

Adams, Tim. 2013. "How to spot a murderer's brain." *The Guardian* May 12.

Agnvall, Beatrix, Johan Bételky, and Per Jensen. 2017. "Brain size is reduced by selection for tameness in Red Junglefowl—correlated effects in vital organs." *Scientific Reports* 7: 3306.

———, Rebecca Katajamaa, Jordi Altimiras, and Per Jensen. 2015. "Is domestication driven by reduced fear of humans? Boldness, metabolism and serotonin levels in divergently selected red junglefowl *(Gallus gallus)*." *Biology Letters* 11: 20150509.

Aguiar, Diego P., Soufien Sghari, and Sophie Creuzet. 2014. "The facial neural crest controls fore- and midbrain patterning by regulating Foxg1 expression through Smad1 activity." *Development* 141: 2494–2505.

Alcock, John. 1987. "Ardent adaptationism." *Natural History* 96: 4.

Alexander, R. D. 1987. *The Biology of Moral Systems.* New York: Aldine de Gruyter.

Alexander, Richard D. 1979. *Darwinism and Human Affairs.* Seattle: University of Washington Press.

Allman, John M. 1999. *Evolving Brains.* New York: Scientific American Library.

Alperson-Afil, Nira. 2008. "Continual fire-making by Hominins at Gesher Benot Ya'aqov, Israel." *Quaternary Science Reviews* 27: 1733–39.

———, and Naama Goren-Inbar. 2016. "Acheulian hafting: Proximal modification of small flint flakes at Gesher Benot Ya'aqov, Israel." *Quaternary International* 411: 34–43.

Amnesty International. 2015. *"Killing in the Name of Justice": The Death Penalty in Saudi Arabia.* London: Amnesty International Ltd.

Anderl, Christine, Tim Hahn, Ann-Kathrin Schmidt, Heike Moldenhauer, Karolien Notebaert, Celina Chantal Clément, and Sabine Windmann. 2016. "Facial width-to-height ratio predicts psychopathic traits in males." *Personality and Individual Differences* 88: 99–101.

Anonymous. 1821. *Account of the short life and ignominious death of Stephen Merrill Clark.* Salem, MA: T. C. Cushing.

Argue, Debbie, Colin P. Groves, Michael S. Y. Lee, and William L. Jungers. 2017. "The affinities of *Homo floresiensis* based on phylogenetic analyses of cranial, dental, and postcranial characters." *Journal of Human Evolution* 107: 107–33.

Arreguín-Toft, Ivan. 2005. *How the Weak Win Wars: A Theory of Asymmetric Conflict.* Cambridge, U.K.: Cambridge University Press.

Asakawa-Haas, Kenji, Martina Schiestl, Thomas Bugnyar, and Jorg J. M. Massen. 2016. "Partner choice in raven *(Corvus corax)* cooperation." *PLoS ONE* 11(6): e0156962.

Atran, Scott. 2003. "Genesis of suicide terrorism." *Science* 299: 1534–39.

Babcock, Julia C., Andra L. T. Tharp, Carla Sharp, Whitney Heppner, and Matthew S. Stanford. 2014. "Similarities and differences in impulsive/premeditated and reactive/proactive bimodal classifications of aggression." *Aggression and Violent Behavior* 19: 251–62.

Bagehot, Walter. 1872. *Physics and Politics: Or Thoughts on the Application of the Principles of 'Natural Selection' and 'Inheritance' to Political Society.* In N. St John-Stevas, ed., *The Collected Works of Walter Bagehot,* vol. VII (Cambridge, MA: Harvard University Press, 1974).

Bagemihl, Bruce. 1999. *Biological Exuberance: Animal Homosexuality and Natural Diversity.* New York: St. Martin's Press.

Bailey, Robert C. 1991. *The Behavioral Ecology of Efe Pygmy Men in the Ituri Forest, Zaire.* Ann Arbor: University of Michigan Museum Press.

Baker, Kate, and Barbara B. Smuts. 1994. "Social relationships of female chimpanzees: Diversity between captive social groups." In *Chimpanzee Cultures,* edited by R. W. Wrangham, W. C. McGrew, F. B. M. de Waal, and P. G. Heltne (Cambridge, Mass.: Harvard University Press), pp. 227–42.

Baker, Laura A., Adrian Raine, Jianghong Liu, and Kristen C. Jacobson. 2008. "Differential genetic and environmental influences on reactive and proactive aggression in children." *Journal of Abnormal Child Psychology* 36: 1265–78.

Banner, Stuart. 2003. *The Death Penalty: An American History.* Cambridge, Mass.: Harvard University Press.

Barash, David. 2003. "Review of 'Tree of Origin.'" *Physiology and Behavior* 78: 513–14.

Barth, Frederik. 1975. *Ritual and Knowledge Among the Baktaman of New Guinea.* New Haven, Conn.: Yale University Press.

Barthes, Julien, Pierre-André Crochet, and Michel Raymond. 2015. "Male homosexual preference: Where, when, why?" *PLoS ONE* 10 (8): e0134817.

Bartlett, Thad Q., Robert W. Sussman, and Jim M. Cheverud. 1993. "Infant killing in primates—a review of observed cases with specific reference to the sexual selection hypothesis." *American Anthropologist* 95: 958–90.

Basedow, Herbert. 1929. *The Australian Aboriginal.* Adelaide: F. W. Preece.

Bateson, Melissa, Daniel Nettle, and G. Roberts. 2006. "Cues of being watched enhance cooperation in a real-world setting." *Biology Letters* 2: 412–14.

Baumard, Nicolas. 2016. *How Evolution Explains Our Moral Nature.* New York: Oxford University Press.

Beard, Mary. 2015. *SPQR: A History of Ancient Rome.* New York: Liveright.

Bedau, Hugo A. 1982. *The Death Penalty in America: Current Controversies.* Oxford, U.K.: Oxford University Press.

Bednarik, Robert G. 2014. "Doing with less: hominin brain atrophy." *HOMO—Journal of Comparative Human Biology* 65: 433–49.

Behncke, Isabel. 2015. "Plan in the Peter Pan ape." *Current Biology* 25: R24–R27.

Behringer, Verena, Tobias Deschner, Róisín Murtagh, Jeroen M. G. Stevens, and Gottfried Hohmann. 2014. "Age-related changes in thyroid hormone levels of bonobos and chimpanzees indicate heterochrony in development." *Journal of Human Evolution* 66: 83–88.

Bellinger Centenary Committee. 1963. *The Bellinger Valley.* Bellingen, Australia: Courier Sun.

Belyaev, Dmitri K., A. O. Ruvinsky, and L. N. Trut. 1981. "Inherited activation-inactivation of the star gene in foxes." *Journal of Heredity* 72: 267–74.

Benedict, Ruth. 1934. *Patterns of Culture.* Boston, Mass.: Houghton Mifflin.

Berger, Lee R., John Hawks, Paul H. G. M. Dirks, Marina Elliott, and Eric M. Roberts. 2017. "*Homo naledi* and Pleistocene hominin evolution in subequatorial Africa." *eLife* 6: e24234.

Berkowitz, Leonard. 1993. *Aggression: Its Causes, Consequences, and Control.* Philadelphia: Temple University Press.

Berndt, Ronald M. 1965. "Law and order in Aboriginal Australia." In *Aboriginal Man in Australia,* edited by Ronald M. Berndt and Catherine H. Berndt (Sydney: Angus and Robertson), pp. 167–206.

———, and Catherine H. Berndt. 1988. *The World of the First Australians.* Canberra: Aboriginal Studies Press.

Bhopal, Raj. 2007. "The beautiful skull and Blumenbach's errors." *British Medical Journal* 335: 1308–9.

Bidau, Claudio J. 2009. "Domestication through the centuries: Darwin's ideas and Dmitry Belyaev's long-term experiment in silver foxes." *Gayana* 73: 55–72.

Birdsell, Joseph B. 1953. "Some environmental and cultural factors influencing the structuring of Australian aboriginal populations." *American Naturalist* 87 (834): 171–207.

Bishop, Charles A., and Victor P. Lytwyn. 2007. "'Barbarism and ardour of war from the tenderest years': Cree-Inuit warfare in the Hudson Bay region." In *North American Indigenous Warfare and Ritual Violence,* edited by Richard J. Chacon and Rubén G. Mendoza (Tucson: University of Arizona Press), pp. 30–57.

Black, M. C., K. C. Basile, M. J. Breiding, S. G. Smith, M. L. Walters, M. T.

Merrick, J. Chen, and M. R. Stevens. 2011. *The National Intimate Partner and Sexual Violence Survey (NISVS): 2010 Summary Report.* Atlanta: National Center for Injury Prevention and Control, Centers for Disease Control and Prevention.

Bloom, Paul. 2012. "Religion, morality, evolution." *Annual Review of Psychology* 63: 179–99.

Blumenbach, Johann Friedrich. 1795. De Generis Humani Varietate Nativa. In *The Anthropological Treatises of Johann Friedrich Blumenbach (1865).* Translated by Thomas Bendyshe. London: Longman, Green, Longman, Roberts, & Green, pp. 153–276.

———. 1806. Contributions to Natural History, Part 1. In *The Anthropological Treatises of Johann Friedrich Blumenbach (1865).* Translated by Thomas Bendyshe. London: Longman, Green, Longman, Roberts, & Green, pp. 277–324.

———. 1811. Contributions to Natural History, Part 2. In *The Anthropological Treatises of Johann Friedrich Blumenbach (1865).* Translated by Thomas Bendyshe. London: Longman, Green, Longman, Roberts, & Green, pp. 325–45.

Boas, Franz. 1888. *The Central Eskimo.* In *Sixth Annual Report of the Bureau of Ethnology, 1884–85,* edited by J. W. Powell (Washington, D.C.: Government Printing Office), pp. 409–675.

———. 1904. "The history of anthropology." *Science* 20: 513–24.

———. 1938. *The Mind of Primitive Man.* New York: Macmillan.

Boehm, Christopher. 1993. "Egalitarian behavior and reverse dominance hierarchy." *Current Anthropology* 34: 227–40.

———. 1999. *Hierarchy in the Forest: The Evolution of Egalitarian Behavior.* Cambridge, Mass.: Harvard University Press.

———. 2011. "Retaliatory violence in human prehistory." *British Journal of Criminology* 51: 518–34.

———. 2012. *Moral Origins: The Evolution of Virtue, Altruism, and Shame.* New York: Basic Books.

———. 2017. "Ancestral precursors, social control, and social selection in the evolution of morals." In *Chimpanzees and Human Evolution,* edited by M. N. Muller, R. W. Wrangham and D. P. Pilbeam (Cambridge, Mass.: Harvard University Press).

Bonanni, Roberto, Simona Cafazzo, Arianna Abis, Emanuela Barillari, Paola Valsecchi, and Eugenia Natoli. 2017. "Age-graded dominance hierarchies and social tolerance in packs of free-ranging dogs." *Behavioral Ecology* 28 (4): 1004–20.

Bonaparte, Napoleon. 1808. "Observations on Spanish affairs." August 27. Written at Saint-Cloud.

Boomsma, Jacobus J. 2016. "Fifty years of illumination about the natural levels of adaptation." *Current Biology* 26: R1247–71.

Borries, Carola, Kristin Launhardt, Cornelia Epplen, Jörg T. Epplen, and Paul Winkler. 1999. "Males as infant protectors in Hanuman langurs *(Presbytis entellus)* living in multimale groups—defence pattern, paternity and sexual behaviour." *Behavioral Ecology and Sociobiology* 46: 350–56.

Bourke, Joanna. 2001. *The Second World War: A People's History.* Oxford, U.K.: Oxford University Press.

Bowles, Samuel. 2009. "Did warfare among ancestral hunter-gatherers affect the evolution of human social behaviors?" *Science* 324: 1293–98.

———, and Herbert Gintis. 2011. *A Cooperative Species: Human Reciprocity and Its Evolution.* Princeton, N.J.: Princeton University Press.

Brace, C. Loring, Karen R. Rosenberg, and Kevin D. Hunt. 1987. "Gradual change in human tooth size in the late Pleistocene and post-Pleistocene." *Evolution* 41: 705–20.

Breggin, Peter R. 2014. *Guilt, Shame and Anxiety: Understanding and Overcoming Negative Emotions.* Amherst, N.Y.: Prometheus Books.

Bridges, E. Lucas. 1948. *Uttermost Part of the Earth.* London: Hodder and Stoughton.

Briggs, Jean. 1970. *Never in Anger: Portrait of an Eskimo Family.* Cambridge, Mass.: Harvard University Press.

Bromhall, Clive. 2003. *The Eternal Child: An Explosive New Theory of Human Origins and Behaviour.* London: Ebury.

Brookman, Fiona. 2003. "Confrontational and revenge homicides among men in England and Wales." *Australian and New Zealand Journal of Criminology* 36: 34–59.

———. 2015. "Killer decisions: The role of cognition, affect and 'expertise' in homicide." *Aggression and Violent Behavior* 20: 42–52.

Brüne, Martin. 2007. "On human self-domestication, psychiatry, and eugenics." *Philosophy, Ethics, and Humanities in Medicine* 2 (21): 1–9.

Brusini, Irene, Miguel Carneiro, Chunliang Wang, Carl-Johan Rubin, Henrik Ring, Sandra Afonso, José A. Blanco-Aguiar et al. 2018. "Changes in brain architecture are consistent with altered fear processing in domestic rabbits." *PNAS* 115: 7380–85.

Buckner, Randy L., and Fenna M. Krienen. 2013. "The evolution of distributed association networks in the human brain." *Trends in Cognitive Sciences* 17 (12): 648–65.

Burch, Ernest S., Jr. 1988. *The Eskimos.* Norman, OK: University of Oklahoma Press.

———. 2007. "Traditional native warfare in western Alaska." In *North American Indigenous Warfare and Ritual Violence,* edited by Richard J. Chacon and Rubén G. Mendoza (Tucson: University of Arizona Press), pp. 11–29.

———. 2005. *Alliance and Conflict: The World System of the Inupiaq Eskimos.* Lincoln: University of Nebraska Press.

Burke, Edmund. 1770. *Thoughts on the Cause of the Present Discontents.* 3rd ed. London: J. Dodsley.

Bushman, Brad J., and Craig A. Anderson. 2001. "Is it time to pull the plug on the hostile versus instrumental aggression dichotomy?" *Psychological Review* 108: 273–79.

Buttner, Alicia Phillips. 2016. "Neurobiological underpinnings of dogs' human-like social competence: How interactions between stress response systems and oxytocin mediate dogs' social skills." *Neuroscience and Biobehavioral Reviews* 71: 198–214.

Butynski, T. M. 1982. "Harem-male replacement and infanticide in the blue monkey *(Cercopithecus mitis stuhlmanni)* in the Kibale Forest, Uganda." *American Journal of Primatology* 3: 1–22.

Byers, John A. 1997. *American Pronghorn: Social Adaptations and the Ghosts of Predators Past.* Chicago: University of Chicago Press.

Cafazzo, Simona, Martina Lazzaroni, and Sarah Marshall-Pescini. 2016. "Dominance relationships in a family pack of captive arctic wolves *(Canis lupus arctos):* The influence of competition for food, age and sex." *PeerJ* 4: e2707; doi 10.7717/peerj.2707.

Candland, Douglas K. 1993. *Feral Children amd Clever Animals: Reflections on Human Nature.* New York: Oxford University Press.

Cardini, Andrea, and Sarah Elton. 2009. "The radiation of red colobus monkeys (Primates, Colobinae): Morphological evolution in a clade of endangered African primates." *Zoological Journal of the Linnean Society* 157: 197–224.

Carnegie Commission on Preventing Deadly Conflict. 1997. *Preventing Deadly Conflict: Final Report.* Washington, D.C.: Carnegie Commission on Preventing Deadly Conflict.

Carneiro, Miguel, Carl-Johan Rubin, Federica Di Palma, Frank W. Albert, Jessica Alföldi, Alvaro Martinez Barrio, Gerli Pielberg, et al. 2014. "Rabbit genome analysis reveals a polygenic basis for phenotypic change during domestication." *Science* 345: 1074–79.

Carré, Justin M., and Cheryl M. McCormick. 2008. "In your face: Facial metrics predict aggressive behavior in the laboratory and in varsity and professional hockey players." *Proceeding of the Royal Society B* 275: 2651–56.

Carré, Justin M., Cheryl M. McCormick, and Ahmad R. Hariri. 2011. "The social neuroendocrinology of human aggression." *Psychoneuroendocrinology* 36: 935–44.

Carré, Justin M., Kelly R. Murphy, and Ahmad R. Hariri. 2013. "What lies beneath the face of aggression?" *SCAN* 8: 224–29.

Cashdan, Elizabeth. 1980. "Egalitarianism among hunter-gatherers." *Current Anthropology* 24: 116–20.

Cassidy, Kira A., and Richard T. McIntyre. 2016. "Do gray wolves *(Canis lupus)* support pack mates during aggressive inter-pack interactions?" *Animal Cognition* 19: 937–47; doi: 10.1007/s10071-016-0994-1.

Chacon, R. J., and R. G. Mendoza, eds. 2007. *North American Indigenous Warfare and Ritual Violence.* Tucson: University of Arizona Press.

Chagnon, Napoleon A. 1997. *Yanomamö.* 5th ed. In *Case Studies in Cultural Anthropology series, ed. George Spindler and Louise Spindler* (Fort Worth, Texas: Harcourt Brace).

Chapais, Bernard. 2015. "Competence and the evolutionary origins of status and power in humans." *Human Nature* 26: 161–83; doi: 10.1007/s12110-015-9227-6.

Chinard, Gilbert. 1931. "Introduction." In *Dialogues curieux entre l'auteur et un sauvage de Bons Sens qui a voyagé, et mémoires de l'Amérique Septentrionale (Baron Louis Armand de Lom d'Arce Lahontan, 1703–1705),* edited by Gilbert Chinard (Baltimore: Johns Hopkins University Press).

Choi, Jung-Kyoo, and Samuel Bowles. 2007. "The coevolution of parochial altruism and war." *Science* 318: 636–40.

Chudasama, Yogita, Alicia Izquierdo, and Elisabeth A. Murray. 2009. "Distinct contributions of the amygdala and hippocampus to fear expression." *European Journal of Neuroscience* 30: 2327–37.

Chudek, Maciej, and Joseph Henrich. 2011. "Culture–gene coevolution, norm-psychology and the emergence of human prosociality." *Trends in Cognitive Sciences* 15 (5): 218–26.

Cieri, Robert L., Steven E. Churchill, Robert G. Franciscus, Jingzhi Tan, and Brian Hare. 2014. "Craniofacial feminization, social tolerance, and the origins of behavioral modernity." *Current Anthropology* 55: 419–43.

Clark, Gregory. 2007. *A Farewell to Alms. A Brief Economic History of the World.* Princeton and Oxford: Princeton University Press.

Clay, Zanna, Takeshi Furuichi, and Frans B. M. de Waal. 2016. "Obstacles and catalysts to peaceful coexistence in chimpanzees and bonobos." *Behaviour* 153 (9–11): 1293–330.

———, and Klaus Zuberbühler. 2012. "Communication during sex among female bonobos: Effects of dominance, solicitation and audience." *Scientific Reports* 2: 291.

Clutton-Brock, Juliet. 1992. "How the wild beasts were tamed." *New Scientist* 133 (1808): 41–43.

Clutton-Brock, Timothy H., Fiona E. Guinness, and Steven D. Albon. 1982. *Red Deer: Behavior and Ecology of Two Sexes.* Chicago: University of Chicago Press.

Cochran, Gregory, and Henry Harpending. 2009. *The 10,000 Year Explosion: How Civilization Accelerated Human Evolution.* New York: Basic Books.

Coid, J., M. Yang, S. Ullrich, A. Roberts, and R. D. Hare. 2009. "Prevalence and correlates of psychopathic traits in the household population of Great Britain." *International Journal of Law and Psychiatry* 32: 65–73.

Coleman, Julia. 1982. "A new look at the north coast: Fish traps and 'villages.'" In *Coastal Archaeology in Eastern Australia,* edited by Sandra Bowdler (Canberra: Australian National University Press), pp. 1–10.

Collier, Jane F., and Michelle Z. Rosaldo. 1981. "Politics and gender in simple societies." In *Sexual Meanings: The Cultural Construction of Gender and Sexuality,* edited by Sherry B. Ortner and Harriet Whitehead (Cambridge, U.K.: Cambridge University Press), pp. 275–329.

Collins, Randall. 2008. *Violence: A Micro-Sociological Theory.* Princeton, N.J.: Princeton University Press.

———. 2009. "Micro and macro causes of violence." *International Journal of Conflict and Violence* 3 (1): 9–22.

Coolidge, Harold J. 1984. "Historical remarks bearing on the discovery of Pan paniscus." In *The Pygmy Chimpanzee,* edited by Randall Susman (New York: Plenum), pp. ix–xiii.

Coppinger, Raymond, and Lorna Coppinger. 2000. *Dogs: A Startling New Understanding of Canine Origin, Behavior, and Evolution.* New York: Scribner.

Corballis, Michael C. 2017. "The evolution of language: Sharing our mental lives." *Journal of Neurolinguistics* 43: 120–32.

Cornell, Dewey G., Janet Warren, Gary Hawk, Ed Stafford, Guy Oram, and Denise Pine. 1996. "Psychopathy in instrumental and reactive violent offenders." *Journal of Consulting and Clinical Psychology* 64: 783–90.

Cowgill, Colleen M., Kimberly Rios, and Ain Simpson. 2017. "Generous heathens? Reputational concerns and atheists' behavior toward Christians in economic games." *Journal of Experimental Social Psychology* 73: 169–79.

Coyne, Jerry A. 2009. *Why Evolution Is True.* New York: Viking.

Craig, Ian W., and Kelly E. Halton. 2009. "Genetics of human aggressive behaviour." *Human Genetics* 126: 101–13.

Creuzet, Sophie E. 2009. "Regulation of pre-otic brain development by the cephalic neural crest." *PNAS* 106: 15774–79.

Crick, N. R., and K. A. Dodge. 1996. "Social information processing mechanisms in reactive and proactive aggression." *Child Development* 67: 993–1002.

Crockett, Caroline M., and Ranka Sekulic. 1984. "Infanticide in red howler monkeys *(Alouatta seniculus).*" In *Infanticide: comparative and evolutionary perspectives,* edited by G. Hausfater and S. B. Hrdy, pp. 173–91. New York: Aldine.

Crockford, Susan J. 2002. "Animal domestication and heterochronic speciation: the role of thyroid hormone." In *Human Evolution Through Developmental Change,* edited by Nancy Minugh-Purvis and Kenneth McNamara (Baltimore: Johns Hopkins University Press).

Crook, D. Paul. 1994. *Darwinism, War and History: The Debate over the Biology of War from the 'Origin of Species' to the First World War.* Cambridge, U.K.: Cambridge University Press.

Cubaynes, Sarah, Daniel R. MacNulty, Daniel R. Stahler, Kira A. Quimby, Douglas W. Smith, and Tim Coulson. 2014. "Density-dependent intraspecific aggression regulates survival in northern Yellowstone wolves *(Canis lupus).*" *Journal of Animal Ecology* 83: 1344–56.

Cushman, Fiery, and Liane Young. 2011. "Patterns of moral judgment derive from nonmoral psychological representations." *Cognitive Science* 35 (6): 1052–75; doi: 10.1111/j.1551-6709.2010.01167.x.

Dagg, Anne I. 2000. "The infanticide hypothesis: A response to the response." *American Anthropologist* 102: 831–34.

Dalberg-Acton, John Emerich Edward. 1949. *Essays on Freedom and Power.* Boston, Mass.: Beacon Press.

Daly, Martin, and Margo Wilson. 1988. *Homicide.* Hawthorne, N.Y.: Aldine.

Damasio, Antonio R. 1995. *Descartes' Error: Emotion, Reason, and the Human Brain.* New York: Avon.

Dambacher, Franziska, Teresa Schuhmann, Jill Lobbestael, Arnoud Arntz, Suzanne Brugman, and Alexander T. Sack. 2015. "Reducing proactive aggression through non-invasive brain stimulation." *SCAN* 10: 1303–9.

Darwin, Charles. 1845 (2005). *Voyage of the Beagle.* In *From So Simple a Beginning: Darwin's Four Great Books* (New York: W. W. Norton).

———. 1868 *The Variation of Animals and Plants Under Domestication.* London: John Murray.

———. 1871 (2005). *The Descent of Man, and Selection in Relation to Sex. In So Simple a Beginning: Darwin's Four Great Books* (New York: W. W. Norton).

———. 1872 (2005). *The Expression of the Emotions in Man and Animals.* In *So Simple a Beginning: Darwin's Four Great Books* (New York: W. W. Norton).

Davidson, Richard J., Katherine M. Putnam, and Christine L. Larson. 2000. "Dysfunction in the neural circuitry of emotion regulation—a possible prelude to violence." *Science* 289: 591–94.

Davie, Maurice R. 1929. *The Evolution of War: A Study of Its Role in Early Societies.* New Haven, Conn.: Yale University Press.

Dawkins, Richard. 1982. *The Extended Phenotype.* Oxford, U.K.: Oxford University Press.

de Almeida, Rosa Maria Martins, João Carlos Centurion Cabral, and Rodrigo Narvaes. 2015. "Behavioural, hormonal and neurobiological mechanisms of aggressive behaviour in human and nonhuman primates." *Physiology & Behavior* 143: 121–35.

de Boer, S. F., B. Olivier, J. Veening, and J. M. Koolhaas. 2015. "The neurobiology of offensive aggression: Revealing a modular view." *Physiology & Behavior* 146: 111–27.

Decety, Jean, Jason M. Cowell, Kang Lee, Randa Mahasneh, Susan Malcolm-Smith, Bilge Selcuk, and Xinyue Zhou. 2015. "The negative association between religiousness and children's altruism across the world." *Current Biology* 25: 2951–55.

Declercq, F., and Kurt Audenaert. 2011. "Predatory violence aiming at relief in a case of mass murder: Meloy's criteria for applied forensic practice." *Behavioral Sciences and the Law* 29: 578–91.

Dediu, Dan, and Stephen C. Levinson. 2013. "On the antiquity of language: The reinterpretation of Neandertal linguistic capacities and its consequences." *Frontiers in Psychology* 4 (307): 1–17.

de Manuel, Marc, Martin Kuhlwilm, Peter Frandsen, Vitor C. Sousa, Tariq Desai, Javier Prado-Martinez, Jessica Hernandez-Rodriguez, et al. 2016. "Chimpanzee genomic diversity reveals ancient admixture with bonobos." *Science* 354: 477–81.

de Quervain, Dominique J.-F., Urs Fischbacher, Valerie Treyer, Melanie Schellhammer, Ulrich Schnyder, Alfred Buck, and Ernst Fehr. 2004. "The neural basis of altruistic punishment." *Science* 305: 1254–58.

DeScioli, Peter, and Robert Kurzban. 2009. "Mysteries of morality." *Cognition* 112: 281–99.

DeScioli, Peter, and Robert Kurzban. 2013. "A solution to the mysteries of morality." *Psychological Bulletin* 139 (2): 477–96.

Des Pres, Terrence. 1976. *The Survivor: An Anatomy of Life in the Death Camps.* New York: Oxford University Press.

de Waal, F. B. M. 2006. *Primates and Philosophers: How Morality Evolved.* Princeton, N.J.: Princeton University Press.

de Waal, Frans. 1996. *Good Natured: The Origins of Right and Wrong in Humans and Other Animals.* Cambridge, Mass.: Harvard University Press.

Dibble, Harold L., Dennis Sandgathe, Paul Goldberg, Shannon McPherron, and Vera Aldeias. 2018. "Were Western European Neandertals able to make fire?" *Journal of Paleolithic Archaeology* 1 (1): 54-79.

Dirks, Paul H. G. M., Eric M. Roberts, Hannah Hilbert-Wolf, Jan D. Kramers, John Hawks, Anthony Dosseto, Mathieu Duval, et al. 2017. "The age of *Homo naledi* and associated sediments in the Rising Star Cave, South Africa." *eLife* 6: e24231.

Dixon, Norman. 1976. *On the Psychology of Military Incompetence.* London: Jonathan Cape.

Dixson, Alan F. 2012. *Primate Sexuality: Comparative Studies of the Prosimians, Monkeys, Apes and Human Beings.* New York: Oxford University Press.

Dobzhansky, T. 1962. *Mankind Evolving.* New Haven, Conn.: Yale University Press.

Dobzhansky, Theodosius. 1973. "Nothing in biology makes sense except in the light of evolution." *The American Biology Teacher* 35 (3): 125–29.

Dodge, Kenneth A., and John D. Coie. 1987. "Social-information-processing factors in reactive and proactive aggression in children's peer groups." *Journal of Personality and Social Psychology* 53 (6): 1146–58.

Domínguez-Rodrigo, Manuel, and Lucía Cobo-Sánchez. 2017. "A spatial analysis of stone tools and fossil bones at FLK Zinj 22 and PTK I (Bed I, Olduvai

Gorge, Tanzania) and its bearing on the social organization of early humans." *Palaeogeography, Palaeoclimatology, Palaeoecology* 488: 21–34.

Drea, Christine M., and A. N. Carter. 2009. "Cooperative problem solving in a social carnivore." *Animal Behaviour* 78: 967–77.

Dreger, Alice. 2011. "Darkness's descent on the American Anthropological Association: A cautionary tale." *Human Nature* 22: 225–46; doi: 10.1007/s12110-011-9103-y.

Duda, Pavel, and Jan Zrzavý. 2013. "Evolution of life history and behavior in Hominidae: Towards phylogenetic reconstruction of the chimpanzee-human last common ancestor." *Journal of Human Evolution* 65: 424–46.

Dugatkin, Lee, and Lyudmila Trut. 2017. *How to Tame a Fox (and Build a Dog): Visionary Scientists and a Siberian Tale of Jump-Started Evolution.* Chicago: University of Chicago Press.

du Picq, Charles Jean Jacques Joseph Ardant. 1921. *Battle Studies: Ancient and Modern Battle.* Translated by John N. Greely and Robert C. Cotton. New York: Macmillan.

Durkheim, Émile. 1902. *The Division of Labor in Society.* New York: Free Press.

Durrleman, Stanley, Xavier Pennec, Alain Trouvé, Nicholas Ayache, and José Braga. 2012. "Comparison of the endocranial ontogenies between chimpanzees and bonobos via temporal regression and spatiotemporal registration." *Journal of Human Evolution* 62: 74–88.

Eliot, T. S., ed. 1941. *A Choice of Kipling's Verse.* London: Faber and Faber.

Elkin, A. P. 1938. *The Australian Aborigines: How to Understand Them.* Sydney: Angus and Robertson.

Ellingson, Ter. 2001. *The Myth of the Noble Savage.* Berkeley: University of California Press.

Engelmann, Jan M., Esther Herrmann, and Michael Tomasello. 2012. "Five-year olds, but not chimpanzees, attempt to manage their reputations." *PLoS ONE* 7(10): e48433.

———. 2016. "The effects of being watched on resource acquisition in chimpanzees and human children." *Animal Cognition* 19: 147–51.

Erdal, David, and Andrew Whiten. 1994. "On human egalitarianism: An evolutionary product of Machiavellian status escalation?" *Current Anthropology* 35: 175–78.

Evin, Allowen, Joseph Owen, Greger Larson, Mélanie Debiais-Thibaud, Thomas Cucchi, Una Strand Vidarsdottir, and Keith Dobney. 2017. "A test for paedomorphism in domestic pig cranial morphology." *Biology Letters* 13: 20170321.

Falk, Dean, and Charles Hildebolt. 2017. "Annual war deaths in small-scale versus state societies scale with population size rather than violence." 58: 805–13.

Fawcett, Katie, and Geresomu Muhumuza. 2000. "Death of a wild chimpanzee community member: Possible outcome of intense sexual competition." *American Journal of Primatology* 51: 243–47.

Feinberg, Matthew, Robb Willer, and Dacher Keltner. 2012. "Flustered and faithful: Embarrassment as a signal of prosociality." *Journal of Personality and Social Psychology* 102 (1): 81–97.

Feinstein, Justin S., Ralph Adolphs, Antonio Damasio, and Daniel Tranel. 2011. "The human amygdala and the induction and experience of fear." *Current Biology* 21: 34–38.

Feldblum, Joseph T., Emily E. Wroblewski, Rebecca S. Rudicell, Beatrice H. Hahn, Thais Paiva, Mine Cetinkaya-Rundel, Anne E. Pusey, and Ian C. Gilby. 2014. "Sexually coercive male chimpanzees sire more offspring." *Current Biology* 24: 2855–60.

Ferguson, R. Brian. 1984. "Introduction." In *Warfare, Culture and Environment,* edited by R. Brian Ferguson (Orlando, Fla.: Academic Press), pp. 1–81.

———. 1997. "Violence and war in prehistory." In *Troubled Times: Violence and Warfare in the Past,* edited by D. Martin and D. Frayer (Amsterdam: Gordon and Breach), pp. 321–55.

———. 2011. "Born to live: Challenging killer myths." In *Origins of Altruism and Cooperation,* edited by R. W. Sussman and C. R. Cloninger (New York: Springer), pp. 249–70.

Ferraro, Joseph V., Thomas W. Plummer, Briana L. Pobiner, James S. Oliver, Laura C. Bishop, David R. Braun, Peter W. Ditchfield, et al. 2013. "Earliest archaeological evidence of persistent hominin carnivory." *PLoS ONE* 8(4): e62174.

Ficks, Courtney A., and Irwin D. Waldman. 2014. "Candidate genes for aggression and antisocial behavior: a meta-analysis of association studies of the 5HTTLPR and MAOA-uVNT." *Behavior Genetics* 44: 427–44.

Fischer, David Hackett. 1992. *Albion's Seed: Four British Folkways in America.* New York: Oxford University Press.

Fiske, Alan Page, and Tage Shakti Rai. 2015. *Virtuous Violence: Hurting and Killing to Create, Sustain, End and Honor Social Relationships.* Cambridge, U.K.: Cambridge University Press.

Fitzgerald, F. Scott. 1936. *The Crack-Up.* New York: New Directions.

Flanagan, James G. 1989. "Hierarchy in simple 'egalitarian' societies." *Annual Review of Anthropology* 18: 245–66.

Flannery, Kent, and Joyce Marcus. 2012. *The Creation of Inequality: How Our Prehistoric Ancestors Set the Stage for Monarchy, Slavery and Empire.* Cambridge, Mass.: Harvard University Press.

Flynn, J. P. 1967. "The neural basis of aggression in cats." In *Neurophysiology and Emotion,* edited by D. C. Glass (New York: Rockefeller University Press), pp. 40–60.

Foucault, Michel. 1994. *Power.* Translated by Robert Hurley et al. Edited by Paul Rabinow. Vol. 3, *Essential Works of Foucault 1954–1984.* New York: New Press.

Francis, Richard. 2015. *Domesticated: Evolution in a Man-Made World.* New York: W. W. Norton.

Fraser, David. 2000. *Frederick the Great: King of Prussia.* London: Allen Lane.

Frayer, David W. 1980. "Sexual dimorphism and cultural evolution in the late Pleistocene and Holocene of Europe." *Journal of Human Evolution* 9: 399–415.

Freuchen, Peter. 1935. *Arctic Adventure—My LIfe in the Frozen North.* New York: Farrar & Rinehart.

Frey, Karin S., Jodi Burrus Newman, and Adaurennaya C. Onyewuenyi. 2014. "Aggressive forms and functions on school playgrounds: profile variations in interaction styles, bystander actions, and victimization." *Journal of Early Adolescence* 34 (3): 285–310.

Fried, Morton H. 1967. *The Evolution of Political Society: An Essay in Political Anthropology.* New York: Random House.

Fromm, Erich. 1973. *The Anatomy of Human Destructiveness.* New York: Holt, Rinehart and Winston.

Frost, Peter, and Henry C. Harpending. 2015. "Western Europe, state formation, and genetic pacification." *Evolutionary Psychology* 13 (1): 230–43.

Fry, Douglas P. 2006. *The Human Potential for Peace: An Anthropological Challenge to Assumptions About War and Violence.* New York: Oxford University Press.

———, and Patrik Söderberg. 2013. "Lethal aggression in mobile forager bands and implications for the origins of war." *Science* 341: 270–74.

Fuentes, Agustin. 2012. *Race, Monogamy, and Other Lies They Told You.* Berkeley: University of California Press.

Furuichi, Takeshi. 1989. "Social interactions and the life history of female Pan paniscus in Wamba." *International Journal of Primatology* 10:173-197.

———. 2009. "Factors underlying party size differences between chimpanzees and bonobos: A review and hypotheses for future study." *Primates* 50: 197–209.

———. 2011. "Female contributions to the peaceful nature of bonobo society." *Evolutionary Anthropology* 20: 131–42.

Futch, Ovid L. 1999. *History of Andersonville Prison.* Gainesville: University Press of Florida.

Garber, P. A., L. M. Porter, J. Spross, and A. Di Fiore. 2016. "Tamarins: Insights into monogamous and non-monogamous single female social and breeding systems." *American Journal of Primatology* 78: 298–314.

García-Moreno, Claudia, Henrica A. F. M. Jansen, Charlotte Watts, and Mary Ellsberg. 2005. *"WHO Multi-Country Study on Women's Health and Domestic Violence Against Women: Summary Report of Initial Results on Prevalence, Health Outcomes and Women's Responses."* Geneva: World Health Organization.

García-Moreno, Claudia, Christina Pallitto, Karen Devries, Heidi Stöckl, Charlotte Watts, and Naeemah Abrahams. 2013. *Global and Regional Estimates of*

Violence Against Women: Prevalence and Health Effects of Intimate Partner Violence and Non-partner Sexual Violence. Geneva, Switzerland: World Health Organization.

Gardner, Brian. 1964. *Up the Line to Death: The War Poets 1914–1918.* London: Methuen.

Gariépy, J., D. Bauer, and R. Cairns. 2001. "Selective breeding for differential aggression in mice provides evidence for heterochrony in social behaviours." *Animal Behaviour* 61: 933–47.

Gat, A. 2006. *War in Human Civilization.* Oxford, U.K.: Oxford University Press.

Gat, Azar. 2015. "Proving communal warfare among hunter-gatherers: The Quasi-Rousseauan Error." *Evolutionary Anthropology* 24: 111–26.

Gavrilets, Sergey, and Laura Fortunato. 2014. "A solution to the collective action problem in between-group conflict with within-group inequality." *Nature Communications* 5: 3256.

Gehlen, Arnold. 1944. *Der Mensch: Seine Natur und seine Stellung in der Welt.* 3rd ed. Berlin: Junker und Dünnhaupt.

Geiger, Madeleine, Allowen Evin, Marcelo R. Sánchez-Villagra, Dominic Gascho, Cornelia Mainini, and Christoph P. E. Zollikofer. 2017. "Neomorphosis and heterochrony of skull shape in dog domestication." *Scientific Reports* 7: 13443.

Gellner, Ernest. 1994. *Conditions of Liberty: Civil Society and Its Rivals.* London: Hamish Hamilton.

Giammarco, Erica A., and Philip A. Vernon. 2015. "Interpersonal guilt and the Dark Triad." *Personality and Individual Differences* 81: 96–101.

Gilby, Ian C., Michael L. Wilson, and Anne E. Pusey. 2013. "Ecology rather than psychology explains co-occurrence of predation and border patrols in male chimpanzees." *Animal Behaviour* 86: 61–74.

Gintis, Herbert, Carel van Schaik, and Christopher Boehm. 2015. "Zoon Politikon: The evolutionary origins of human political systems." *Current Anthropology* 56: 327–53.

Giraud, Tatiana, Jes S. Pedersen, and Laurent Keller. 2002. "Evolution of supercolonies: The Argentine ants of southern Europe." *PNAS* 99 (9): 6075–79.

Glasse, Robert M. 1968. *Huli of Papua: A Cognatic Descent System.* The Hague, Netherlands: Mouton.

Glowacki, L., and R. Wrangham. 2013. "The role of rewards in motivating participation in simple warfare: A test of the cultural rewards war-risk hypothesis." *Human Nature* 24: 444–60.

Glowacki, Luke, and Richard W. Wrangham. 2015. "Warfare and reproductive success in a tribal population." *PNAS* 112: 348–53; doi: 10.1073/pnas.1412287112.

Goffman, Erving. 1956. "Embarrassment and social organization." *American Journal of Sociology* 62: 264–71.

Golden, Sam A., Conor Heins, Marco Venniro, Daniele Caprioli, Michelle

Zhang, David H. Epstein, and Yavin Shaham. 2017. "Compulsive addiction-like aggressive behavior in mice." *Biological Psychiatry* 82 (4): 239–48.

Goldhagen, Daniel J. 1996. *Hitler's Willing Executioners: Ordinary Germans and the Holocaust.* New York: Alfred A. Knopf.

Goldizen, Anne W. 1989. "Social relationships in a cooperatively polyandrous group of tamarins *(Saguinus fuscicollis).*" *Behavioral Ecology and Sociobiology* 24: 79–89.

Goldstein, Joshua S. 2012. *Winning the War on War: The Decline of Armed Conflict Worldwide.* New York: Plume Books.

Goodall, Jane. 1986. *The Chimpanzees of Gombe: Patterns of Behavior.* Cambridge, Mass.: Harvard University Press.

———. 1999 *Reason For Hope: A Spiritual Journey.* New York: Warner.

Goren-Inbar, N., A. Alperson, M. E. Kislev, O. Simchoni, Y. Melamed, A. Ben-Nun, and E. Werker. 2004. "Evidence of Hominin Control of Fire at Gesher Benot Ya'aqov, Israel." *Science* 304: 725–27.

Goren-Inbar, Naama, Yoel Melamed, Irit Zohar, Kumar Akhilesh, and Shanti Pappu. 2014. "Beneath still waters—multistage aquatic exploitation of *Euryale ferox* (Salisb.) during the Acheulian." *Internet Archaeology* 37; doi: 10.11141/ia.37.1.

———, Gonen Sharon, Nira Alperson-Afil, and Gadi Herzlinger. 2015. "A new type of anvil in the Acheulian of Gesher Benot Ya'aqov, Israel." *Philosophical Transactions of the Royal Society B* 370: 20140353.

Gould, Stephen Jay. 1977. *Ontogeny and Phylogeny.* Cambridge, Mass.: Harvard University Press.

———. 1987. "Freudian slip." *Natural History* 96: 14–21.

———. 1996. *The Mismeasure of Man.* New York: W. W. Norton.

———. 1997. "The exaptive excellence of spandrels as a term and prototype." *PNAS* 94: 10750–55.

———, and Richard Lewontin. 1979. "The spandrels of San Marco and the Panglossian paradigm: A critique of the adaptationist programme." *Proceedings of the Royal Society B* 205: 581–98.

Gourevitch, P. 1998. *We Wish to Inform You That Tomorrow We Will Be Killed with Our Families.* New York: Farrar, Straus and Giroux.

Grainger, John D. 2007. *Alexander the Great Failure: The Collapse of the Macedonian Empire.* London: Hambledon Continuum.

Graves, Robert. 1929 (1960). *Goodbye to All That.* London: Penguin.

Greene, Joshua. 2013. *Moral Tribes: Emotion, Reason, and the Gap Between Us and Them.* New York: Penguin.

Grinker, Roy Richard. 1994. *Houses in the Rainforest: Ethnicity and Inequality Among Farmers and Foragers in Central Africa.* Berkeley: University of California Press.

Grossman, Dave. 1995. *On Killing: The Psychological Costs of Learning to Kill in War and Society.* New York: Little, Brown.

Groves, Colin P. 1989. "Feral mammals of the Mediterranean islands: Documents of early domestication." In *The Walking Larder: Patterns of Domestication, Pastoralism, and Predation,* edited by J. Clutton-Brock (London: Unwin Hyman), pp. 46–58.

Gunz, Philipp, Simon Neubauer, Lubov Golovanova, Vladimir Doronichev, Bruno Maureille, and Jean-Jacques Hublin. 2012. "A uniquely modern human pattern of endocranial development: Insights from a new cranial reconstruction of the Neandertal newborn from Mezmaiskaya." *Journal of Human Evolution* 62: 300–313.

———, Bruno Maureille, and Jean-Jacques Hublin. 2010. "Brain development after birth differs between Neanderthals and modern humans." *Current Biology* 20: R921–22.

Gurven, Michael, W. Allen-Arave, Kim Hill, and A. Magdalena Hurtado. 2000. " 'It's a wonderful life': Signaling generosity among the Ache of Paraguay." *Evolution and Human Behavior* 21: 263–82.

Haidt, Jonathan. 2007. "The new synthesis in moral psychology." *Science* 316: 998–1002.

———. 2012. *The Righteous Mind: Why Good People Are Divided by Politics and Religion.* New York: Pantheon.

Haldane, J. B. S. 1956. "The argument from animals to men: An examination of its validity for anthropology." *Journal of the Royal Anthropological Institute of Great Britain and Ireland* 86: 1–14.

Hamburg, D. A., and M. B. Trudeau, eds. 1981. *Biobehavioral Aspects of Aggression.* New York: Alan R. Liss.

Hamburg, David A. 2002. *No More Killing Fields: Preventing Deadly Conflict.* New York: Rowman and Littlefield.

———. 2010. *Preventing Genocide: Practical Steps Toward Early Detection and Effective Action.* Boulder, Colo.: Paradigm.

Hamlin, J. Kiley, and Karen Wynn. 2011. "Young infants prefer prosocial to antisocial others." *Cognitive Development* 26 (1): 30–39.

Hansen, Thomas Blom, and Finn Stepputat. 2006. "Sovereignty revisited." *Annual Review of Anthropology* 35: 295–315.

Hanson, Victor Davis. 2001. *Carnage and Culture: Landmark Battles in the Rise of Western Power.* New York: Doubleday.

Hardy, Karen, Stephen Buckley, Matthew J. Collins, Almudena Estalrrich, Don Brothwell, Les Copeland, Antonio García-Tabernero, et al. 2012. "Neanderthal medics? Evidence for food, cooking, and medicinal plants entrapped in dental calculus." *Naturwissenschaften* 99: 617–26.

Hare, B., and S. Kwetuenda. 2010. "Bonobos voluntarily share their own food with others." *Current Biology* 20: 230–31.

Hare, Brian. 2017. "Survival of the friendliest: *Homo sapiens* evolved via selection for prosociality." *Annual Review of Psychology* 68: 155–86.

———, Michelle Brown, Christina Williamson, and Michael Tomasello. 2002. "The domestication of social cognition in dogs." *Science* 298: 1634–36.

———, Alicia P. Melis, Vanessa Woods, Sara Hastings, and Richard Wrangham. 2007. "Tolerance allows bonobos to outperform chimpanzees on a cooperative task." *Current Biology* 17: 619–23.

———, Irene Plyusnina, Natalie Ignacio, Oleysa Schepina, Anna Stepika, Richard Wrangham, and Lyudmila Trut. 2005. "Social cognitive evolution in captive foxes is a correlated by-product of experimental domestication." *Current Biology* 15: 1–20.

———, Victoria Wobber, and Richard W. Wrangham. 2012. "The self-domestication hypothesis: Bonobos evolved due to selection against male aggression." *Animal Behaviour* 83: 573–85.

Harvati, Katerina. 2007. "100 years of *Homo heidelbergensis*—life and times of a controversial taxon." *Mitteilungen der Gesellschaft für Urgeschichte* 16: 85–94.

Haselhuhn, Michael P., Margaret E. Ormiston, and Elaine M. Wong. 2015. "Men's facial width-to-height ratio predicts aggression: a meta-analysis." *PLoS ONE* 10 (4): e0122637.

Hathaway, Oona A., and Scott J. Shapiro. 2017. "Outlawing war? It actually worked." *New York Times* September 2.

Hauser, Marc D. 2006. *Moral Minds: How Nature Designed Our Universal Sense of Right and Wrong.* New York: HarperCollins.

———, and Jeffrey Watumull. 2017. "The Universal Generative Faculty: The source of our expressive power in language, mathematics, morality, and music." *Journal of Neurolinguistics* 43: 78–94.

Hayden, Brian. 1993. "The cultural capacities of Neandertals: A review and re-evaluation." *Journal of Human Evolution* 24: 113–46.

———. 2012. "Neandertal social structure?" *Oxford Journal of Archaeology* 31 (1): 1–26.

———. 2014. *The Power of Feasts: From Prehistory to the Present.* Cambridge, U.K.: Cambridge University Press.

Hearne, Vicki. 1986. *Adam's Task: Calling Animals by Name.* New York: Alfred A. Knopf.

Heider, Karl. 1972. *The Dani of West Irian: An Ethnographic Companion to the Film "Dead Birds."* New York: Warner Modular Publication.

Heider, Karl G. 1997. *Grand Valley Dani: Peaceful Warriors.* New York: Holt, Rinehart and Winston.

Hemmer, H. 1990. *Domestication: The Decline of Environmental Appreciation.* Cambridge, U.K.: Cambridge University Press.

Henneberg, Maciej. 1998. "Evolution of the human brain: Is bigger better?" *Clinical and Experimental Pharmacology and Physiology* 25: 745–49.

———, and M. Steyn. 1993. "Trends in cranial capacity and cranial index in Subsaharan Africa during the Holocene." *American Journal of Human Biology* 5: 473–79.

Henrich, Joseph. 2016. *The Secret of Our Success: How Culture Is Driving Human Evolution, Domesticating Our Species, and Making Us Smarter.* Princeton, N.J.: Princeton University Press.

Henzi, Peter, and Louise Barrett. 2003. "Evolutionary ecology, sexual conflict, and behavioral differentiation among baboon populations." *Evolutionary Anthropology* 12: 217–30.

Herculano-Houzel, Suzana. 2016. *The Human Advantage: A New Understanding of How Our Brain Became Remarkable.* Boston: MIT Press.

Herdt, Gilbert. 1987. *The Sambia: Ritual and Gender in New Guinea.* Fort Worth, Texas: Harcourt Brace Jovanovich.

Herrera, Ana M., Patricia L. R. Brennan, and Martin J. Cohn. 2015. "Development of avian external genitalia: Interspecific differences and sexual differentiation of the male and female phallus." *Sexual Development* 9 (1): 43–52.

Hess, Nicole, Courtney Helfrecht, Edward Hagen, Aaron Sell, and Barry Hewlett. 2010. "Interpersonal aggression among Aka hunter-gatherers of the Central African Republic: Assessing the effects of sex, strength, and anger." *Human Nature* 21: 330–54.

Hiatt, Les R. 1996. *Arguments About Aborigines: Australia and the Evolution of Social Anthropology.* New York: Cambridge University Press.

Higham, Tom, Katerina Douka, Rachel Wood, Christopher Bronk Ramsey, Fiona Brock, Laura Basell, Marta Camps, et al. 2014. "The timing and spatiotemporal patterning of Neanderthal disappearance." *Nature* 512: 306–9.

Hinde, Robert, and Joseph Rotblat. 2003. *War No More: Eliminating Conflict in the Nuclear Age.* London: Pluto Press.

Hinde, Robert A. 2008. *Ending War: A Recipe.* Nottingham, U.K.: Spokesman.

Hines, Melissa. 2011. "Prenatal endocrine influences on sexual orientation and on sexually differentiated childhood behavior." *Frontiers in Neuroendocrinology* 32: 170–82.

Hinton, Alexander L. 2005. *Why Did They Kill? Cambodia in the Shadow of Genocide.* Berkeley: University of California Press.

Hoebel, E. Adamson. 1954. *The Law of Primitive Man: A Study in Comparative Legal Dynamics.* Cambridge, Mass.: Harvard University Press.

Hoffman, Carl. 2014. *Savage Harvest: A Tale of Cannibals, Colonialism, and Michael Rockefeller's Tragic Quest for Primitive Art.* New York: William Morrow.

Hoffman, Moshe, Erez Yoeli, and Carlos David Navarrete. 2016. "Game theory and morality." In *The Evolution of Morality,* edited by T. K. Shackelford and R. D. Hansen, 289-316. New York: Springer.

Hood, Bruce. 2014. *The Domesticated Brain.* London: Penguin.

Hrdy, S. B. 1977. *The Langurs of Abu.* Cambridge, Mass.: Harvard University Press.

Hrdy, Sarah B. 2014. "Development plus social selection in the emergence of 'emotionally modern' humans." In *Origins and Implications of the Evolution of*

Childhood, edited by C. L. Meehan and A. N. Crittenden (Santa Fe, N.Mex.: SAR Press).

Hrdy, Sarah Blaffer. 2009. *Mothers and Others: The Evolutionary Origins of Mutual Understanding.* Cambridge, Mass.: Harvard University Press.

Hublin, Jean-Jacques, Abdelouahed Ben-Ncer, Shara E. Bailey, Sarah E. Freidline, Simon Neubauer, Matthew M. Skinner, Inga Bergmann, et al. 2017. "New fossils from Jebel Irhoud, Morocco and the pan-African origin of *Homo sapiens.*" *Nature* 546: 289–92.

———, Simon Neubauer, and Philipp Gunz. 2015. "Brain ontogeny and life history in Pleistocene hominins." *Philosophical Transactions of the Royal Society B* 370: 20140062.

Huck, Maren, Petra Löttker, Uta-Regina Böhle, and Eckhard W. Heymann. 2005. "Paternity and kinship patterns in polyandrous moustached tamarins *(Saguinus mystax).*" *American Journal of Physical Anthropology* 127: 449–64.

Hughes, Joelene, and David W. Macdonald. 2013. "A review of the interactions between free-roaming domestic dogs and wildlife." *Biological Conservation* 157: 341–51.

Hutchinson, J. Robert. 1898. *The Romance of a Regiment: Being the True and Diverting Story of the Giant Grenadiers of Potsdam, How They Were Caught and Held in Captivity 1713–1740.* London: Sampson Low.

Huxley, Aldous. 1939. *After Many a Summer Dies the Swan.* New York: Harper and Row.

Huxley, Thomas Henry. 1863 (1959). *Man's Place in Nature.* Ann Arbor, Mich.: Ann Arbor Paperbacks.

Hylander, William L. 2013. "Functional links between canine height and jaw gape in catarrhines with special reference to early hominins." *American Journal of Physical Anthropology* 150: 247–59.

Isen, Joshua D., Matthew K. McGue, and William G. Iacono. 2015. "Aggressive-antisocial boys develop into physically strong young men." *Psychological Science* 26 (4): 444–55.

Ishikawa, S. S., A. Raine, T. Lencz, S. Bihrle, and L. LaCasse. 2001. "Increased height and bulk in antisocial personality disorder and its subtypes." *Psychiatry Research* 105 (3): 211–19.

Jensen, Keith, Josep Call, and Michael Tomasello. 2007. "Chimpanzees are rational maximizers in an ultimatum game." *Science* 318: 107–9.

———. 2013. "Chimpanzee responders still behave like rational maximizers." *PNAS* 110 (20): E1837.

Johnson, Dominic D. P. 2004. *Overconfidence and War: The Havoc and Glory of Positive Illusions.* Cambridge, Mass.: Harvard University Press.

———, and Niall J. MacKay. 2015. "Fight the power: Lanchester's laws of combat in human evolution." *Evolution and Human Behavior* 36 (2): 152–63.

———, and Bradley Thayer. 2016. "The evolution of offensive realism: Survival under anarchy from the Pleistocene to the present." *Politics and the Life Sciences* 35 (1): 1–26.

———, and Monica Duffy Toft. 2014. "Grounds for war: The evolution of territorial conflict." *International Security* 38: 7–38.

Johnson, Samuel. 1771 (1913). "Thoughts on the late transactions respecting Falkland's Islands." In *The Works of Samuel Johnson.* Troy, N.Y.: Pafraets & Co.

Joly, Marine, Jerome Micheletta, Arianna De Marco, Jan A. Langermans, Elisabeth H. M. Sterck, and Bridget M. Waller. 2017. "Comparing physical and social cognitive skills in macaque species with different degrees of social tolerance." *Proceedings of the Royal Society B* 284: 20162738.

Junger, Sebastian. 2016. *Tribe: On Homecoming and Belonging.* New York: Grand Central Publishing.

Kaburu, S. S. K., S. Inoue, and N. E. Newton-Fisher. 2013. "Death of the alpha: Within-community lethal violence among chimpanzees of the Mahale Mountains National Park." *American Journal of Primatology* 75: 789–97.

Kagan, Jerome. 1994. *Galen's Prophecy: Temperament in Human Nature.* New York: Westview Press.

Kaiser, Sylvia, Michael B. Hennessy, and Norbert Sachser. 2015. "Domestication affects the structure, development and stability of biobehavioural profiles." *Frontiers in Zoology* 12 (suppl. 1): S19.

Kalikow, Theodora J. 1983. "Konrad Lorenz's ethological theory: Explanation and ideology, 1938–1943." *Journal of the History of Biology* 16 (1): 39–73.

Kano, Takayoshi. 1992. *The Last Ape: Pygmy Chimpanzee Behavior and Ecology.* Stanford, Calif.: Stanford University Press.

Keedy, Edwin R. 1949. "History of the Pennsylvania Statute creating degrees of murder." *University of Pennsylvania Law Review* 97 (6): 759–77.

Keegan, John. 1976. *The Face of Battle: A Study of Agincourt, Waterloo and the Somme.* London: Jonathan Cape.

———. 1999. *The First World War.* New York: Alfred A. Knopf.

Keeley, Lawrence H. 1996. *War Before Civilization.* New York: Oxford University Press.

Kelley, Jay. 1995. "Sexual dimorphism in canine shape among extant great apes." *American Journal of Physical Anthropology* 96 (4): 365–89.

Kelly, Raymond C. 1993. *Constructing Inequality: The Fabrication of a Hierarchy of Virtue Among the Etoro.* Ann Arbor: University of Michigan Press.

———. 2000. *Warless Societies and the Origins of War.* Ann Arbor: University of Michigan Press.

———. 2005. "The evolution of lethal intergroup violence." *PNAS* 102 (43): 15294–98.

Kelly, Robert L. 1995. *The Foraging Spectrum: Diversity in Hunter-Gatherer Lifeways.* Washington, D.C.: Smithsonian Institution.

Keltner, Dacher. 2009. *Born to Be Good: The Science of a Meaningful Life.* New York: W. W. Norton.

Kim, Tong-Hyung. 2015. "South Korea says Kim Jong Un has executed 70 officials in 'Reign of Terror.'" *Huffington Post,* July 9.

Klein, Richard G. 2017. "Language and human evolution." *Journal of Neurolinguistics* 43: 204–21.

Knauft, Bruce M. 1985. *Good Company and Violence: Sorcery and Social Action in a Lowland New Guinea Society.* Berkeley: University of California Press.

———. 1991. "Violence and sociality in human evolution." *Current Anthropology* 32:391-428.

———. 2009. *The Gebusi: Lives Transformed in a Rainforest World.* New York: McGraw-Hill.

Kornbluh, Peter, ed. 1998. *Bay of Pigs Declassified: The Secret Report on the Invasion of Cuba.* New York: New Press.

Kruska, D. 1988. "Mammalian domestication and its effects on brain structure and behavior." In *The Evolutionary Biology of Intelligence,* edited by H. J. Jerison and I. Jerison (Berlin, Heidelberg: Springer), pp. 211–50.

Kruska, D. C. T., and V. E. Sidorovich. 2003. "Comparative allometric skull morphometrics in mink (*Mustela vison* Schreber, 1777) of Canadian and Belarus origin: Taxonomic status." *Mammalian Biology* 68: 257–76.

Kruska, Dieter. 2014. "Comparative quantitative investigations on brains of wild cavies *(Cavia aperea)* and guinea pigs (*Cavia aperea* f. porcellus): A contribution to size changes of CNS structures due to domestication." *Mammalian Biology* 79: 230–39.

Kruska, Dieter C. T., and Katja Steffen. 2013. "Comparative allometric investigations on the skulls of wild cavies *(Cavia aperea)* versus domesticated guinea pigs (*C. aperea* f. porcellus) with comments on the domestication of this species." *Mammalian Biology* 78: 178–86.

Kübler, Simon, Peter Owenga, Sally C. Reynolds, Stephen M. Rucina, and Geoffrey C. P. King. 2015. "Animal movements in the Kenya Rift and evidence for the earliest ambush hunting by hominins." *Scientific Reports* 5: 14011.

Künzl, Christine, Sylvia Kaiser, Edda Meier, and Norbert Sachser. 2003. "Is a wild mammal kept and reared in captivity still a wild animal?" *Hormones and Behavior* 43: 187–96.

———, and Norbert Sachser. 1999. "The behavioral endocrinology of domestication: A comparison between the domestic guinea pig (*Cavia aperea f. porcellus)* and its wild ancestor, the cavy *(Cavia aperea).*" *Hormones and Behavior* 35: 28–37.

Kurzban, Robert, and Mark R. Leary. 2001. "Evolutionary origins of stigmatization: The functions of social exclusion." *Psychological Bulletin* 127 (2): 187–208.

LaFave, Wayne R., and Austin W. Scott, Jr. 1986. *Criminal Law.* 2nd ed. St. Paul, Minn.: West.

Lahr, Marta Mirazón, F. Rivera, R. K. Power, A. Mounier, B. Copsey, F. Crivellaro, J. E. Edung, et al. 2016. "Inter-group violence among early Holocene hunter-gatherers of West Turkana, Kenya." *Nature* 529 (7586): 394–98.

Langergraber, Kevin E., Kay Prüfer, Carolyn Rowney, Christophe Boesch, Catherine Crockford, Katie Fawcett, Eiji Inouef, et al. 2012. "Generation times in wild chimpanzees and gorillas suggest earlier divergence times in great ape and human evolution." *PNAS* 109 (39): 15716–21; doi: 10.1073/pnas.1211740109.

———, Grit Schubert, Carolyn Rowney, R. Wrangham, Zinta Zommers, and Linda Vigilant. 2011. "Genetic differentiation and the evolution of cooperation in chimpanzees and humans." *Proceedings of the Royal Society B* 278: 2546–52.

Leach, Helen. 2003. "Human domestication reconsidered." *Current Anthropology* 44: 349–68.

Lee, Richard B. 1969. "Eating Christmas in the Kalahari." *Natural History* 78: 14–22; 60–63.

———. 1979. *The !Kung San: Men, Women and Work in a Foraging Society.* Cambridge, U.K.: Cambridge University Press.

———. 1984. *The Dobe !Kung.* New York: Holt, Rinehart and Winston.

———. 2014. "Hunter-gatherers on the best-seller list: Steven Pinker and the 'Bellicose School's' treatment of forager violence." *Journal of Aggression, Conflict and Peace Research* 6 (4): 216–28.

Lefevre, C. E., G. J. Lewis, D. I. Perrett, and L. Penke. 2013. "Telling facial metrics: Facial width is associated with testosterone levels in men." *Evolution and Human Behavior* 34 (4): 273–79.

Lescarbot, Marc. 1609. *Nova Francia: A Description of Arcadia.* Translated by P. Erondelle. London: Routledge, 1928.

Lewejohann, Lars, Thorsten Pickel, Norbert Sachser, and Sylvia Kaiser. 2010. "Wild genius—domestic fool? Spatial learning abilities of wild and domestic guinea pigs." *Frontiers in Zoology* 7 (9): 1–8.

Li, Caixia, Manhong Jia, Yanling Ma, Hongbing Luo, Qi Li, Yumiao Wang, Zhenhui Li, et al. 2016. "The relationship between digit ratio and sexual orientation in a Chinese Yunnan Han population." *Personality and Individual Differences* 101: 26–29.

Liberman, Kenneth. 1985. *Understanding Interaction in Central Australia: An Ethnomethodological Study of Australian Aboriginal People.* Boston: Routledge and Kegan Paul.

Librado, Pablo, Cristina Gamba, Charleen Gaunitz, Clio Der Sarkissian, Mélanie Pruvost, Anders Albrechtsen, Antoine Fages, et al. 2017. "Ancient genomic changes associated with domestication of the horse." *Science* 356: 442–45.

Lieberman, Daniel E. 2008. "Speculations about the selective basis for modern human craniofacial form." *Current Anthropology* 17: 55–68.

———. 2013. *The Story of the Human Body: Evolution, Health, and Disease.* New York: Pantheon.

———, Brandeis M. McBratney, and Gail Krovitz. 2002. "The evolution and development of cranial form in *Homo sapiens*." *PNAS* 99: 1134–39.

Lieberman, D. E., J. Carlo, M. Ponce de León, and C. Zollikofer. 2007. "A geometric morphometric analysis of heterochrony in the cranium of chimpanzees and bonobos." *Journal of Human Evolution* 52: 647–62.

Limolino, Mark V. 2005. "Body size evolution in insular vertebrates: Generality of the island rule." *Journal of Biogeography* 32: 1683–99.

Liu, Xiling, Dingding Han, Mehmet Somel, Xi Jiang, Haiyang Hu, Patricia Guijarro, Ning Zhang, et al. 2016. "Disruption of an evolutionarily novel synaptic expression pattern in autism." *PLoS Biology* 14 (9): e1002558.

———, Mehmet Somel, Lin Tang, Zheng Yan, Xi Jiang, Song Guo, Yuan Yuan, et al. 2012. "Extension of cortical synaptic development distinguishes humans from chimpanzees and macaques." *Genome Research* 22 (4): 611–22.

Lopez, Anthony C. 2016. "The evolution of war: Theory and controversy." *International Theory* 8 (1): 97–139.

———. 2017. "The evolutionary psychology of war: Offense and defense in the adapted mind." *Evolutionary Psychology* 15 (4): 1–23; doi: 10.1177/1474704917742720.

———, Rose McDermott, and Michael Bang Petersen. 2011. "States in mind: Evolution, coalitional psychology, and international politics." *International Security* 36: 48–83.

Lord, C. G., L. Ross, and M. R. Lepper. 1979. "Biased assimilation and attitude polarization: The effects of prior theories on subsequently considered evidence." *Journal of Personality and Social Psychology* 37 (11): 2098–109.

Lord, Kathryn. 2013. "A comparison of the sensory development of wolves *(Canis lupus lupus)* and dogs *(Canis lupus familiaris)*." *Ethology* 119: 110–20.

———, Mark Feinstein, Bradley Smith, and Raymond Coppinger. 2013. "Variation in reproductive traits of members of the genus Canis with special attention to the domestic dog *(Canis familiaris)*." *Behavioral Processes* 92: 131–42.

Lorenz, Konrad. 1966. *On Aggression*. New York: Harcourt Brace.

Lorenz, Konrad Z. 1940. "Durch Domestikation verursachte Störungen arteigener Verhalten." *Zeitschrift für angewandte Psychologie und Charakterkunde* 59: 1–81.

———. 1943. "Die angeborenen Formen möglicher Erfahrung." *Zeitschrift für Tierpsychologie* 5: 235–409.

Losos, Jonathan B., and Robert E. Ricklefs. 2009. "Adaptation and diversification on islands." *Nature* 457: 830–37.

Lothrop, S. K. 1928. *The Indians of Tierra del Fuego*. New York: Museum of the American Indian, Heye Foundation.

Lourandos, Harry. 1997. *Continent of Hunter-Gatherers*. Cambridge, U.K.: Cambridge University Press.

Lovejoy, C. Owen. 2009. "Reexamining human origins in light of *Ardipithecus ramidus*." *Science* 326: 74e1–74e8.

Lovisek, Joan A. 2007. "Aboriginal warfare on the Northwest Coast: Did the potlatch replace warfare?" In *North American Indigenous Warfare and Ritual Violence,* edited by Richard J. Chacon and Rubén G. Mendoza (Tucson: University of Arizona Press), pp. 58–73.

Lukas, Dieter, and Elise Huchard. 2014. "The evolution of infanticide by males in mammalian societies." *Science* 346: 841–44.

MacHugh, David E., Greger Larson, and Ludovic Orlando. 2017. "Taming the past: Ancient DNA and the study of animal domestication." *Annual Review of Animal Biosciences* 5: 329–51.

MacLean, Evan L., Brian Hare, Charles L. Nunn, Elsa Addessi, Federica Amici, Rindy C. Anderson, Filippo Aureli, et al. 2014. "The evolution of self-control." *PNAS* 111: E2140–48.

Malmkvist, Jens, and Steffen W. Hansen. 2002. "Generalization of fear in farm mink, *Mustela vison,* genetically selected for behaviour towards humans." *Animal Behaviour* 64: 487–501.

Malone, Paul. 2014. *The Peaceful People: The Penan and Their Fight for the Forest.* Petaling Jaya, Malaysia: Strategic Information and Research Development Centre.

Manjila, Sunil, Gagandeep Singh, Ayham M. Alkhachroum, and Ciro Ramos-Estebanez. 2015. "Understanding Edward Muybridge: Historical review of behavioral alterations after a 19th-century head injury and their multifactorial influence on human life and culture." *Neurosurgical Focus* 39 (1): E4.

Manson, Joseph H., Julie Gros-Louis, and Susan Perry. 2004. "Three apparent cases of infanticide by males in wild white-faced capuchins *(Cebus capucinus).*" *Folia primatologica* 75: 104–6.

Marean, Curtis W. 2015. "An evolutionary anthropological perspective on modern human origins." *Annual Review of Anthropology* 44: 533–56.

Marks, Jonathan. 2002. *What It Means to Be 98% Chimpanzee.* Berkeley: University of California Press.

Marlowe, Frank. 2002. "Why the Hadza are still hunter-gatherers." In *Ethnicity, Hunter-Gatherers, and the "Other": Association or Assimilation in Africa,* edited by Sue Kent (Washington, D.C.: Smithsonian Institution Press), pp. 247–75.

Marlowe, Frank W. 2004. "What explains Hadza food sharing?" *Economic Anthropology* 23: 69–88.

———. 2005. "Hunter-gatherers and human evolution." *Evolutionary Anthropology* 14: 54–67.

Mashour, George A., Erin E. Walker, and Robert L. Martuza. 2005. "Psychosurgery: Past, present, and future." *Brain Research Reviews* 48: 409–19.

Mathew, Sarah, and Robert Boyd. 2011. "Punishment sustains large-scale cooperation in prestate warfare." *PNAS* 108: 11375–80.

McCrae, Robert R., and Paul T. Costa. 1997. "Personality trait structure as a human universal." *American Psychologist* 52 (5): 509–16.

McDermott, Rose, Dustin Tingley, Jonathan Cowden, Giovanni Frazzetto, and Dominic D. P. Johnson. 2009. "Monoamine oxidase A gene (MAOA) predicts behavioral aggression following provocation." *PNAS* 106 (7): 2118–23.

McDonald, Melissa M., Carlos David Navarrete, and Mark Van Vugt. 2012. "Evolution and the psychology of intergroup conflict: The male warrior hypothesis." *Philosophical Transactions of the Royal Society B* 367: 670–79.

McIntyre, Matthew H., Esther Herrmann, Victoria Wobber, Michel Halbwax, Crispin Mohamba, Nick de Sousa, Rebeca Atencia, Debby Cox, and Brian Hare. 2009. "Bonobos have a more human-like second-to-fourth finger length ratio (2D:4D) than chimpanzees: A hypothesized indication of lower prenatal androgens." *Journal of Human Evolution* 56: 361–65.

Mead, Margaret. 1954. "Some theoretical considerations on the problem of mother-child separation." *American Journal of Orthopsychiatry* 24: 471–83.

Mech, L. David, L. G. Adams, T. J. Meier, J. W. Burch, and B. W. Dale. 1998. *The Wolves of Denali.* Minneapolis: University of Minnesota Press.

———, Shannon M. Barber-Meyer, and John Erb. 2016. "Wolf *(Canis lupus)* generation time and proportion of current breeding females by age." *PLoS ONE* 11(6): e0156682.

Meggitt, M. J. 1965. "Marriage among the Walbiri of central Australia: A statistical examination." In *Aboriginal Man in Australia,* edited by Ronald M. Berndt and Catherine H. Berndt (Sydney: Angus and Robertson), pp. 146–66.

Melamed, Yoel, Mordechai E. Kisleva, Eli Geffen, Simcha Lev-Yadunc, and Naama Goren-Inbar. 2016. "The plant component of an Acheulian diet at Gesher Benot Ya'aqov, Israel." *PNAS* 113 (51): 14674–79.

Melis, Alicia P., Brian Hare, and Michael Tomasello. 2006. "Engineering cooperation in chimpanzees: Tolerance constraints on cooperation." *Animal Behaviour* 72: 275–86.

Meloy, J. Reid. 2006. "Empirical basis and forensic application of affective and predatory violence." *Australian and New Zealand Journal of Psychiatry* 40: 539–47.

Mencken, H. L. 1949. *A Mencken Chrestomathy.* New York: Alfred A. Knopf.

Mighall, Robert. 2002. "Introduction." In *Robert Louis Stevenson: The Strange Case of Dr. Jekyll and Mr. Hyde, and Other Tales of Terror,* edited by Robert Mighall (London: Penguin), pp. ix–xlii.

Miller, Daniel J., Tetyana Duka, Cheryl D. Stimpson, Steven J. Schapiro, Wallace B. Bazeb, Mark J. McArthur, Archibald J. Fobbs, et al. 2012. "Prolonged myelination in human neocortical evolution." *PNAS* 109 (41): 16480–85.

Mitani, J. C., D. P. Watts, and S. J. Amsler. 2010. "Lethal intergroup aggression leads to territorial expansion in wild chimpanzees." *Current Biology* 20: R507–8.

Mitteroecker, Philipp, Philipp Gunz, Markus Bernhard, Katrin Schaefer, and Fred L. Bookstein. 2004. "Comparison of cranial ontogenetic trajectories among great apes and humans." *Journal of Human Evolution* 46: 679–98.

Mohnot, S.M. 1971. "Some aspects of social changes and infant-killing in the Hanuman langur, *Presbytis entellus* (Primates: Cercopithecidae), in Western India." *Mammalia* 35: 175–98.

Montague, Michael J., Gang Li, Barbara Gandolfi, Razib Khan, Bronwen L. Aken, Steven M. J. Searle, Patrick Minx, et al. 2014. "Comparative analysis of the domestic cat genome reveals genetic signatures underlying feline biology and domestication." *PNAS* 111 (48): 17230–325.

Moorjani, Priya, Carlos Eduardo G. Amorim, Peter F. Arndt, and Molly Przeworski. 2016. "Variation in the molecular clock of primates." *PNAS* 113 (38): 10607–12.

Moran, John. 2007. *The Anatomy of Courage.* London: Robinson.

Morgenthau, Hans. 1973. *Politics Among Nations.* New York: Alfred A. Knopf.

Morris, N., and D. J. Rothman, eds. 1995. *Oxford History of the Prison: The Practice of Punishment in Western Society.* New York: Oxford University Press.

Morrison, Rachel, and Diana Reiss. 2018. "Precocious development of self-awareness in dolphins." *PLoS ONE* 13(1): e0189813; doi: 10.1371/journal.pone.0189813.

Muller, Martin N. 2002. "Agonistic relations among Kanyawara chimpanzees." In *Behavioural Diversity in Chimpanzees and Bonobos,* edited by Christophe Boesch, Gottfried Hohmann, and Linda Marchant (Cambridge, U.K.: Cambridge University Press), pp. 112–24.

———, and David R. Pilbeam. 2017. "Evolution of the human mating system." In *Chimpanzees and Human Evolution,* edited by M. N. Muller, R. W. Wrangham, and D. R. Pilbeam (Cambridge, Mass.: Harvard University Press), pp. 328–426.

———, M. Emery Thompson, Sonya M. Kahlenberg, and Richard W. Wrangham. 2011. "Sexual coercion by male chimpanzees shows that female choice may be more apparent than real." *Behavioral Ecology and Sociobiology* 65: 921–33.

———, and Richard W. Wrangham. 2004. "Dominance, aggression and testosterone in wild chimpanzees: A test of the 'challenge hypothesis.'" *Animal Behaviour* 67: 113–23.

Murphy, R. F. 1957. "Intergroup hostility and social cohesion." *American Anthropologist* 59: 1018–35.

Muscarella, Frank. 2000. "The evolution of homoerotic behavior in humans." *Journal of Homosexuality* 40 (1): 51–77.

Myers Thompson, Jo A. 2001. "On the nomenclature of *Pan paniscus.*" *Primates* 42 (2): 101–11.

Naef, Albert. 1926. "Über die Urformen der Anthropomorphen und die Stammesgeschichte des Menschenschädels" ("The prototype of anthropomorphen and the phylogeny of human impairment.") *Naturwissenschaften* 14: 472–77. (Article in German.)

Nash, George. 2005. "Assessing rank and warfare strategy in prehistoric hunter-gatherer society: A study of representational warrior figures in rock-art from the Spanish Levant." In *Warfare, Violence and Slavery in Prehistory,* edited by M. Parker Pearson and I. J. N. Thorpe (Oxford, U.K.: Archaeopress), pp. 75–86.

Nelson, Walter H. 1970. *The Soldier Kings: The House of Hohenzollern.* New York: G. P. Putnam.

Nesse, Randolph M. 2007. "Runaway social selection for displays of partner value and altruism." *Biological Theory* 2 (2): 143–55.

———. 2010. "Social selection and the origins of culture." In *Evolution, Culture, and the Mind,* edited by Mark Schaller, Ara Norenzayan, Steven J. Heine, Toshio Yamagishi, and Tatsuya Kameda (New York: Psychology Press), pp. 137–50.

Nettle, Daniel, Katherine A. Cronin, and Melissa Bateson. 2013. "Responses of chimpanzees to cues of conspecific observation." *Animal Behaviour* 86: 595–602.

Neumann, Craig S., Robert D. Hare, and Dustin A. Pardini. 2015. "Antisociality and the Construct of Psychopathy: Data from Across the Globe." *Journal of Personality* 83 (66): 678–92.

Nielsen, Rasmus, Joshua M. Akey, Mattias Jakobsson, Jonathan K. Pritchard, Sarah Tishkoff, and Eske Willerslev. 2017. "Tracing the peopling of the world through genomics." *Nature* 541: 302–10.

Nikulina, E. M. 1991. "Neural control of predatory aggression in wild and domesticated animals." *Neuroscience & Biobehavioral Reviews* 15: 545–47.

Nisbett, Alec. 1976. *Konrad Lorenz.* New York: Harcourt Brace Jovanovich.

Nishida, Toshisada. 1996. "The death of Ntologi, the unparalleled leader of M group." *Pan Africa News* 3: 4.

———. 2012. *Chimpanzees of the Lakeshore: Natural History and Culture at Mahale.* New York: Cambridge University Press.

———, K. Hosaka, Michio Nakamura, and M. Hamai. 1995. "A within-group gang attack on a young adult male chimpanzee: Ostracism of an ill-mannered member?" *Primates* 36: 207–11.

Nowak, Katarzyna, Andrea Cardini, and Sarah Elton. 2008. "Evolutionary acceleration and divergence in *Procolobus kirkii.*" *International Journal of Primatology* 29: 1313–39.

Nowak, Martin A., Corina E. Tarnita, and Edward O. Wilson. 2010. "The evolution of eusociality." *Nature* 466: 1057–62.

Oftedal, Olav Y. 2012. "The evolution of milk secretion and its ancient origins." *Animal* 6 (3): 355–68.

Oka, Rahul C., Marc Kissel, Mark Golitko, Susan Guise Sheridan, Nam C. Kim, and Agustín Fuentes. 2017. "Population is the main driver of war group size and conflict casualties." *PNAS* 114 (52): E11101–10.

Okada, Daijiro, and Paul M. Bingham. 2008. "Human uniqueness—self-interest and social cooperation." *Journal of Theoretical Biology* 253: 261–70.

Orwell, George. 1938. *Homage to Catalonia*. London: Secker and Warburg.

Otterbein, Keith F. 1986. *The Ultimate Coercive Sanction: A Cross-Cultural Study of Capital Punishment*. New Haven, Conn.: HRAF Press.

———. 2004. *How War Began*. College Station: Texas A&M University Press.

Overy, Richard. 2009. *The Twilight Years: The Paradox of Britain Between the Wars*. New York: Viking.

Painter, Nell Irvin. 2010. *The History of White People*. New York: W. W. Norton.

Palagi, Elisabetta. 2006. "Social play in bonobos *(Pan paniscus)* and chimpanzees *(Pan troglodytes):* Implications for natural social systems and interindividual relationships." *American Journal of Physical Anthropology* 129: 418–26.

Pallitto, Christina, and Claudia García-Moreno. 2013. "Intimate partner violence and its measurement: Global considerations." In *Family Problems and Family Violence,* edited by Steven R. H. Beach, Richard E. Heyman, Amy Smith Slep, and Heather M. Foran (New York: Springer), pp. 15–32.

Palombit, Ryne A. 2012. "Infanticide: Male strategies and female counterstrategies." In *The Evolution of Primate Societies,* edited by John C. Mitani, Josep Call, Peter M. Kappeler, Ryne A. Palombit, and Joan B. Silk (Chicago: University of Chicago Press), pp. 432–68.

Paquin, Stephane, Eric Lacourse, Mara Brendgen, Frank Vitaro, Ginette Dionne, Richard E. Tremblay, and Michel Boivin. 2014. "The genetic-environmental architecture of proactive and reactive aggression throughout childhood." *Monatsschrift für Kriminologie und Strafrechtsreform* 97 (5–6): 398–420.

Parish, Amy M. 1994. "Sex and food control in the 'Uncommon Chimpanzee': How bonobo females overcome a phylogenetic legacy of male dominance." *Ethology and Sociobiology* 15: 157–79.

Parry, Richard Lloyd. 1998. "What young men do." *Granta* 62: 83–124.

Payn, Graham, and Sheridan Morley, eds. 1982. *The Nöel Coward Diaries*. London: Papermac.

Pearce, Eiluned, Chris Stringer, and R. I. M. Dunbar. 2013. "New insights into differences in brain organization between Neanderthals and anatomically modern humans." *Proceedings of the Royal Society B* 280: 20130168.

Pendleton, Amanda L., Feichen Shen, Angela M. Taravella, Sarah Emery, Krishna R. Veeramah, Adam R. Boyko, and Jeffrey M. Kidd. 2018. "Comparison of village dog and wolf genomes highlights therole of the neural crest in dog domestication." *BMC Biology, in press. doi:* 10.1186/s12915-018-0535-2.

Peplau, Letitia Anne, and Adam W. Fingerhut. 2007. "The close relationships of lesbians and gay men." *Annual Review of Psychology* 58: 405–24.

Perry, Susan, and Joseph H. Manson. 2008. *Manipulative Monkeys: The Capuchins of Lomas Barbudal*. Cambridge, Mass.: Harvard University Press.

Peterson, Dale. 2011. *The Moral Lives of Animals*. New York: Bloomsbury.

Phillips, Herbert P. 1965. *Thai Peasant Personality: The Patterning of Interpersonal Behavior in the Village of Bang Chan.* Berkeley: University of California Press.

Phillips, Tim, Jiawei Li, and Graham Kendall. 2014. "The effects of extra-somatic weapons on the evolution of human cooperation towards non-kin." *PLoS ONE* 9(5): e95742.

Pickering, Travis R. 2013. *Rough and Tumble: Aggression, Hunting and Human Evolution.* Berkeley: University of California Press.

Pilbeam, David R., and Daniel E. Lieberman. 2017. "Reconstructing the Last Common Ancestor of chimpanzees and humans." In *Chimpanzees and Human Evolution,* edited by M. N. Muller, D. R. Pilbeam, and R. W. Wrangham (Cambridge, Mass.: Harvard University Press), pp. 22–141.

Pilot, Małgorzata, Tadeusz Malewski, Andre E. Moura, Tomasz Grzybowski, Kamil Olenski, Stanisław Kaminski, Fernanda Ruiz Fadel, et al. 2016. "Diversifying selection between pure-breed and free-breeding dogs inferred from genome-wide SNP analysis." *Genes, Genomes and Genetics* 6 (8): 2285–98.

Pinker, Steven. 2011. *The Better Angels of Our Nature: Why Violence Has Declined.* New York: Penguin.

———. 2012. "The false allure of group selection." *https://www.edge.org/conversation/the-false-allure-of-group-selection;* doi: 10.1002/9781119125563.evpsych236.

Plomin, Robert. 2014. "Genotype-environment correlation in the era of DNA." *Behavior Genetics* 44: 629–38.

Plyusnina, Irina Z., Maria Yu Solov'eva, and Irina N. Oskina. 2011. "Effect of domestication on aggression in gray norway rats." *Behavior Genetics* 41: 583–92.

Polk, Kenneth. 1995. "Lethal violence as a form of masculine conflict resolution." *Australian and New Zealand Journal of Criminology* 28: 93–115.

Power, Margaret. 1991. *The Egalitarians—Human and Chimpanzee: An Anthropological View of Social Organization.* Cambridge, U.K.: Cambridge University Press.

Price, Edward O. 1999. "Behavioral development in animals undergoing domestication." *Applied Animal Behaviour Science* 65: 245–71.

Pruetz, Jill D., Kelly Boyer Ontl, Elizabeth Cleaveland, Stacy Lindshield, Joshua Marshack, and Erin G. Wessling. 2017. "Intragroup lethal aggression in West African chimpanzees *(Pan troglodytes verus):* Inferred killing of a former alpha male at Fongoli, Senegal." *International Journal of Primatology* 38: 31–57.

Prüfer, K., F. Racimo, N. Patterson, F. Jay, S. Sankararaman, S. Sawyer, A. Heinze, et al. 2014. "The complete genome sequence of a Neanderthal from the Altai Mountains." *Nature* 505: 43–49.

Prüfer, Kay, Kasper Munch, Ines Hellmann, Keiko Akagi, Jason R. Miller, Brian Walenz, Sergey Koren, et al. 2012. "The bonobo genome compared with the chimpanzee and human genomes." *Nature* 486: 527–31.

Pusey, Anne, Carson Murray, William Wallauer, Michael Wilson, Emily

Wroblewski, and Jane Goodall. 2008. “Severe aggression among female *Pan troglodytes schweinfurthii* at Gombe National Park, Tanzania.” *International Journal of Primatology* 29: 949–73.
———, and Craig Packer. 1994. “Infanticide in lions: Consequences and counterstrategies.” In *Infanticide and Parental Care,* edited by S. Parmigiani and F. von Saal (London: Harwood Academic Publishers), pp. 277–330.

Rabett, Ryan J. 2018. “The success of failed *Homo sapiens* dispersals out of Africa and into Asia.” *Nature Ecology and Evolution* 2: 212–19.
Radcliffe-Brown, A. 1922. *The Andaman Islanders: A Study in Social Anthropology.* Cambridge, U.K.: Cambridge University Press.
Raia, Pasquale, Fabio M. Guarino, Mimmo Turano, Gianluca Polese, Daniela Rippa, Francesco Carotenuto, Daria M. Monti, Manuela Cardi, and Domenico Fulgione. 2010. “The blue lizard spandrel and the island syndrome.” *BMC Evolutionary Biology* 10 (289): 1–16.
Raine, A., J. R. Meloy, S. Bihrle, J. Stoddard, L. LaCasse, and M. S. Buchsbaum. 1998a. “Reduced prefrontal and increased subcortical brain functioning assessed using positron emission tomography in predatory and affective murderers.” *Behavioral Sciences and the Law* 16 (3): 319–32.
Raine, Adrian. 2013. *The Anatomy of Violence: The Biological Roots of Crime.* London: Allen Lane.
———, Chandra Reynolds, Peter H. Venables, Sarnoff A. Mednick, and David P. Farrington. 1998b. “Fearlessness, stimulation-seeking, and large body size at age 3 years as early predispositions to childhood aggression at age 11 years.” *Archives of General Psychiatry* 55 (8): 745–51.
Ramm, Steven A., L. Schärer, J. Ehmcke, and J. Wistuba. 2014. “Sperm competition and the evolution of spermatogenesis.” *Molecular Human Reproduction* 20 (12): 1169–79.
Range, Friederike, Caroline Ritter, and Zsófia Virányi. 2015. “Testing the Myth: Tolerant Dogs and Aggressive Wolves.” *Proceedings of the Royal Society B* 282: 20150220.
Ridley, Matthew. 1996. *The Origins of Virtue.* London: Viking.
Roebroeks, Will, and Marie Soressi. 2016. “Neandertals revised.” *PNAS* 113 (23): 6372–79.
Rosaldo, Michelle Z. 1974. “Women, culture and society: A theoretical overview.” In *Woman, Culture and Society,* edited by Michelle Z. Rosaldo and Louise Lamphere (Stanford, Calif.: Stanford University Press), pp. 17–42.
Roselli, Charles E., Kay Larkin, John A. Resko, John N. Stellflug, and Fred Stormshak. 2004. “The volume of a sexually dimorphic nucleus in the ovine medial preoptic area/anterior hypothalamus varies with sexual partner preference.” *Endocrinology* 145 (2): 478–83.
———, Radhika C. Reddy, and Katherine R. Kaufman. 2011. “The development of male-oriented behavior in rams.” *Frontiers in Neuroendocrinology* 32: 164–69.

Roth, H. Ling. 1890. *The Aborigines of Tasmania.* London: Kegan Paul, Trench.

Rowson, B., B. H. Warren, and C. F. Ngereza. 2010. "Terrestrial molluscs of Pemba Island, Zanzibar, Tanzania, and its status as an 'oceanic' island." *ZooKeys* 70: 1–39.

Rudolf von Rohr, Claudia, Carel P. van Schaik, Alexandra Kissling, and Judith M. Burkart. 2015. "Chimpanzees' bystander reactions to infanticide: An evolutionary precursor of social norms?" *Human Nature* 26: 143–60.

Ruff, C. B., E. Trinkaus, and T. W. Holliday. 1997. "Body mass and encephalization in Pleistocene *Homo.*" *Nature* 387: 173–76.

Ruff, Christopher B., Eric Trinkaus, Alan Walker, and Clark Spencer Larsen. 1993. "Postcranial Robusticity in *Homo.* I: Temporal trends and mechanical interpretation." *American Journal of Physical Anthropology* 91: 21–53.

Saey, Tina Hesman. 2017. "DNA evidence is rewriting domestication origin stories." *Science News* 191 (13): 20–36.

Sánchez-Villagra, Marcelo R., Madeleine Geiger, and Richard A. Schneider. 2016. "The taming of the neural crest: A developmental perspective on the origins of morphological covariation in domesticated mammals." *Royal Society Open Science* 3: 160107.

———, Valentina Segura, Madeleine Geiger, Laura Heck, Kristof Veitschegger, and David Flores. 2017. "On the lack of a universal pattern associated with mammalian domestication: Differences in skull growth trajectories across phylogeny." *Royal Society Open Science* 4: 170876; doi: 10.1098/rsos.170876.

Sanislow, Charles A., D. S. Pine, K. J. Quinn, M. J. Kozak, M. A. Garvey, R. K. Heinssen, P. S. Wang, and B. N. Cuthbert. 2010. "Developing constructs for psychopathology research: Research domain criteria." *Journal of Abnormal Psychology* 119 (4): 631–39.

Sapolsky, Robert M. 2017. *Behave: The Biology of Humans at Our Best and Worst.* New York: Penguin.

Saucier, Gerard. 2018. "Culture, morality and individual differences: Comparability and incomparability across species." *Philosophical Transactions of the Royal Society B* 373: 20170170.

Schlesinger, Louis B. 2007. "Sexual homicide: Differentiating catathymic and compulsive murders." *Aggression and Violent Behavior* 12: 242–56.

Schrire, Carmel. 1980. "An inquiry into the evolutionary status and apparent identity of San hunter-gatherers." *Human Ecology* 8 (1): 9–32.

Schultz, Ted R., and Seán G. Brady. 2008. "Major evolutionary transitions in ant agriculture." *PNAS* 105: 5435–40.

Schwing, Raoul, Élodie Jocteur, Amelia Wein, Ronald Noë, and Jorg J. M. Massen. 2016. "Kea cooperate better with sharing affiliates." *Animal Cognition* 19: 1093–102.

Scollon, Christie N., Ed Diener, Shigehiro Oishi, and Robert Biswas-Diener. 2004. "Emotions across cultures and methods." *Journal of Cross-Cultural Psychology* 35 (3): 304–26.

Segal, Nancy. 2012. *Born Together—Reared Apart: The Landmark Minnesota Twins Study*. Cambridge, Mass.: Harvard University Press.

Shea, Brian T. 1989. "Heterochrony in human evolution: The case for neoteny reconsidered." *Yearbook of Physical Anthropology* 32: 69–104.

Shea, John J., and Matthew L. Sisk. 2010. "Complex projectile technology and *Homo sapiens* dispersal into Western Eurasia." *PaleoAnthropology* 2010: 100–122; doi: 10.4207/PA.2010.ART36.

Shermer, Michael. 2004. *The Science of Good and Evil: Why People Cheat, Gossip, Care, Share, and Follow the Golden Rule*. New York: Henry Holt.

Shimamura, Arthur P. 2002. "Muybridge in motion: Travels in art, psychology and neurology." *History of Photography* 26 (4): 341–50.

Shostak, Marjorie. 1981. *Nisa: The Life and Words of a !Kung Woman*. New York: Random House.

Shumny, V. K. 1987. "In memory of Dmitri Konstantinovich Belyaev." *Theoretical and Applied Genetics* 73: 932–33.

Sidorovich, V., and D. W. Macdonald. 2001. "Density dynamics and changes in habitat use by the European mink and other native mustelids in connection with the American mink expansion in Belarus." *Netherlands Journal of Zoology* 51 (1): 107–26.

Siegel, A., and J. Victoroff. 2009. "Understanding human aggression: New insights from neuroscience." *International Journal of Law and Psychiatry* 32: 209–15.

Siever, Larry J. 2008. "Neurobiology of aggression and violence." *American Journal of Psychiatry* 165: 429–42.

Simões-Costa, Marcos, and Marianne E. Bronner. 2015. "Establishing neural crest identity: A gene regulatory recipe." *Development* 142: 242–57.

Singh, J. A. L., and Robert M. Zingg. 1942. *Wolf-Children and Feral Man*. Hamden, Conn.: Archon.

Singh, Manvir, Richard W. Wrangham, and Luke Glowacki. 2017. "Self-interest and the design of rules." *Human Nature* 28: 457–80.

Singh, Nandini, Frank W. Albert, Irina Plyusnina, Lyudmila Trut, Svante Pääbo, and Katerina Harvati. 2017. "Facial shape differences between rats selected for tame and aggressive behaviors." *PLoS ONE* 12(4): e0175043.

Skorska, Malvina N., and Anthony F. Bogaert. 2015. "Sexual orientation: Biological influences." *International Encyclopedia of the Social & Behavioral Sciences* 21: 773–78.

Slon, Viviane, Bence Viola, Gabriel Renaud, Marie-Theres Gansauge, Stefano Benazzi, Susanna Sawyer, Jean-Jacques Hublin, et al. 2017. "A fourth Denisovan individual." *Science Advances* 3: e1700186.

Smith, Richard J., and William L. Jungers. 1997. "Body mass in comparative primatology." *Journal of Human Evolution* 32: 523–59.

Snyder, Timothy. 2010. *Bloodlands: Europe Between Hitler and Stalin*. New York: Basic Books.

Sober, Elliott, and David S. Wilson. 1998. *Unto Others: The Evolution and Psychology of Unselfish Behavior.* Cambridge, Mass.: Harvard University Press.

Sommer, Volker. 2000. "The holy wars about infanticide: Which side are you on?" In *Infanticide by Males and Its Implications,* edited by C. P. van Schaik and C. Janson (Cambridge, U.K.: Cambridge University Press), pp. 9–26.

Sorensen, Andrew C. 2017. "On the relationship between climate and Neandertal fire use during the Last Glacial in south-west France." *Quaternary International* 436 114-128.

Stamps, J. A., and M. Buechner. 1985. "The territorial defense hypothesis and the ecology of insular vertebrates." *Quarterly Review of Biology* 60: 155–81.

Starks, Philip T. 2003. "Selection for uniformity: Xenophobia and invasion success." *Trends in Ecology and Evolution* 18 (4): 159–62.

Statham, Mark J., Lyudmila N. Trut, Ben N. Sacks, Anastasiya V. Kharlamova, Irina N. Oskina, Rimma G. Gulevich, Jennifer L. Johnson, et al. 2011. "On the origin of a domesticated species: Identifying the parent population of Russian silver foxes *(Vulpes vulpes).*" *Biological Journal of the Linnean Society* 103: 168–75.

Stearns, Jason K. 2011. *Dancing in the Glory of Monsters: The Collapse of the Congo and the Great War of Africa.* New York: PublicAffairs, Perseus.

Stevenson, Robert Louis. 1991 (1886). *The Strange Case of Dr. Jekyll and Mr. Hyde.* New York: Dover.

Stimpson, Cheryl D., Nicole Barger, Jared P. Taglialatela, Annette Gendron-Fitzpatrick, Patrick R. Hof, William D. Hopkins, and Chet C. Sherwood. 2016. "Differential serotonergic innervation of the amygdala in bonobos and chimpanzees." *Social Cognitive and Affective Neuroscience* 11 (3): 413–22.

Stirrat, Michael, Gert Stulp, and Thomas V. Pollet. 2012. "Male facial width is associated with death by contact violence: Narrow-faced males are more likely to die from contact violence." *Evolution and Human Behavior* 33: 551–56.

Stringer, Chris. 2016. "The origin and evolution of *Homo sapiens.*" *Philosophical Transactions of the Royal Society B* 371: 20150237.

Stringer, Christopher B. 2012. *The Origin of Our Species.* London: Penguin.

Surbeck, Martin, Tobias Deschner, Verena Behringer, and Gottfried Hohmann. 2015. "Urinary C-peptide levels in male bonobos *(Pan paniscus)* are related to party size and rank but not to mate competition." *Hormones and Behavior* 71: 22–30.

———, Tobias Deschner, Grit Schubert, Anja Weltring, and Gottfried Hohmann. 2012. "Mate competition, testosterone and intersexual relationships in bonobos, *Pan paniscus.*" *Animal Behaviour* 83: 659–69.

Surbeck, Martin, Cédric Girard-Buttoz, Christophe Boesch, Catherine Crockford, Barbara Fruth, Gottfried Hohmann, Kevin E. Langergraber, Klaus Zuberbühler, Roman M. Wittig, and Roger Mundry. 2017. "Sex-specific association patterns in bonobos and chimpanzees reflect species differences in cooperation." *Royal Society Open Science* 4: 161081.

———, and Gottfried Hohmann. 2013. "Intersexual dominance relationships and the influence of leverage on the outcome of conflicts in wild bonobos (Pan paniscus)." *Behavioral Ecology and Sociobiology* 67: 1767–80.

Sussman, Robert W., and Joshua Marshack. 2010. "Are humans inherently killers?" *Global Non-Killing Working Papers* 1: 7–28.

Sussman, Robert W., ed. 1998. *The Biological Basis of Human Behavior: A Critical Review.* New York: Prentice Hall.

Suzuki, Kenta, Maki Ikebuchi, Hans-Joachim Bischof, and Kazuo Okanoya. 2014. "Behavioral and neural trade-offs between song complexity and stress reaction in a wild and a domesticated finch strain." *Neuroscience and Biobehavioral Reviews* 46: 547–56.

Swedell, Larissa, and Thomas W. Plummer. 2012. "A papionin multilevel society as a model for hominin social evolution." *International Journal of Primatology* 33 (5): 1165–93.

Sznycer, Daniel, John Tooby, Leda Cosmides, Roni Porat, Shaul Shalvie, and Eran Halperin. 2016. "Shame closely tracks the threat of devaluation by others, even across cultures." *PNAS* 113 (10): 2625–30.

Takemoto, Hiroyuki, Yoshi Kawamoto, and Takeshi Furuichi. 2015. "How did bonobos come to range south of the Congo River? Reconsideration of the divergence of *Pan paniscus* from other *Pan* populations." *Evolutionary Anthropology* 24: 170–84.

Tattersall, Ian. 2015. *The Strange Case of the Rickety Cossack, and Other Cautionary Tales from Human Evolution.* New York: Palgrave Macmillan.

———. 2016. "A tentative framework for the acquisition of language and modern human cognition." *Journal of Anthropological Sciences* 94: 157–66.

Terburg, David, and Jack van Honk. 2013. "Approach–avoidance versus dominance–submissiveness: A multilevel neural framework on how testosterone promotes social status." *Emotion Review* 5 (3): 296–302.

Teten Tharp, Andra L., Carla Sharp, Matthew S. Stanford, Sarah L. Lake, Adrian Raine, and Thomas A. Kent. 2011. "Correspondence of aggressive behavior classifications among young adults using the Impulsive Premeditated Aggression Scale and the Reactive Proactive Questionnaire." *Personality and Individual Differences* 50: 279–85.

Theofanopoulou, Constantina, Simone Gastaldon, Thomas O'Rourke, Bridget D. Samuels, Angela Messner, Pedro Tiago Martins, Francesco Delogu, Saleh Alamri, and Cedric Boeckx. 2017. "Comparative genomic evidence for self-domestication in *Homo sapiens.*" *PLoS ONE* 12(10): e0185306.

Thomas, Elizabeth Marshall. 1959. *The Harmless People.* New York: Vintage.

Tindale, Norman B. 1940. "Distribution of Australian aboriginal tribes: A field survey." *Transactions of the Royal Society of South Australia* 64 (1): 140–231.

———. 1974. *Aboriginal Tribes of Australia: Their Terrain, Environmental Controls, Distribution, Limits, and Proper Names.* Berkeley: University of California Press.

Tokuyama, Nahoko, and Takeshi Furuichi. 2016. "Do friends help each other? Patterns of female coalition formation in wild bonobos at Wamba." *Animal Behaviour* 119: 27–35.

Tomasello, Michael. 2016. *A Natural History of Human Morality.* Cambridge, Mass.: Harvard University Press.

———, and Malinda Carpenter. 2007. "Shared intentionality." *Developmental Science* 10 (1): 121–25.

Tooby, J., and L. Cosmides. 1988. "The evolution of war and its cognitive foundations." *Technical Report, No. 88–1.* Santa Barbara: Institute for Evolutionary Studies, University of California, Santa Barbara.

———. 1990. "On the universality of human-nature and the uniqueness of the individual—the role of genetics and adaptation." *Journal of Personality* 58 (1): 17–67.

Treves, Adrian, and Lisa Naughton-Treves. 1997. "Case study of a chimpanzee recovered from poachers and temporarily released with wild conspecifics." *Primates* 38 (3): 315–24.

Trut, L. N. 1999. "Early canid domestication: The farm-fox experiment." *American Scientist* 87: 160–69.

———, F. Ya Dzerzhinskii, and V. S. Nikol'skii. 1991. "Intracranial allometry and craniological changes during domestication of silver foxes." *Genetika* 27 (9): 1605–11. (Article in Russian.)

Trut, Lyudmila N., Irina Oskina, and Anastasiya Kharlamova. 2009. "Animal evolution during domestication: The domesticated fox as a model." *BioEssays* 31: 349–60.

Tuchman, Barbara W. 1985. *The March of Folly: From Troy to Vietnam.* New York: Random House.

Tulogdi, A., M. Toth, J. Halasz, E. Mikics, T. Fuzesi, and J. Haller. 2010. "Brain mechanisms involved in predatory aggression are activated in a laboratory model of violent intra-specific aggression." *European Journal of Neuroscience* 32: 1744–53.

Tulogdi, Aron, Laszlo Biro, Beata Barsvari, Mona Stankovic, Jozsef Haller, and Mate Toth. 2015. "Neural mechanisms of predatory aggression in rats—Implications for abnormal intraspecific aggression." *Behavioural Brain Research* 283: 108–15.

Turney-High, H. H. 1949. *Primitive War: Its Practice and Concepts.* Columbia: University of South Carolina Press.

Tuvblad, Catherine, and Laura A. Baker. 2011. "Human aggression across the lifespan: Genetic propensities and environmental moderators." *Advances in Genetics* 75: 171–214.

———, Adrian Raine, Mo Zheng, and Laura A. Baker. 2009. "Genetic and environmental stability differs in reactive and proactive aggression." *Aggressive Behavior* 35: 437–52.

Twain, Mark. 1917. *What Is Man?* New York: Harper.

Umbach, Rebecca, Colleen M. Berryessa, and Adrian Raine. 2015. "Brain imaging research on psychopathy: Implications for punishment, prediction, and treatment in youth and adults." *Journal of Criminal Justice* 43: 295–306.

Valentova, Jaroslava Varella, Karel Kleisner, Jan Havlícek, and Jirí Neustupa. 2014. "Shape differences between the faces of homosexual and heterosexual men." *Archives of Sexual Behavior* 43: 353–61.

van den Audenaerde, D. F. E. 1984. "The Tervuren Museum and the pygmy chimpanzee." In *The Pygmy Chimpanzee: Evolutionary Biology and Behavior,* edited by R. L. Susman (New York: Plenum), pp. 3–11.

van der Dennen, Johann M. G. 2006. "Review essay: Buss, D. M. (2005), the Murderer Next Door: Why the Mind Is Designed to Kill." *Homicide Studies* 10 (4): 320–35.

VanderLaan, Doug P., Lanna J. Petterson, and Paul L. Vasey. 2016. "Femininity and kin-directed altruism in androphilic men: A test of an evolutionary developmental model." *Archives of Sexual Behavior* 45: 619–33.

van der Plas, Ellen A. A., Aaron D. Boes, John A. Wemmie, Daniel Tranel, and Peg Nopoulos. 2010. "Amygdala volume correlates positively with fearfulness in normal healthy girls." *SCAN* 5: 424–31.

van Schaik, Carel P. 2016. *The Primate Origins of Human Nature.* Hoboken, N.J.: John Wiley.

Vasey, Paul L. 1995. "Homosexual behavior in primates: A review of evidence and theory." *International Journal of Primatology* 16: 173–204.

Veroude, Kim, Yanli Zhang-James, Noelia Fernandez-Castillo, Mireille J. Bakker, Bru Cormand, and Stephen V. Faraone. 2015. "Genetics of aggressive behavior: An overview." *American Journal of Medical Genetics Part B* 171B: 3–43.

Villa, Paola, and Will Roebroeks. 2014. "Neandertal demise: An archaeological analysis of the modern human superiority complex." *PLoS ONE* 9(4): e96424.

Vrba, Rudolf. 1964. *I Cannot Forgive.* New York: Grove.

Wadley, Lyn. 2010. "Were snares and traps used in the Middle Stone Age and does it matter? A review and a case study from Sibudu, South Africa." *Journal of Human Evolution* 58: 179–92.

Wallen, Kim. 2001. "Sex and context: Hormones and primate sexual motivation." *Hormones and Behavior* 40: 339–57.

Wang, Xu, Lenore Pipes, Lyudmila N. Trut, Yury Herbeck, Anastasiya V. Vladimirova, Rimma G. Gulevich, Anastasiya V. Kharlamova, et al. 2017. "Genomic responses to selection for tame/aggressive behaviors in the silver fox *(Vulpes vulpes).*" *bioRxiv;* doi: 10.1101/228544.

Warneken, Felix. 2018. "How children solve the two challenges of cooperation." *Annual Review of Psychology* 69: 205–29.

Warner, W. Lloyd. 1958. *A Black Civilization: A Social Study of an Australian Tribe.* Revised ed. New York: Harper.

Watts, David P. 1989. "Infanticide in mountain gorillas: new cases and a reconsideration of the evidence." *Ethology* 81: 1–18.

———. 2004. "Intracommunity coalitionary killing of an adult male chimpanzee at Ngogo, Kibale National Park, Uganda." *International Journal of Primatology* 25 (3): 507–21.

Weaver, T. D., C. C. Roseman, and C. B. Stringer. 2008. "Close correspondence between quantitative- and molecular-genetic divergence times for Neandertals and modern humans." *PNAS* 105: 4645–49.

Weiner, Tim. 1998. "C.I.A. bares own bungling in Bay of Pigs report." *New York Times,* February 22, 1998.

Weinshenker, N. J., and A. Siegel. 2002. "Bimodal classification of aggression: Affective defense and predatory attack." *Aggression and Violent Behavior* 7: 237–50.

Weinstock, Eugene. 1947. *Beyond the Last Path.* Translated by Clara Ryan. New York: Boni and Gaer.

West, S. A., A. S. Griffin, and A. Gardner. 2007. "Social semantics: Altruism, cooperation, mutualism, strong reciprocity and group selection." *Journal of Evolutionary Biology* 20: 415–32.

Wheeler, G. 1910. *The Tribe and Intertribal Relations in Australia.* London: John Murray.

Whitam, Frederick L. 1983. "Culturally invariable properties of male homosexuality: Tentative conclusions from cross-cultural research." *Archives of Sexual Behavior* 12 (3): 207–26.

White, Isobel, ed. 1985. *Daisy Bates: The Native Tribes of Western Australia.* Canberra: National Library of Australia.

Wiessner, Polly. 2005. "Norm enforcement among the Ju/'hoansi Bushmen: A case of strong reciprocity?" *Human Nature* 16: 115–45.

———. 2006. "From spears to M-16s: Testing the imbalance of power hypothesis among the Enga." *Journal of Anthropological Research* 62: 165–91.

Wilkins, Adam S., Richard W. Wrangham, and W. Tecumseh Fitch. 2014. "The 'domestication syndrome' in mammals: A unified explanation based on neural crest cell behavior and genetics." *Genetics* 197: 795–808.

Williams, Frank L. 2013. "Neandertal craniofacial growth and development and its relevance for modern human origins." In *The Origins of Modern Humans: Biology Reconsidered,* edited by Fred H. Smith and James C. M. Ahern (New York: John Wiley), pp. 253–84.

Williams, J., G. Oehlert, J. Carlis, and A. Pusey. 2004. "Why do male chimpanzees defend a group range?" *Animal Behaviour* 68: 523–32.

Williams, Kipling D., and Blair Jarvis. 2006. "Cyberball: A program for use in research on interpersonal ostracism and acceptance." *Behavior Research Methods* 38 (1): 174–80.

———, and Steve A. Nida. 2011. "Ostracism: Consequences and coping." *Current Directions in Psychology* 20: 71–75.

Wilmsen, E. 1989. *Land Filled with Flies: A Political Economy of the Kalahari.* Chicago: University of Chicago Press.

Wilson, Edward O. 2012. *The Social Conquest of Earth.* New York: Liveright.

Wilson, Margo, and Martin Daly. 1985. "Competitiveness, risk taking, and violence: The young male syndrome." *Ethology and Sociobiology* 6: 59–73.

Wilson, Michael L., Christophe Boesch, Barbara Fruth, Takeshi Furuichi, I. C. Gilby, Chie Hashimoto, Catherine Hobaiter, et al. 2014. "Lethal aggression in *Pan* is better explained by adaptive strategies than human impacts." *Nature* 513: 414–17.

Wittig, Roman M., and Christophe Boesch. 2003. "Food competition and linear dominance hierarchy among female chimpanzees of the Taï National Park." *International Journal of Primatology* 24 (4): 847–67.

Wobber, Victoria, Richard Wrangham, and Brian Hare. 2010. "Bonobos exhibit delayed development of social behavior and cognition relative to chimpanzees." *Current Biology* 20: 226–30.

Wolfgang, Marvin. 1958. *Patterns of Criminal Homicide.* Philadelphia: University of Pennsylvania Press.

Won, Y., and J. Hey. 2005. "Divergence population genetics of chimpanzees." *Molecular Biology and Evolution* 22: 297–307.

Woodburn, James. 1982. "Egalitarian societies." *Man* 17 (3): 431–51.

Workman, B. K. 1964. *They Saw It Happen in Classical Times: An Anthology of Eye-Witnesses' Accounts of Events in the Histories of Greece and Rome, 1400 B.C.–A.D. 540.* New York: Blackwell.

Wrangham, Richard W. 1999. "Evolution of coalitionary killing." *Yearbook of Physical Anthropology* 42: 1–39.

———. 2018. "Two types of aggression in human evolution." *PNAS* 115 (2): 245–53; doi: 10.1073/pnas.1713611115.

———, and Dale Peterson. 1996. *Demonic Males: Apes and the Origins of Human Violence.* Boston: Houghton Mifflin.

———, Michael L. Wilson, and Martin N. Muller. 2006. "Comparative rates of aggression in chimpanzees and humans." *Primates* 47: 14–26.

Wrangham, R. W., and L. Glowacki. 2012. "War in chimpanzees and nomadic hunter-gatherers: Evaluating the chimpanzee model." *Human Nature* 23: 5–29.

Wright, Robert. 1994. *The Moral Animal.* New York: Pantheon.

Wrinch, Pamela N. 1951. "Science and politics in the U.S.S.R.: The genetics debate." *World Politics* 3 (4): 486–519.

Wyden, Peter. 1979. *Bay of Pigs: The Untold Story.* New York: Simon and Schuster.

Wynn, Thomas, Karenleigh A. Overmann, and Frederick L. Coolidge. 2016. "The false dichotomy: A refutation of the Neandertal indistinguishability claim." *Journal of Anthropological Sciences* 94: 201–21.

Xu, Jiaquan, Sherry L. Murphy, Kenneth D. Kochanek, and Brigham A. Bastian. 2016. "Deaths: Final data for 2013." *National Vital Statistics Reports* 64 (2): 1–118.

Yamakoshi, Gen. 2004. "Food seasonality and socioecology in *Pan:* Are West African chimpanzees another bonobo?" *African Study Monographs* 25 (1): 45–60.

Young, Lindsay C., and Eric A. VanderWerf. 2014. "Adaptive value of same-sex pairing in Laysan albatross." *Proceedings of the Royal Society B* 281: 20132473.

Zammito, John H. 2006. "Policing polygeneticism in Germany, 1775: (Kames,) Kant, and Blumenbach." In *The German Invention of Race,* edited by Sara Eigen and Mark Larrimore (Albany: State University of New York Press), pp. 35–54.

Zefferman, Matthew R., and Sarah Mathew. 2015. "An evolutionary theory of large-scale human warfare: Group-structured cultural selection." *Evolutionary Anthropology* 24: 50–61.

Zegwaard, G. A. 1959. "Head-hunting practices of the Asmat of Netherlands New Guinea." *American Anthropologist* 61: 1020–41.

Zollikofer, Christoph P. E. 2012. "Evolution of hominin cranial ontogeny." In *Progress in Brain Research,* edited by M. A. Hofman and Dean Falk (Amsterdam: Elsevier), pp. 273–92.

Index

A NOTE ABOUT THE AUTHOR

Richard Wrangham is Ruth B. Moore Professor of Biological Anthropology, Harvard University. He is the author of *Catching Fire: How Cooking Made Us Human,* and *Demonic Males: Apes and the Origins of Human Violence* (with Dale Peterson). Professor Wrangham is a primate behavioral ecologist and President Emeritus of the International Primatological Society. He is the recipient of the Rivers Memorial Medal from the Royal Anthropological Institute and a MacArthur Foundation Fellowship. He is a fellow of the American Academy of Arts and Sciences and of the British Academy.

A NOTE ON THE TYPE

This book was set in Adobe Garamond. Designed for the Adobe Corporation by Robert Slimbach, the fonts are based on types first cut by Claude Garamond (ca. 1480–1561). Garamond was a pupil of Geoffroy Tory and is believed to have followed the Venetian models, although he introduced a number of important differences, and it is to him that we owe the letter we now know as "old style." He gave to his letters a certain elegance and feeling of movement that won their creator an immediate reputation and the patronage of Francis I of France.

Typeset by Scribe,
Philadelphia, Pennsylvania

Printed and bound by Berryville Graphics,
Berryville, Virginia

Designed by Michael Collica